Signals and Communication Technology

More information about this series at http://www.springer.com/series/4748

Mladen Božanić · Saurabh Sinha

Power Amplifiers for the S-, C-, X- and Ku-bands

An EDA Perspective

Springer

Mladen Božanić
Engineering and Built Environment
University of Johannesburg
Johannesburg
South Africa

Saurabh Sinha
Engineering and Built Environment
University of Johannesburg
Johannesburg
South Africa

ISSN 1860-4862 ISSN 1860-4870 (electronic)
Signals and Communication Technology
ISBN 978-3-319-28375-3 ISBN 978-3-319-28376-0 (eBook)
DOI 10.1007/978-3-319-28376-0

Library of Congress Control Number: 2015958906

Printed on acid-free paper

This Springer imprint is published by SpringerNature
The registered company is Springer International Publishing AG Switzerland

Preface

In the age where telecommunication has become a standard, almost every portable device has some kind of transmitter and receiver allowing it to connect to a cellular network or available Wi-Fi networks. We are also driving cars that are smarter and equipped with new technologies, such as radars for collision detection. Other types of radars are used in both civilian and military applications. Nowadays, we even receive signals from satellites on our phones from Global Positioning Systems. Radio frequency (RF) identification devices are becoming more and more common and are being used in many applications, from access control to medical applications. In other words, the spectrum around us is full of transmitted signals waiting to be received. Each signal is transmitted by some kind of power amplifier. As a result, all researchers are likely to face the challenge of designing an RF or microwave power amplifier at some stage of their careers.

Design of power amplifiers, however, is not an easy task. Even the great number of power amplifier classes suggests that no single configuration is capable of delivering acceptable performance for several frequency bands and for several applications at once. Thus, the aim of this book is twofold. First, the idea is to provide researchers with enough power amplifier theory to gain sufficient knowledge to choose the best power amplifier stage for the specific application and to understand the most important defining equations and parameters. Second, the design equations to achieve this are very complex, and if they are used to design by hand, they tend to put off researchers and designers. Thus, this book also aims to provide its readers with some ideas on how to simplify the design process by introducing their own software-based procedures or, in other words, by developing their own electronic design automation (EDA). Although MATLAB is used throughout the book to illustrate the concept of EDA (sometimes also termed computer-aided design or CAD), the exact programming language is not important. The accent is on how to identify what is needed as the end result of the power amplifier design, and how to develop custom EDA to reach this result; essentially, this book focuses on the methodology of power amplifier design.

This methodology is conceptualized so that it “trivializes” the approach to power amplifier design by removing the “black magic” approach typically used in the process. This advances research by allowing the readers to shift the focus of research from power amplifiers onto other less-explored components of the system or even on specification refinement. It can be used equally well by researchers focusing on integrated design, or researchers focusing on discrete implementations, which are typically used for high power or higher frequencies. The researchers focusing on the thin line between integrated circuits and discrete implementations, sometimes termed systems-on-package (SOP), are not excluded. Even though different approaches sometimes need to be followed for design in different frequency bands up to and including the Ku-band, similar principles of EDA apply.

This book is organized in two parts. Part I focuses on the main concepts of power amplification, and this part can be used like any reference book. It first presents a review of transmission bands and their implications for transceiver system design. The feasibility of different passive component implementations in each frequency range is investigated, and power amplifiers are placed into the context of the transceiver system. The theory behind transistor operation at large signal levels is included, and various semiconductor fabrication technologies are discussed for full system integration or power transistor fabrication, together with substrates for the implementation of discrete passives and SOP packaging. Other basic aspects of communication theory, such as *S*-parameters, *Y*-parameters, Smith charts, resonance, loaded quality factor (Q-factor), insertion loss impedance transformation, and Fourier theory, are reviewed, which allows for better understanding of power amplifier concepts. This is followed by an in-depth analysis of power amplifier stages. Most of the commonly used power amplification classes (among others A, AB, C, J, D, E, E^{-1}, F, and F^{-1}) are discussed, and the defining equations are included. Passives and their Q-factors are also covered in Part I. This includes resistors, capacitors, integrated inductors, solenoids, toroids, RF chokes, and transformers. Special focus is placed on both discrete and integrated inductor designs, as they tend to exhibit low Q-factors and are therefore paramount to power amplifier design. Micro-electro-mechanical systems are discussed as a promising technology for the design of passives. Lumped and transmission line impedance matching, which is important if the power amplifier is to be connected to the rest of the transceiver system with minimum losses, is also discussed from an analytical, graphical, and EDA perspective, for both real and complex sources and loads.

Part II of the book focuses on developing CAD procedures to aid practical power amplifier design based on the theory reviewed in Part I. Although this book is not intended for computer engineers, it is assumed that every RF researcher has some basic programming skills. The automation and intelligent design ideas for inductor design are presented in this part, which is followed by automation and intelligent design ideas of various on- and off-chip power amplifier classes described in Part I. Previously described design of passive elements and matching are incorporated into this methodology. Real-life power amplifier design examples using the proposed methodology are explored by means of examples, and developed algorithms are illustrated both graphically and in MATLAB. IC layout and fabrication are considered as one alternative for practical implementations, and discrete implementations are considered as a second alternative. A practical aspect of packaging of discrete systems is also explored. Other practical aspects that are not necessarily covered by the EDA flow are also discussed in this part, and a formal procedure for practical power amplifier design is presented. Part II of the book concludes with the ideas for future research.

Acknowledgment

The authors would like to recognize the research-capacity grant of the Department of Higher Education and Training, South Africa, for sponsoring the work covered in this book. Furthermore, the authors would like to recognise Dr. Riëtte de Lange, Postgraduate School, University of Johannesburg, South Africa, for her effective administration of this grant.

Contents

About the Authors

Mladen Božanić SMIEE, obtained his B.Eng. (with distinction), B.Eng. (Hons) (with distinction) and Ph.D. degrees in Electronic Engineering from the University of Pretoria (UP) in 2006, 2008 and 2011 respectively. In 2008, he joined Azoteq, a fabless IC design company originating in South Africa where he was responsible for the silicon-level design, simulation characterization design for testability (DFT) of various analog, RF, digital and mixed-mode circuits. While actively working in the industry, he also participates in research activities, currently with the University of Johannesburg (UJ) where he is serving as a Senior Research Fellow. Since 2011, Dr. Božanić has been fulfilling the role of a Specialist Editor of the South African Institute of Electrical Engineers (SAIEE). He is a recipient of SAMES Award and CEFIM Fellowship Award, and an author or co-author of over 10 peer-reviewed journal and conference articles, one book chapter and one book.

Saurabh Sinha SMIEEE, FSAIEE, FSAAE, obtained his B.Eng., M.Eng. and Ph.D. degrees in Electronic Engineering from the University of Pretoria (UP), South Africa. He achieved both his B.Eng. and M.Eng. with distinction. As a published researcher, he has authored or co-authored over 85 publications in peer-reviewed journals and at international conferences. In addition, he is the managing editor of the South African Institute of Electrical Engineers (SAIEE) Africa Research Journal. Prof. Sinha served the UP for over a decade, his last service being as Director of the Carl and Emily Fuchs Institute for Microelectronics, Department of Electrical, Electronic and Computer Engineering. On 1 October 2013, Prof. Sinha was appointed Executive Dean of the Faculty of Engineering and the Built Environment (FEBE) at the University of Johannesburg (UJ). Professor Saurabh Sinha is the 2014–2015 Vice-President, IEEE Educational Activities and serves on the IEEE Board of Directors.

Chapter 1
Introduction

In today's communication age, almost every portable device has some sort of transmitter—be it a radio for third generation (3G), long-term evolution (LTE) or Worldwide Interoperability for Microwave Access (WiMAX) networks, Bluetooth or WiFi [1–4]. We are driving cars that are smarter, equipped with new technologies, such as radars for collision detection. Other types of radars are used in both civilian and military applications. We receive signals from satellites in our cars and on our phones from global positioning systems (GPS), as well as at home (satellite TV receivers). Radio-frequency identification (RFID) devices are becoming more and more common and are finding use even in medical applications [5, 6].

Essentially, the spectrum around us is full of transmitted signals waiting to be received. Each signal was transmitted by some sort of power amplifier (sometimes abbreviated as PA). Thus, every circuit designer is likely, sooner or later in his or her career, to face the challenge of designing a radio-frequency (RF) or microwave (sometimes abbreviated MW) transmitter, and inherently, a power amplifier for one of the following bands: S-, C-, X- or Ku-bands [7] operating at ultra-high frequencies (UHF) and super-high frequencies (SHF) from 2 to 18 GHz.

The great number of power amplifier types (termed classes) suggests that no single configuration is capable of delivering acceptable performance in all frequency bands and for all applications. One of the aims of this book is to provide its readers with enough power amplifier theory to gain sufficient knowledge to choose the best power amplifier stage for the specific application and to understand the most important defining equations and parameters. The power amplifier increases the power level of the input signal, resulting in a signal with a higher output power level. Therefore an important focus of power amplifiers is output power as well as power gain. The design equations and process to achieve this are very complex and if they are used to design by hand, they tend to almost frighten the designers. Thus another aim of the book is to provide readers with some ideas on how to simplify the design process by introducing software-based routines in a programming language of their choice, and provide enough examples to make this task easier.

The software-aided methodologies presented this book are conceptionalized so that they can be used equally well for designing increasingly popular integrated circuits (ICs), or well-established discrete implementations (typically used for high

M. Božanić and S. Sinha, *Power Amplifiers for the S-, C-, X- and Ku-bands*,
Signals and Communication Technology, DOI 10.1007/978-3-319-28376-0_1

power), or combinations of these two (e.g. integrated power amplifier with matching or any other components off-chip). Although different approaches need to be followed for design in different bands (albeit without an exactly defined border where one approach stops and the next one begins) up to and including the Ku-band, similar principles of computer-aided design (CAD) and electronic design automation (EDA) will apply.

This introductory chapter touches on basic principles of signal transmission and reception, active and passive devices, basic operation of power amplifiers and their classification, basic design parameters and characterization. It also presents the justification for the use of a computer-aided approach in power amplifier design. The purpose of this chapter is to touch on the basic topics of this book in order to interest the reader in later chapters.

1.1 Power Amplifier as Part of a Transceiver System

A typical transceiver system consists of a transmitter and receiver [8], as shown in Fig. 1.1. In order for a signal to be transmitted over a channel, the signal is first processed. This usually means that some sort of digital encoding is performed. The signal is thereafter modulated onto a carrier frequency using one of the modulation schemes suitable for a particular band of operation, for example phase-shift keying (PSK), quadrature PSK (QPSK), direct sequence spread spectrum (DSSS), quadrature amplitude modulation (QAM), on-off keying (OOK) or orthogonal frequency division multiplexing (OFDM) [9–13]. The code-division multiple

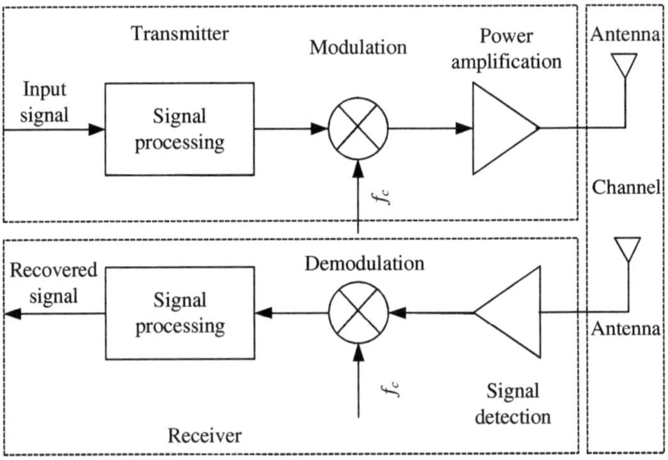

Fig. 1.1 Power amplifier as part of a simple telecommunication system [8]

access (CDMA) technique, a type of channel access method where several transmitters can send information simultaneously, is often confused with some modulation schemes.

The first two stages will set the correct signal voltage levels, carrier frequency and bandwidth. However, this signal is still unsuitable for transmission. The amount of power needs to be increased in order to drive the antenna. Power amplification is normally the third stage of the transmitter and last before the antenna, through which the power amplifier converts the direct current (DC) input power from the supply rails into a significant amount of RF or microwave power [2].

On the receiver side, a similar process occurs, but in reverse order. First, the signal is detected from a channel using a low-noise amplifier (LNA) [14]. A carrier recovery scheme may be employed. Thereafter, the signal is demodulated, and reverse signal processing and filtering are used to reproduce the original signal.

The power amplifier, marked bold in Fig. 1.1, needs to deliver high efficiency, high linearity, high power gain and large dynamic range simultaneously [15]. Consequently, it consumes the largest amount of DC power. The increasing demand for a higher data rate and increasing modulation complexity, complemented by the need to keep the transmitter costs low, calls for innovative art of transmitter design [16, 17]. The power amplifier therefore remains a bottleneck in the design of wireless transceivers.

Inclusion of a power amplifier is a particular problem in integrated devices, especially if integration is done in pure silicon complementary metal-oxide semiconductor (CMOS) processes, mainly owing to the amount of power that needs to be generated on chip and the size of passive components. For this reason, most commercial wireless devices use an external power amplifier using discrete components to drive an antenna. The driving transistor device is usually fabricated in semiconductor technologies superior to the silicon (Si) CMOS, such as silicon-germanium (SiGe), gallium-arsenide (GaAs), indium-phosphate (InP), gallium-nitride (GaN), silicon carbide (SiC) and others [18–24]. However, in discrete power amplifier implementations, other factors can introduce limitations, such as the design and material of the printed-circuit board (PCB) used, and insertion loss of lumped devices and discrete designs are not straightforward either.

1.2 Active and Passive Devices for Power Amplifier Design

A basic power amplifier is designed around a minimum of one active device. This could be metal-oxide semiconductor field-effect transistors (MOSFETs), bipolar junction transistors (BJTs), heterojunction bipolar transistors (HBTs), high-electron-mobility transistors (HEMTS), or another type. Vacuum-tube power amplifiers are still used [9].

MOSFET devices have generally been considered less suitable for the power amplification task because they require more current to achieve the same amount of power amplification than their bipolar counterparts (HBTs) [25], but this difference is becoming smaller as superior MOSFET technologies emerge. Apart from differences based on fundamental device properties, transistors (both integrated and discrete) will yield different performance when fabricated in different technologies. Several factors, including the transistor transition frequency f_T (frequency at which transistor gain-bandwidth product becomes zero), the breakdown voltage of the transistor and the driving capability of the transistor, need to be taken into consideration when choosing the best technology or a transistor for power amplifier implementation [26]. It is worthwhile noting that the performance of active devices is severely affected by the trend in device scaling, but technologies capable of reaching even mm-wave frequencies have been reported [19]. There are several figures of merit that can be used to quantify the suitability of semiconductor material for power transistor fabrication, and Johnson's figure of merit (JFOM) and Baliga's figure of merit (BFOM) will be mentioned later in this book [27, 28].

For illustration purposes, two power transistors (a layout of an HBT transistor in IBM 7WL technology and a photograph of a power Darlington pair) are shown in Fig. 1.2. Active devices will be discussed in more detail in Chap. 2.

Additional to the active devices, a power amplifier contains a number of passive components, such as inductors and capacitors used for filtering and matching. Other passive components, including among others transformers that are used for power combining [9], and transmission lines can also be found in the power amplifier. At RF, designing with ideal devices seldom generates good results even on the first design iteration. Real devices and their parasitic effects need to be considered. The

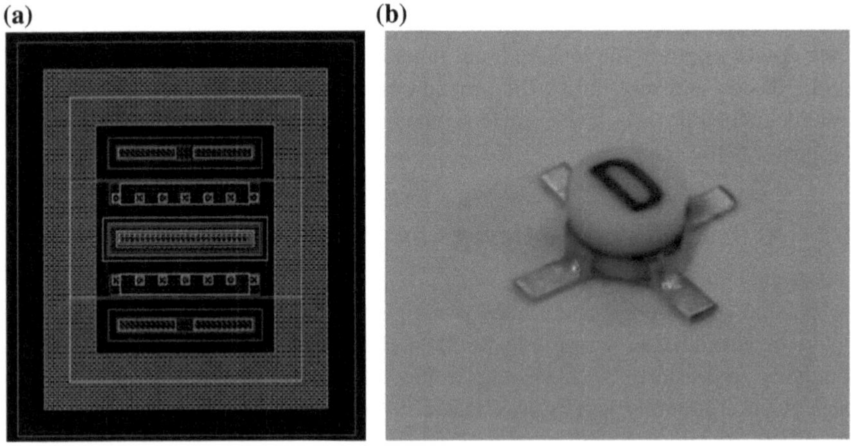

Fig. 1.2 An example of the HBT transistor layout for integrated power amplifier implementations (**a**), and a photograph of a Darlington power transistor for discrete power amplifier implementations (**b**)

greater the frequency, the more difficult it is to find a device with the expected performance. This principle applies particularly to inductors, which tend to have inferior performance, both on- and off-chip. Here, the substrate on which the passive component is fabricated plays a major role, and a quality factor (Q-factor) is used as a measure of quality. Instead of using lumped passives, passives implemented using transmission lines (i.e. open and short-circuited stubs) can be used at RF, but they are mostly practical off-chip. At mm-wave frequencies, transmission lines can be used on-chip; however, mm-wave frequencies are not one of the main focuses of this book but they will be discussed in Chap. 10 when dealing with future directions.

Figure 1.3 shows a photograph of a wire-wound inductor, an integrated spiral octagonal inductor in IBM 7WL technology and a 2:1 wire-wound transformer.

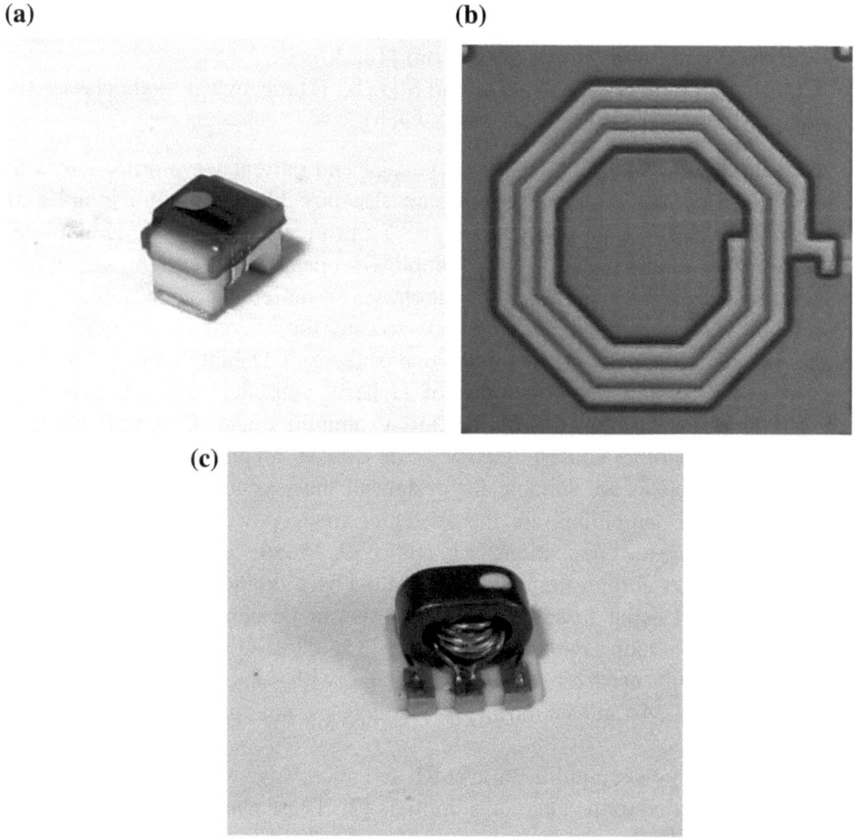

Fig. 1.3 Photographs of different inductors and transformers: a wire-wound inductor (**a**), an integrated spiral inductor (**b**), and a 2:1 transformer (**c**)

1.3 Classification of Power Amplifiers

Several groupings of power amplifiers are possible and all groupings are used interchangeably. Power amplifiers are commonly grouped into broadband and narrowband amplifiers. Sometimes, they are grouped depending on whether they are intended for linear or constant envelope operation [29]. Finally, the most common grouping of power amplifiers is grouping into classes according to the nature of their voltage and current waveforms. The variety of power amplifier classes reflects the inability of any single circuit to satisfy stringent requirements for linearity, power gain, output power and efficiency, all described later in this chapter.

A letter or combinations of letters of the alphabet are used to define different power amplifier classes. This classification is based on the shape of the voltage (v_D) and current (i_D) waveforms of the driving transistor. The following classes are commonly used for different applications:

- Classes A, B, AB and C are classes exhibiting continuous mode of operation (i.e. the driving transistor is always on) [13, 30].
- Classes D, DE, E, F, FE, G, H, J and S [7, 8, 31] are switch-mode classes (i.e. the driving transistor functions as a switch).

Inverse classes, where the shape of voltage and current waveforms across the power transistors are swapped around, are also possible. Common examples are inverse Class-C (C^{-1}), inverse Class-E (E^{-1}) and inverse Class-F (F^{-1}) amplifiers [32, 33]. Most of the real-life power amplifiers operate with current and voltage waveforms that lie between two different classes. If more than one power amplifier of different classes are combined in parallel to cater for different modes of operation of the transmitter (usually one main and one peaking), a Doherty power amplifier is created [34]. Common combinations of Doherty amplifiers are a Class-AB or Class-B amplifier combined with a Class-C amplifier and a Class-F amplifier combined with another Class-F, Class-F^{-1} or Class-C amplifier.

Not all the classes are suitable for design all the way up to the Ku-band. For example, Class-D amplifiers are the switching-mode power amplifiers generally used in low-frequency applications (e.g. audio) [9, 35, 36], and the use of this class of power amplifier at high frequencies is limited by prominent parasitic reactances that lead to substantial losses. However, they can be considered at higher frequencies when operating in the current mode [35]. Class-G and Class-H amplifiers are also commonly used for audio applications, with some limited use in digital telephony and CDMA at low megahertz frequencies, not applicable to the topic of this book.

Traditionally, power amplification at RF and microwave was done with amplifier classes A to C, often termed classic amplifiers [7]. These classes (with exception of Class C) generally have high linearity but suffer from low efficiencies. Class-E,

Class-F amplifiers and other switchmode classes are considered modern amplifiers, since they can be used in many high-end applications. They suffer from low linearity, but their efficiencies can reach 100 % in theory.

Because of their importance, all amplifier classes mentioned will be presented in separate sections in this book.

1.4 Basic Principles of Operation of Power Amplifiers

1.4.1 Power Amplifier Block Diagram

Figure 1.4 shows a block diagram general single-ended power amplifier [9]. In this model, V_{DD} is the voltage supply, R_L is the load, *RFC* is the RF choke—ideally an inductor with infinite reactance and zero series resistance. RFC is large enough to ensure the substantially constant current through the drain. In some designs, RFC can be replaced by a finite inductor, if the output filter can be designed to resonate with it. The output filter mentioned is also shown in this figure [37]. It can include harmonic tuning and wave shaping, impedance matching or any other passive circuitry. The transistor T_1 is shown as an n-channel MOS (NMOS) transistor, but it can be any power transistor (MOS, HBT, BJT, HEMT or other) used in a particular power amplifier application.

Note that throughout this book, terms for device terminals associated with MOS transistors (gate, source, drain) and terms for device terminals of BJTs (such as base, collector, emitter) are used interchangeably.

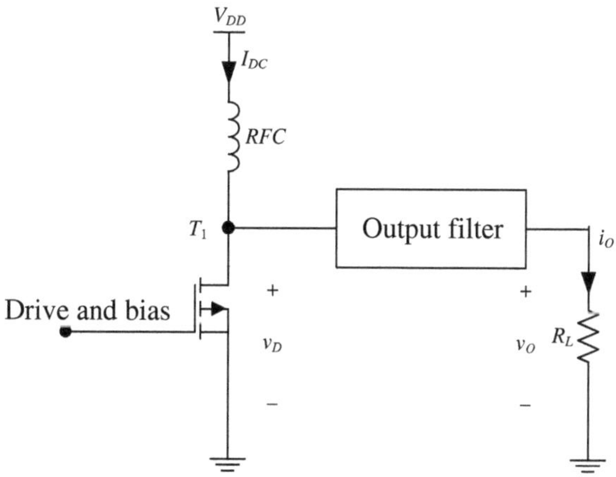

Fig. 1.4 General model of a power amplifier [9]

1.4.2 Output Power and Gain

The task of a power amplifier is to deliver a given power into the load [8]. This power is determined by the power supply voltage V_{DD} and the load R_L. The maximum power that can be delivered is

$$P = \frac{V_{DD}^2}{2R_L}. \tag{1.1}$$

Equation (1.1) is applicable to sinusoidal waveforms. Depending on the shape of waveforms generated for a specific power amplifier stage, it may be possible to deliver more power to the same size load. Integrated power amplifiers are generally designed for low values of R_L.

Power gain is defined as a ratio of output power to input power:

$$G = \frac{P_{out}}{P_{in}}. \tag{1.2}$$

1.4.3 Power Consumption

Total DC power consumption is an important quantity in a power amplifier design, especially for battery-powered portable devices. The DC input power of a power amplifier is the current drawn from the voltage supply over a period of time, or

$$P_{dc} = \frac{1}{T}\int_0^T V_{DD} i_D dt = \frac{V_{DD}}{T}\int_0^T i_D dt = V_{DD} I_{DC}, \tag{1.3}$$

where I_{DC} is the DC component of the current waveform.

1.4.4 Power Efficiency

Efficiency is a measure of performance of a power amplifier. The performance of a power amplifier will be better if its efficiency is higher, irrespective of its definition. Several definitions of efficiency are commonly used with power amplifiers [7, 9, 13]. In order to define the efficiency, the RF output power needs to be defined first.

Assuming sinusoidal voltage and current, the RF output power is given by

$$P_{out} = v_{eff} i_{eff} = \frac{i_1 v_1}{2} = \frac{i_1^2 R_L}{2}. \tag{1.4}$$

Drain (or collector) **efficiency** η is defined as the ratio of RF output power (P_{out}) to DC input power (P_{dc}), or

$$\eta = \frac{P_{out}}{P_{dc}}, \tag{1.5}$$

where i_{eff} and v_{eff} are effective and i_1 and v_1 are the peak fundamental components of current and voltage respectively, and the DC input power is given by Eq. (1.3).

Power added efficiency (PAE) takes into account the input power (P_{in}) by subtracting it from the output power:

$$PAE = \frac{P_{out} - P_{in}}{P_{dc}} = \frac{P_{out} - \frac{P_{out}}{G}}{P_{dc}} = \eta\left(1 - \frac{1}{G}\right). \tag{1.6}$$

The PAE will give a good indication of the performance of a power amplifier for high amplifier gains but it can even become negative for low gains. This relationship is shown in Fig. 1.5. As a consequence, power amplifiers achieve their greatest efficiency only in saturated operation [10] and during non-saturated or idle cycles, a large amount of power is wasted. This is why many commercial devices use Doherty power amplifiers, with different driving stages activated depending on the intensity of the transmitted signal [38].

Overall efficiency is the ratio of output power to the sum of input power and DC input power:

$$OE = \frac{P_{out}}{P_{dc} + P_{in}}. \tag{1.7}$$

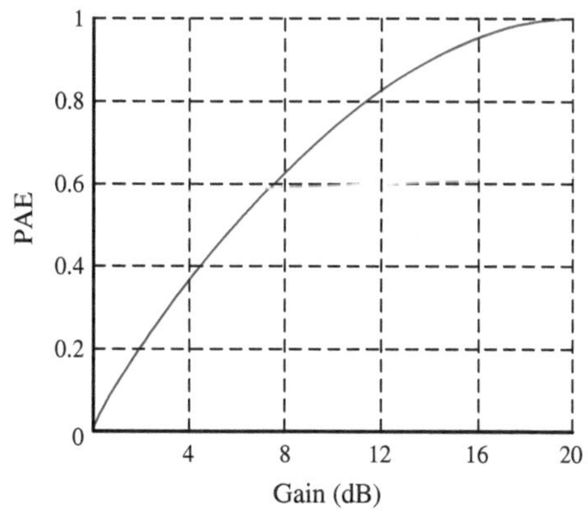

Fig. 1.5 Normalized PAE versus power amplifier gain [8]

Instead of instantaneous efficiencies defined above, **average efficiency** may be more applicable to signals with time-varying amplitudes:

$$\eta_{AVG} = \frac{\overline{P}_{out}}{\overline{P}_{dc}}. \tag{1.8}$$

To achieve maximum drain efficiency, no power can be dissipated over the drain (collector) of the driving transistor [7]. Thus, the power in the drain must be zero, or

$$P_D = \frac{1}{T}\int_0^T v_D i_D dt = 0. \tag{1.9}$$

This implies that the product of the current and voltage needs to be zero at any moment:

$$i_D v_D = 0. \tag{1.10}$$

Therefore, for 100 % efficiency, at any time, either $i_D = 0$ or $v_D = 0$, or in other words, the ideal power amplifier, must have non-overlapping current and voltage waveforms.

Efficiencies in practical power amplifier implementations are generally much lower than ideal values calculated for each stage because of a number of factors: low quality factor of passives, saturation voltage in the transistors and transistor parasitics, tuning errors and temperature variations.

1.4.5 Output Power Capability

Output power capability is defined as ratio of the maximum power delivered to the load and the product of maximum values of i_D and v_D:

$$c_p = \frac{P_{out(\max)}}{I_{DM(\max)} V_{DS(\max)}}. \tag{1.11}$$

1.4.6 Maximum Operating Frequency of Power Amplifiers

Another limiting factor in power amplifier design is the maximum operating frequency for a predetermined power and supply voltage. It is dependent on the transistor output capacitance C_{OUT}, and for a Class-E power amplifier it can be expressed as [39]:

$$f_{MAX} = \frac{1}{2\pi^2} \cdot \frac{P_{out}}{C_{OUT} V_{DD}^2}. \quad (1.12)$$

This relation shows that the greater the amount of power that needs to be delivered, the more limiting the driving transistor will be in reaching higher frequencies.

1.4.7 Temperature Aspects of Power Amplifiers

As discussed in previous sections, efficiency is the ability of the power amplifier to convert the electrical energy into output power. Excess power is converted to heat, which can limit the performance of an amplifier [40]. With increased efficiency, the amount of heat generated decreases. However, even with high efficiency, high output power configurations [41] will dissipate significant amounts of heat. The amount of heat generated and the way in which that heat is dissipated in a power amplifier system depends on the technology in which the IC is fabricated or type of a substrate used for discrete implementation, as well as on the type of active device (transistor) used for power amplification.

Typically, if heating poses a problem, the excess heat can be removed by means of heat sinks, heat slugs or heat spreaders [42–44] Any of the three mentioned components is basically a piece of metal used to dissipate heat away from the chip better than the substrate or the package is able to. Typically, they are made of aluminum or copper.

1.4.8 Matching for Desired Power

A power amplifier needs to be inserted between the modulator and the antenna with minimum insertion loss. This calls for careful impedance matching.

Figure 1.6 shows a block diagram of a power amplifier illustrating matching on the input and output side. At the input side of the power amplifier, care needs to be taken so that the correct current and voltage waveforms are delivered at the gate or

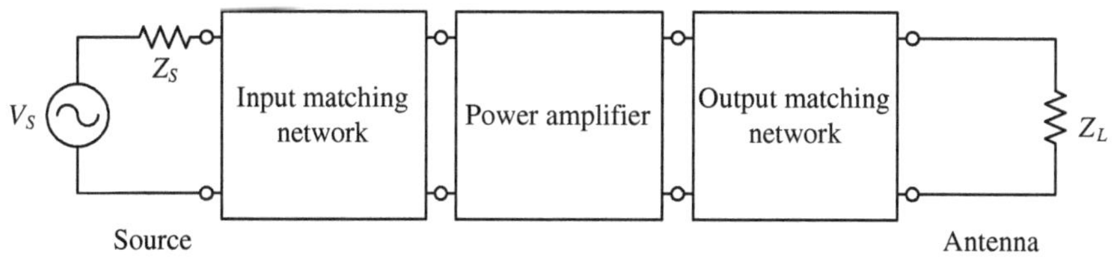

Fig. 1.6 Block diagram of a power amplifier showing input and output matching networks

base of the transistor to achieve a particular class of operation, thus matching needs to be performed simultaneously with biasing described later.

On the other hand, at the output side load has to be chosen correctly. Searching for the optimum impedance for maximum power output, PAE and gain for the power amplifier is usually achieved using load pull.

From Eq. (1.1), it is obvious that the only two parameters influencing the output power are the voltage supply, V_{DD} or V_{CC}, and the load impedance, R_L. The supply is normally fixed for a given application, so that the only degree of freedom left to the designer is the impedance of the load. This impedance will often differ from standard impedances of 50 or 75 Ω, and in IC impedances of less than 10 Ω are not uncommon. Impedance matching networks are used to convert standard impedances to required load impedances as defined by amplifier design equations or obtained by load pull. At mm-wave frequencies, where wavelengths are correspondingly small, this matching can be accomplished with transmission lines [45]. At UHF and SHF, the transmission lines are impractically long to be used on a chip but they can be implemented on a PCB. Matching using discrete or integrated passive components can be deployed both on- and off-chip, provided that sufficiently high-Q circuit elements of required value can be achieved at the matching frequency.

Two-component networks (L networks) and three-component networks (T and Π networks) are commonly used. Eight L-network configurations are possible, as shown in Fig. 1.7a, b, where X_1 and X_2 can be any combination of inductors and capacitors, Z_S is the source impedance and Z_L is the load impedance. Such an L network is a broadband (high-pass or low-pass) network. Conversely, the T and Π networks with passives X_1, X_2 and X_3, shown in Fig. 1.8a, b, are narrowband networks.

1.4.9 Biasing

Input impedance matching, mentioned in the previous section, is used to ensure that correct amplitudes of alternating current (AC) signals appear at input. Biasing,

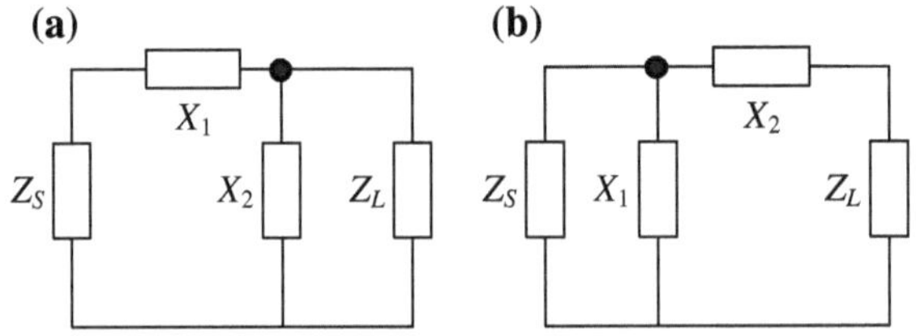

Fig. 1.7 Two-component matching networks where passive component is parallel to **a** load and **b** source [45]

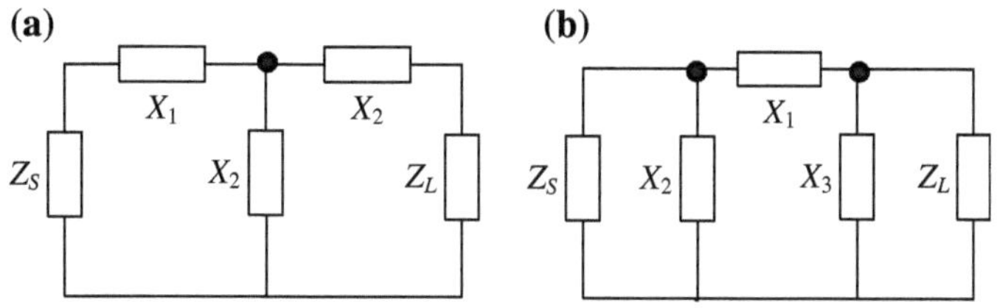

Fig. 1.8 Three-component matching networks: **a** T network and **b** Π network [45]

described in this section, provides the appropriate quiescent point for the power amplifier [45].

The biasing point should remain constant irrespective of transistor parameter variations or temperature fluctuations. Active and passive biasing networks are possible. Figure 1.9 shows one-resistor and three-resistor biasing networks commonly used with BJT power amplifiers.

Adaptive bias techniques can be used with power amplifiers in order to avoid too-large or too-small current, as well as to improve linearity and efficiency [13].

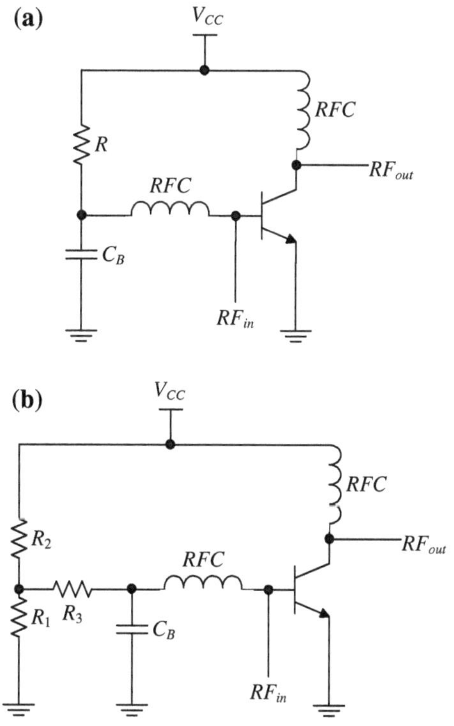

Fig. 1.9 Passive biasing networks for a BJT power amplifier: **a** one-resistor configuration, **b** three-resistor configuration [45]

1.4.10 Conduction Angle

Conduction angle, 2θ, is another important parameter of power amplifiers. It represents a portion of the cycle during which an amplifier is conducting current. In Class A, AB, B and C amplifiers, it directly determines the shape of current and voltage waveforms and thus it is commonly used to differentiate between Class A, AB, B and C amplifiers. In switch-mode power amplifiers (e.g. Class E or F), the conduction angle is limited to describing the biasing level at which the input waveform is applied to the active device. Thus, a Class-E power amplifier can be biased at Class-AB, B or C level based on the portion of the cycle during which the switch is activated.

1.4.11 Distortion, Linearization and Increase of Power Output

High linearity is one of the main requirements of each power amplifier. Distortion is manifested either by the harmonics of the carrier frequency (harmonic distortion, where the nth harmonic is designated as nf_c) or by intermodulation products (intermodulation distortion IMD, designated by $f_{IMD} = nf_1 \pm mf_2$) [7]. In practice, the IMD is tested by a two-tone test or a two-tone test, whereby two or more sinusoidal waveforms connected in series are applied to the amplifier. Carrier-to-intermodulation ratio (C/I) should be higher than 30 dBc, where dBc indicates the number of decibels below the carrier.

A prominent IMD type is third-order intermodulation distortion (IMD3). If a system with at least a third-order non-linearity can be approximated by a polynomial series

$$v_{out}(t) = a + bv_{in}(t) + cv_{in}^2(t) + dv_{in}^3(t) \tag{1.13}$$

and a two-tone signal

$$v_{in} = A\cos\omega_1 t + B\cos\omega_2 t \tag{1.14}$$

is used as an input of the amplifier, then the output of the nonlinear amplifier is

$$\begin{aligned} v_{out} = {} & a + b(A\cos\omega_1 t + B\cos\omega_2 t) \\ & + c(A^2\cos^2\omega_1 t + B^2\cos^2\omega_2 t + 2AB\cos\omega_1 t\cos\omega_2 t) \\ & + d(A^3\cos^3\omega_1 t + A^2B\cos^2\omega_1 t\cos\omega_2 t \\ & + AB^2\cos\omega_1 t\cos^2\omega_2 t + B^3\cos^3\omega_2 t). \end{aligned} \tag{1.15}$$

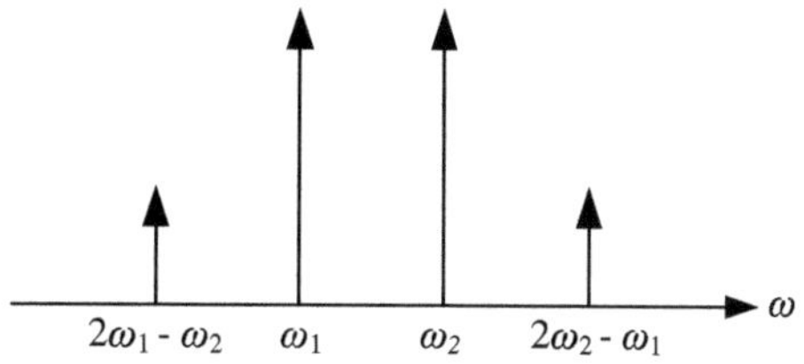

Fig. 1.10 Intermodulation tones

In this equation, $dA^2B\cos^2\omega_1t\cos\omega_2t$ and $dAB^2\cos\omega_1t\cos^2\omega_2t$ are the third-order intermodulation terms at frequencies $2\omega_1 - \omega_2$ and $2\omega_2 - \omega_1$, illustrated in Fig. 1.10.

Total harmonic distortion (THD) is the ratio of the sum of the power in all harmonic components to the power contained in the fundamental frequency, expressed as [46]

$$THD = \frac{\sqrt{\sum_{n=2}^{\infty} V_{on}^2}}{V_1}, \tag{1.16}$$

where V_{on} is the root-mean square (RMS) value of the voltage of the nth harmonic and V_1 is the RMS value of the voltage of the signal at fundamental frequency.

Distortion can usually be improved by various linearization techniques [47–49]. These include feedback and feedforward techniques, various analogue predistortion (APD) and digital predistortion (DPD) techniques, as well as pulse-width modulation (PWM) and supply modulation.

To improve the output power, the efficiency can be boosted by means of adaptive bias and the already mentioned Doherty techniques [50]. Power combining is usually used to increase the total output power of a power amplifier system. This can be performed on- and off-chip [8, 41, 51] and is typically done with the aid of transformers.

Chireix outphasing is another technique that is gaining popularity because of advances of low-power and high-speed digital processing [15]. In this system, an amplitude-modulated (AM) signal is split into two phase-modulated signals with constant amplitude, which are amplified separately and combined. This leads to increased efficiency [52]. Outphasing systems have been implemented successfully for Class-B, D, F and E amplifiers.

1.4.12 Impact of Power Amplifier Turn-on Characteristics

In many power amplifier applications, the power amplifier is switched off between transmissions to save power [1]. In the time period during which the amplifier is transitioning on or off, it will operate outside design specifications and the amount of power consumed will increase. If the delay time the amplifier takes to turn on is defined as T_d and time during the ON cycle is marked as T_{on}, then the total power consumption of the power amplifier during the ON and turn-on part of the cycle is given by

$$P_{total} = \frac{T_d P_{turn-\mathrm{on}}}{T_{on}} + P_{static}, \tag{1.17}$$

where $P_{turn\text{-}on}$ is the power it consumes during turn-on, and P_{static} is the power it consumes during normal operation. It can be deduced from this equation that unless $T_{on} \gg T_d$, power during turn-on can become prominent and should thus not be ignored.

In addition to increased power consumption, intermodulation and harmonic distortion will also increase during the turn-on phase.

1.4.13 Noise in Power Amplifiers

Voltage and current waveforms driving the power amplifier are generated in the modulation block of the transmitter. Three types of noise are applicable to these waveforms: AM noise, frequency-modulated (FM) noise, and phase noise (PM noise). AM noise arises from amplitude variations inside the oscillator producing the carrier frequency. FM and PM noise are due to frequency spreading around the carrier frequency. Around the carrier, the PM noise is most prominent. Noise is measured in units of dBc/Hz, or the number of decibels below the carrier per hertz. Over a bandwidth of one Hz in single sideband, noise power is defined as the noise-to-carrier power ratio

$$NCP = 10\log\frac{N}{C}, \tag{1.18}$$

where N is noise power and C is the carrier power. In addition to carrier-induced noise, power amplifiers can introduce thermal noise, which should be kept below −130 dBm.

1.4.14 Measuring Large-Signal Power Amplifier Performance

A practical means of large-signal performance measurement of the power amplifier is experimentally accomplished by large-signal scattering parameters (S-parameters

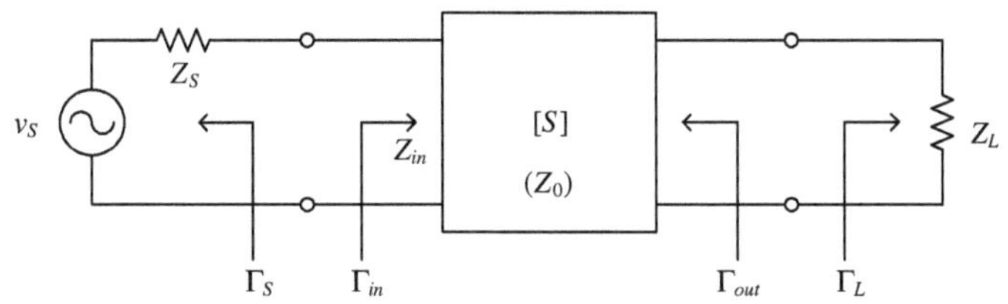

Fig. 1.11 Two-port model of a power amplifier showing the scattering matrix and reflection coefficients

for short). Two-port S-parameters are normally used, where S_{11} and S_{22} indicate the quality of input and output matching respectively and S_{21} and S_{12} indicate the forward and reverse gain. All four two-port scattering parameters are normally termed scattering matrix and denoted [*S*]. A two-port power amplifier model showing the scattering matrix and reflection coefficients defined later is shown in Fig. 1.11.

1.4.15 Measuring Amplifier Power Gain and Stability

Amplifier power gain and stability are usually defined in terms of reflection coefficients and are treated together [45].

Gain of the amplifier between the source and the load is defined as transducer gain and is the ratio between the power delivered to the load and power supplied from the source:

$$G_T = \frac{P_L}{P_A}. \tag{1.19}$$

In Fig. 1.11, the reflection coefficient seen looking to the source is defined as:

$$\Gamma_S = \frac{Z_S - Z_0}{Z_S + Z_0}. \tag{1.20}$$

Similarly, the reflection coefficient seen looking to the load is defined as

$$\Gamma_L = \frac{Z_L - Z_0}{Z_L + Z_0}. \tag{1.21}$$

The input impedance of an antenna is affected by objects in its vicinity [53], therefore Γ_L will not be zero even in almost perfectly matched systems. The transducer gain can be expressed as a relation of S-parameters and reflection coefficients as

$$G_T = \frac{(1 - |\Gamma_L|^2)|S_{21}|^2(1 - |\Gamma_S|^2)}{|(1 - S_{11}\Gamma_S)(1 - S_{22}\Gamma_L) - S_{21}S_{12}\Gamma_L\Gamma_S|^2}. \tag{1.22}$$

If the model of the driving transistor is known, it is possible to derive simpler gain equations that are not derived from S-parameters [54].

Depending on frequency and termination, an amplifier can become unstable and begin to oscillate. Therefore, any amplifier must also meet stability conditions in the frequency range of interest. For stability, two more reflection coefficients, also shown in Fig. 1.11, have to be defined:

$$\Gamma_{in} = S_{11} + \frac{S_{12}S_{21}\Gamma_L}{1 - S_{22}\Gamma_L} \tag{1.23}$$

and

$$\Gamma_{out} = S_{22} + \frac{S_{12}S_{21}\Gamma_S}{1 - S_{11}\Gamma_S}. \tag{1.24}$$

If we define a new quantity Δ as

$$\Delta = S_{11}S_{22} - S_{12}S_{21} \tag{1.25}$$

then we can express Γ_{in} and Γ_{out} as

$$\Gamma_{in} = \frac{S_{11} - \Gamma_L\Delta}{1 - S_{22}\Gamma_L} \tag{1.26}$$

and

$$\Gamma_{in} = \frac{S_{11} - \Gamma_L\Delta}{1 - S_{22}\Gamma_L} \tag{1.27}$$

Stability implies that the magnitudes of all reflection coefficients are less than unity. In other words,

$$|\Gamma_L| < 1,\ |\Gamma_S| < 1,\ |\Gamma_{in}| < 1,\ |\Gamma_{out}| < 1. \tag{1.28}$$

In can be shown, using the theory of stability circles, that the system will be unconditionally stable if

$$k = \frac{1 - |S_{11}|^2 - |S_{22}|^2 + |\Delta|^2}{2|S_{12}||S_{21}|} > 1 \tag{1.29}$$

and

$$|\Delta| < 1. \tag{1.30}$$

Quantity k introduced in Eq. (1.29) is called the stability or Rollett factor.

In a matched power amplifier system, the output capacitance of the driving transistor (C_{OUT}) changes with the drain voltage, and as a result, the amplifier stability suffers at low voltages [55].

It should be noted that in packaged devices, the package parasitics need to be included in the stability measurement or calculation.

1.5 Justification for Computer-Aided Design

Computer-aided design is not uncommon in circuit design, and as a matter of fact, is used in many stages of circuit design.

A typical design flow for an electronic circuit could look as follows: Firstly, a conceptual design or modeling is performed in a mathematical package. Hand design of subsystems is then done before the design options are drawn in a schematic editor. Performance of the circuit is then normally simulated in a SPICE[1] or RF SPICE [56] based simulator. Digital circuitry can be designed using a hardware description language (e.g. VHDL[2] [57]) and simulated in digital simulators (such as Modelsim [58]). Furthermore, a synthesizer can be used to convert digital blocks designed in hardware description languages into digital gates. Automatic place-and-route tools can be used to connect synthesized digital cells automatically with minimum involvement of a design engineer in order to create functional ICs. Different place-and-route tools can be used for route tracks on PCBs for discrete implementations. Automated tools can also be used for parasitic extraction, and electromagnetic (EM) simulators or other specialist software can be used to cater for transmission lines or other passive devices.

It is clear, however, that even with the increased computing power and new innovative EDA ideas, an amount of hand design still needs to be performed, and several hand-design steps (gaps), marked with bold blocks in the flow chart in Fig. 1.12, will remain. This is particularly applicable to analog design, specifically to the circuits operating at increased frequencies. A power amplifier is a very common example of a circuit that requires a lot of hand design work, and therefore fits into one of the gaps in the flow diagram. In the case of rapid design (where a number of different devices need to be designed in a short span of time), even a small amount of hand design work could quickly turn into a very tedious and time-consuming task.

---

[1]Simulation Program with Integrated Circuit Emphasis.

[2]Very High Speed Integrated Circuit Hardware Description Language.

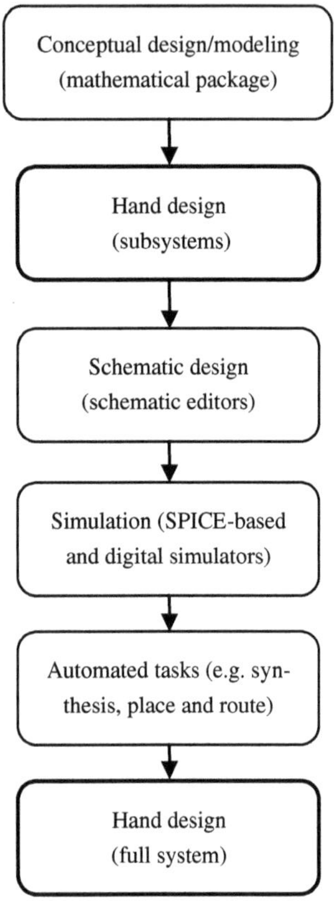

Fig. 1.12 Role of CAD/EDA in circuit design, where hand design stages are termed gaps and are marked in bold

In this book, we describe how power amplifier design equations can be used as a starting point to develop a set of software routines that will aid the design process. Furthermore, we describe and give examples of CAD design of passives, particularly inductors, which have been identified as being traditionally difficult to implement because of low-quality factors and their indeterministic behavior at high frequencies. Because of the strong influence of substrates and many degrees of freedom that need to be considered in inductor design, we also demonstrate intelligent search procedures for inductors that replace iterated procedures commonly used. Finally, it will be shown how the complex task of matching can be simplified by introducing matching algorithms. We also try to identify the basic parameters each designer needs to take into consideration when performing the design. Some of the parameters (e.g. the carrier frequency and antenna impedance) may be more obvious and easier to determine than others (e.g. process parameters such as substrate resistivity). There are parameters over which the designer typically has no control (e.g. thickness of a metal for inductor implementation or a carrier frequency). Other parameters can be treated as design parameters (e.g. output power). Therefore, throughout this book we help the reader to identify and isolate the needed information on design parameters.

Algorithms presented in this book are coded in MATLAB from Mathworks [59]. This package is a scripting programming language that supports a great number of mathematical functions that add to the simplicity of the code. The authors are of the opinion that most of the readers of this book would have at least a basic knowledge of MATLAB to understand the examples provided.

The authors verified that the MATLAB scripts provided throughout the book as examples work as expected in at least two versions of MATLAB: version 2007b and version 2014b. It is thus likely that they will work correctly in any version released between version 2007b and 2014b, and also in any newer version, but it is impossible to verify this. The examples in MATLAB, however, are just for illustration purposes and any other programming or scripting language can be used to accomplish the same task (e.g. python, C#, Delphi). Licenses for certain languages may be free of charge but may still have good mathematical libraries.

1.6 Organization of the Book

This book is organized in two parts. The first part focuses on the main concepts of power amplification and this part can be used like any reference book. The second part focuses on developing CAD routines to aid power amplifier design practically.

This chapter summarized the basic reasoning behind introducing custom EDA into the design flow. Also, the basic principles of power amplifiers are discussed in some detail, where some information serves as background information to the reader and will not be discussed further in this book, but many topics will be expanded in later chapters.

Chapter 2 will present a review of communication systems as applicable to power amplifiers. The chapter will include a review of transmission bands and their implications for transceiver system design. The feasibility of different passive component implementations in each frequency range will be investigated. Power amplifiers will be placed into the context of the transceiver system, and different modulation schemes suitable for a particular band of operation will also be introduced (including PSK, QPSK, DSSS, QAM, OOK and OFDM) The chapter will also include the theory behind transistor operation as applicable to transceiver theory. Various semiconductor fabrication technologies will be discussed for full system integration or power transistor fabrication (SiGe, Si, GaAs, GaN). Substrates for the implementation of discrete passives and their packaging will also be discussed. Furthermore, the chapter will focus on the *S*-parameters and *Y*-parameters review, Smith charts and some other aspects of RF and microwave engineering. The concepts of resonance and resonant tank, loaded quality factor, insertion loss and impedance transformation will also be introduced.

Chapters 3 and 4 will describe power amplifier stages in great detail. Most of the commonly used power amplification classes (among others A, AB, C, D, E, E^{-1}, F and F^{-1}) will be discussed and the defining equations will be included. Power-combining methods and methods for improving the efficiency of amplifiers (e.g. Doherty) will be discussed. The two chapters will also include the physics of the amplifier operation and examples of both integrated and non-integrated state-of-the-art designs found in the literature. Chapter 3 will focus on continuous (classic) stages, while switch-mode classes will be covered in Chap. 4.

In Chap. 5, passive components will be discussed. The chapter will cover resistors, capacitors, integrated inductors, solenoids, toroidal inductors, RF-chokes and transformers, among others. Q-factors of these devices will be investigated in detail. Special focus will be placed on both discrete and integrated inductor designs, as they tend to exhibit low Q-factors and are therefore paramount to power amplifier design. Micro-electro-mechanical systems (MEMS) will be discussed as a promising technology for the design of passives.

Chapter 6 will be the last chapter of Part 1 and will deal with impedance matching, which is important if the power amplifier is to be connected to the rest of the transceiver system with minimum losses. Impedance matching with lumped elements and transmission lines will be discussed, together with aspects of matching both on- and off-chip. Analytical, graphical, and EDA matching solutions will be presented, both for real and complex sources and loads.

Chapter 7 will be the first chapter in Part 2 and will present inductor design automation and intelligent design ideas. The chapter will try to cover all inductors needed to design a stand-alone system; this will include both filtering and matching inductors. The Q-factor and its dependence on various inductor design parameters will be described in detail. Together with Chap. 5, this chapter will contain information beneficial not only for the design of power amplifiers but also for the design of other devices that require high-quality passives, such as LNAs and DC-to-DC converters.

Chapter 8 will introduce automation and intelligent design of various on- and off-chip power amplifier classes in step-by-step manner. The algorithms will be illustrated by means of flow charts and their development will be demonstrated in MATLAB, with various examples demonstrating the use of each procedure. With ideas for inductor design presented in Chap. 7, the automation of the quarter-wave transformer and impedance matching networks will be considered in Chap. 8. Finally, the ideas of both chapters will be merged to present the development of a fully functional power amplifier design program as a proof-of-concept to the readers.

Chapter 9 will be dedicated to practical power amplifier considerations for discrete, integrated, and hybrid power amplifier implementations. Packaging will be discussed in some detail for both system-on-chip (SOC) and system-on-package (SOP) architectures. Layout of integrated circuits will be reconsidered here with a few additional useful subroutines for rapid layout design. Finally, a suggestion on how to execute a practical design of the power amplifier will be shared with the readers.

Chapter 10 will cover future power amplifier directions. Topics in this chapter will include mm-wave and transmission line theory, as well an introduction to near terahertz (THz) transmissions. Other EDA opportunities will also be discussed. Finally, this chapter will include some concluding remarks.

References

1. Watkins G, Wang S. The impact of power amplifier turn-on characteristics in cognitive radio networks. Microwave J. 2014;57(2):86–92.
2. Raab FH, Asbeck P, Kenington PB, Cripps S, Popovic ZB, Pothecary N, Sevic JF, Sokal NO. RF and microwave power amplifier and transmitter technologies—part 1. High Freq Electron. 2003;2:22–36.
3. Fang J, Moreno J, Quaglia R, Camarchia V, Pirola M, Guerrieri SD, Ramella C, Ghione G. 3.5 GHz WiMAX GaN Doherty power amplifier with second harmonic tuning. Microwave Opt Technol Lett. 2012;54(11):2601–5.
4. Wagner R. 3G power ampliifiers: moving from GaAs to CMOS. Microwave J. 2012;55(4):44–8.
5. Khannur PB, Chen X, Yan DL, Shen D, Zhao B, Raja MK, Sindunata R, Yeoh GW, Singh R. Universal UHF RFID reader IC in 0.18-μm CMOS technology. IEEE J Solid-State Circuits. 2008;43(5):1146–55.
6. Kaynak M, Tekin I, Gurbuz Y. Fully integrated low-power SiGe power amplifier for biomedical applications. IET Microwaves Antennas Propag. 2011;5(2):214–9.
7. Kazimierczuk MK. RF power amplifiers. 2nd ed. Chiechester: Wiley; 2015.
8. Rogers J, Plett C. Radio frequency integrated circuit design. 2nd ed. Boston: Artech House; 2010.
9. Raab FH, Asbeck P, Cripps S, Kenington PB, Popovic ZB, Pothecary N, Sevic JF, Sokal NO. Power amplifiers and transmitters for RF and microwave. IEEE Trans Microw Theory Tech. 2002;50(3):814–26.
10. Barton TW, Perreault DJ. Four-way microstrip-based power combining for microwave outphasing power amplifiers. IEEE Trans Circuits Syst—I: Regul Pap. 2014;61(10):2987–98.
11. Naude N, Bozanic M, Sinha S. Anlogue direct sequence spread spectrum transceiver with carrier recovery employing complex spreading sequences. In: Malaga P, editor. IEEE mediterranean electrotechnical conference; 2006. pp. 1227–30.
12. Iguchi S, Saito A, Watanabe K, Sakurai T, Takamiya M. Design method of class-F power amplifier with output power of 20 dBm and efficient dual supply voltage transmitter. IEEE Trans Circuits Syst—I: Regul Pap. 2014;61(10):2978–86.
13. Lee ML, Liou CY, Tsai WT, Lou CY, Hsu HL, Mao SS. Fully monolithic BiCMOS reconfigurable power amplifier for multi-mode and multi-band applications. IEEE Trans Microw Theory Tech. 2015;63(2):614–24.
14. Szczepkowski G, Farrell R. Study of linearity and power consumption requirements of CMOS low noise amplifiers in context of LTE systems and beyond. ISRN Electron. 2014;2014:1–11.
15. Calvillo-Cortes DA, van der Heijden MP, Acar M, de Langen M, Wesson R, van Rijs F, de Vreede LCN. A package-integrated chireix outphasing RF switch-mode high-power amplifier. IEEE Trans Microw Theory Tech. 2013;61(10):3721–32.
16. Eswaran U, Ramiah H, Kanesan J, Reza AW. Class-E GaAs HBT power amplifier with passive linearization scheme for mobile wireless communications. Turkish J Electr Eng Comput Sci. 2014;22:1210–8.
17. Anastasijevic A, Coja D, Neskovic N, Neskovic A, Budimir D. Joint power amplifier and I/Q modulator impairments modelling and compensation for LTE transmitters using artificial neural networks. Int J Electron Commun (AEÜ). 2014;69(1):529–38.
18. Poulin D. The III-V vs silicon battle. Microwave J. 2009;52(4):22–38.
19. Avenier G, Diop M, Chevalier P, Troillard G, Loubet N, Bouvier J, Depoyan L. 0.13 m SiGe BiCMOS technology fully dedicated to mm-wave applications. IEEE J Solid-State Circuits. 2009;44(9):2312–21.
20. Nikandish G, Medi A. A design procedure for high-efficiency and compact-Size 5–10-W MMIC power amplifiers in GaAs pHEMT technology. IEEE Trans Microw Theory Tech. 2013;61(8):2922–33.

21. Sheppard ST, Doverspike K, Pribble WL, Allen ST, Palmour JW, Kehias LT, Jenkins TJ. High-power microwave GaN/AlGaN HEMTs on semi-insulating silicon carbide substrates. IEEE Electron Device Lett. 1999;20(4):161–3.
22. Nguyen C, Micovic M. The State-of-the-art of GaAs and InP power devices and amplifiers. IEEE Trans Electron Device. 2001;48(3):472–8.
23. Pribble WL, Palmour W, Sheppard ST, Smith RP, Allen ST, Smith TJ, Ring Z, Sumakeris JJ, Saxler AW, Milligan JW. Applications of SiC MESFETs and GaN HEMTs in power amplifier design. In: 2002 IEEE MTT-S international microwave symposium digest. Seattle; 2002. pp. 1819–22.
24. Nellis K, Zampardi P. A comparison of linear handset power amplifier in different bipolar technologies. IEEE J Solid-State Circuits. 2004;39(10):1746–54.
25. Chiou HK, Liao HY, Chen CC, Wang SM, Chen CC. A 2.6-GHz fully integrated CMOS power amplifier using power combining transformer. Microwave Opt Technol Lett. 2010;52 (2):299–302.
26. Pierco R, Torfs G, De Keulenaer T, Vandecasteele B, Missinne J, Bauwelinck J. A Ka-band SiGe BiCMOS power amplifier With 24 dBm output power. Microwave Opt Technol Lett. 2015;57(3):718–22.
27. Johnson EO. Physical limitations on frequency and power parameters of transistors. RCA Rev. 1965;26:163–77.
28. Baliga BJ. Power semiconductor device figure of merit for high-frequency applications. Electron Device Lett. 1989;10(10):455–7.
29. Lee TH. The design of CMOS radio-frequency integrated circuits. 2nd ed. Cambridge: Cambridge University Press; 2004.
30. Joo S, Han S, Song M, Kim H, Ham J, Seo M, Kim M, Park C, Yang Y. 1-Watt highly linear HBT power amplifier driver integrated circuit with low current consumption. Microwave Opt Technol Lett. 2013;55(7):1680–3.
31. Grebennikov A. High-frequency class-FE tuned power amplifiers. IEEE Trans Circuits Syst-I: Regul Pap. 2008;55(10):3284–92.
32. Stameroff AN, Ta HH, Pham AV, Leoni RE. Wide-bandwidth power-combining and inverse class-F GaN power amplifier at X-band. IEEE Trans Microw Theory Tech. 2013;61(3):1291–300.
33. Raab FH. Class-E, class-C, and class-F power amplifiers based upon a finite number of harmonics. IEEE Trans Microw Theory Tech. 2001;49(12):1462–8.
34. Eccleston KW, Smith KJI, Gough PT. A compact class-F/class-C Doherty amplifier. Microwave Opt Technol Lett. 2011;53(7):1606–10.
35. Aflaki P, Negra R, Ghannouchi FM. Enhanced architecture for microwave currentmode class-D amplifiers applied to the design of an S-band GaN-based power amplifier. IET Microwave Antennas Prop. 2009;3(6):997–1006.
36. Grebennikov A, Sokal NO. Switchmode RF power amplifiers. 1st ed. Burlington: Elsevier; 2007.
37. Van JH, Kim MS, Jung SC, Park HC, Ahn G, Park CS, Kim BS, Yang Y. A high frequency and high power quasi-class-E amplifier design using a finite bias feed inductor. Microwave Opt Technol Lett. 2007;49(5):1114–8.
38. Yang LY, Chen HS, Chen YJE. A 2.4 Ghz fully integrated cascode-cascade CMOS Doherty power amplifier. IEEE Microwave Wirel Compon Lett. 2008;18(3):197–9.
39. Thian M, Fusco VF. Transmission-line class-E power amplifier with extended maximum operating frequency. IEEE Trans Circuits Syst—II: Express Briefs. 2011;58(4):195–9.
40. Dongyue J, Wanrong Z, Pei S, Hongyun X, Yang W, Wei Z, Lijian H, Yongping S, Jia L, Junning G. Multi-finger power SiGe HBTs for thermal stability enhancement over a wide biasing range. Solid-State Electron. 2008;52(6):937–40.
41. Aoki I, Kee S, Magoon R, Aparicio R, Bohn F, Zachan J, Hatcher G, McClymont D, Hajimiri A. A fully-integrated quad-band GSM/GPRS CMOS power amplifier. IEEE J Solid-State Circuits. 2008;43(12):2747–58.

42. Brama R, Larcher L, Mazzanti A, Svelto F. A 30.5 dBm 48 % PAE CMOS class-E PA with integrated balun for RF applications. IEEE J Solid-State Circuits. 2008;43(8):1755–62.
43. Gaw C, Arnold T, Martin R, Zupac D. Evaluation of SiGe: C HBT intrinsic reliability using conventional and step stress methodologies. Microelectron Reliab. 2006;46(8):1272–8.
44. Chen A, Lo HY. Semiconductor packaging: materials interaction and reliability. 1st ed. Boca Ranton: CRC Press; 2012.
45. Ludwig Bretchko. RF circuit design: theory and applications. 1st ed. Upper Saddle River: Prentice Hall; 2000.
46. Chen FY, Chen JF, Lin RL. Low-harmonic push-pull class-E power amplifier with a pair of LC resonant networks. IEEE Trans Circuits Systems-I: Regul Pap. 2007;54(3):579–89.
47. Kim D, Choi D, Moon Y, Baek D, Baek K, Choi Y. The design and realization of a highly linear power amplifier module for a WiMAX/WiBro (802.16e) base station. Microwave Opt Technol Lett. 2010;52(9):1952–5.
48. Jiang H, Wilford A. Digital predistortion for power amplifiers using separable functions. IEEE Trans Signal Process. 2010;58(8):4121–30.
49. Walling S, Allstot DJ. Linearizing CMOS switching power amplifiers using supply regulators. IEEE Trans Circuits Systems II: Express Briefs. 2010;57(7):497–501.
50. Nam J, Kim B. The Doherty power amplifier with on-chip dynamic bias control circuit for handset application. IEEE Trans Microw Theory Tech. 2007;55(4):633–42.
51. An KH, Lee O, Kim H, Lee DH, Han J, Yang KS, Kim Y. Power-combining transformer techniques for fully-integrated CMOS power amplifiers. IEEE J Solid-State Circuits. 2008;43 (5):1064–75.
52. Zhang R, Acar M, van der Heijden MP, Apostolidou M, Leenaerts DMW. Generalized semi-analytical design methodology of class-E outphasing power amplifier. IEEE Trans Circuits Syst-I: Regul Pap. 2014;61(10):2951–60.
53. Keerti A, Pham AVH. RF characterization of SiGe HBT power amplifiers under load mismatch. IEEE Trans Microw Theory Tech. 2007;55(2):207–14.
54. Komijani A, Natarajan A, Hajimiri A. A 24-GHz, + 14.5-dBm fully integrated power amplifier in 0.18-μm CMOS. IEEE J Solid-State Circuits. 2005;40(9):1901–8.
55. Eroglu A. Stabilization of class E amplifiers with a diode network. AEU-Int J Electron Commun. 2010;64(3):224–30.
56. Roberts GW, Sedra AS. SPICE. 2nd ed. Oxford: Oxford University Press; 1997.
57. Navabi Z. VHDL: Analysis and Modeling of Digital Systems. 2nd ed. Boston: McGraw-Hill; 1998.
58. Granberg T. Handbook of digital techniques for high-speed design. 1st ed. Upper Saddle River: Prentice Hall; 2004.
59. Attaway S. Matlab: a practical introduction to programming and problem solving. 3rd ed. Oxford: Butterworth-Heinemann; 2013.

Part I
Power Amplifier Theory

Chapter 2
Review of Telecommunication Aspects for Power Amplifier Design

In this chapter RF and microwave telecommunication theory, together with semiconductor and substrate theory, as applicable to power amplifier design, is presented to complement the basic theory of power amplifiers presented in Chap. 1. Several topics are addressed.

The chapter starts with a review of the frequency spectrum and various transmission bands and their implications for transceiver system design. The feasibility of passive component implementations in each frequency range is investigated and power amplifiers are placed into the context of the transceiver system. A review of various digital modulation schemes, commonly used with power amplifiers, is also presented.

Secondly, the theory behind transistor operation under large-signal and high power is discussed. Different transistor types are discussed and their advantages and disadvantages compared to other types are listed. In addition, various semiconductor fabrication technologies are compared for complete system integration or just for active device (transistor) integration. This is complemented by a discussion of different substrates for discrete implementations, PCBs and packaging. Heating presents a typical problem when large amounts of power are dissipated, thus temperature aspects of transistors and semiconductor materials are also discussed.

Last part of this chapter focuses on the scattering (*S*-) parameters and admittance (*Y*-) parameters review, Smith charts and some other aspects of RF and microwave engineering, such as resonance, loaded quality factor, bandwidth, insertion loss and impedance transformation. Fourier analysis of periodic signals is also discussed.

2.1 Wavelength and Transmission Bands

Frequency of operation has a major influence on the behavior of passive and active devices. Below 30 GHz, transceivers constructed by lumped elements can be more compact than designs based on transmission lines. Above 30 GHz, transceivers and their elements require accurate modeling and high-precision manufacturing. Above 60 GHz, transmission lines and waveguides are more practical. As described in Chap. 1, this book presents a design methodology for power amplifiers up to

M. Božanić and S. Sinha, *Power Amplifiers for the S-, C-, X- and Ku-bands*,
Signals and Communication Technology, DOI 10.1007/978-3-319-28376-0_2

Ku-band, the top end of which is located at 18 GHz [1–3]. Thus, the classification of the different bands in the frequency spectrum is beneficial for power amplifier design and is included in this section. Frequency bands are defined by the international telecommunication union [4].

The frequency is related to wavelength according to relation

$$\lambda = \frac{v}{f}, \tag{2.1}$$

where v is the phase speed of the wave and f is the wave frequency. The phase speed of the electromagnetic wave is the speed of light, which is about 3×10^8 m/s. At lower frequencies, the wavelengths of signals are very large, so the size of the electrical components has little impact on these signals. At 2.4 GHz, the wavelength is 12.5 cm. This means that any component or a connection should not be greater than a tenth of the wavelength (12.5 mm) for a system to behave with minimal loss. This can still be accomplished on a PCB. At 18 GHz, the wavelength is 1.7 cm and transmission lines can only be avoided on chip without incurring mismatches due to connections longer than about 1/10 of the wavelength. In all other cases, careful matching is paramount.

The extremely low frequency (ELF), voice frequency (VF) and very-low-frequency (VLF) ranges span from 30 to 30 kHz and contain audible frequencies and are thus not suitable for radio transmission. Low frequencies (LF) span from 30 to 300 kHz and are used for long-range navigation, submarine communication and telegraphy. Medium frequencies (MF) or medium waves span from 300 to 3 MHz and are used for commercial radio. The high-frequency (HF) range with frequencies from 3 to 30 MHz is used for military tactical radios and for amateur radio operators because of the long-distance propagation properties of the waves with 30-m-long waves.

The very-high-frequency (VHF) range with frequencies from 30 to 300 MHz and the UHF range with frequencies from 300 to 3 GHz are used for television broadcast, cordless and cellular telephone transmission, as well as for other wireless applications, such as wireless local area networks (WLANs) and Bluetooth®. This is also suitable for industrial heating and microwave ovens.

The SHF range includes frequencies from 3 to 30 GHz and the extremely-high-frequency (EHF) range includes frequencies from 30 to 300 GHz. These two ranges are mostly used for satellite communication and radar applications.

UHF, SHF and EHF frequency ranges are further divided into *L*-band (1–2 GHz), *S*-band (2–4 GHz), *C*-band (4–8 GHz), *X*-band (8–12.4 GHz), *K*-band (18–26.5 GHz), *Ku*-band (26.5–40 GHz), *V*-band (40–75 GHz) and *W*-band (75–110 GHz). Informally, the spectrum is also divided into RF, microwave and mm-waves, with the boundaries between the three bands somewhat loosely defined. The frequency spectrum is illustrated in Table 2.1. This table also shows the feasibility of passives as applicable to each frequency range.

Table 2.1 The frequency spectrum and feasibility of passives

Frequency range	Wavelength range	Range name	UHF/SHF/EHF band name	Other names/feasibility of passives
30–300 Hz	10,000–1000 km	ELF	–	No transmission
300–3000 Hz	1000–100 km	VF		
3–30 kHz	100–10 km	VLF		RF/lumped passives
30–300 kHz	10–1 km	LF		
300–3000 kHz	1000–100 m	MF		
3–30 MHz	100–10 m	HF		
30–300 MHz	10–1 m	VHF		
300–1000 MHz	100–30 cm	UHF		
1–2 GHz	30–15 cm		*L*-band	
2–3 GHz	15–10 cm		*S*-band	
3–4 GHz	10–7.5 cm	SHF		Microwave/lumped passives, transmission lines off-chip
4–8 GHz	7.5–3.75 cm		*C*-band	
8–12.4 GHz	37.5–24 mm		*X*-band	
12.4–18 GHz	24–17 mm		*Ku*-band	
18–26.5 GHz	17–1.1 mm		*K*-band	
26.5–30 GHz	1.1–1 mm		*Ka*-band	
30–40 GHz	<1 mm	EHF		mm-wave/transmission lines on and off chip
40–75 GHz			*V*-band	
75–110 GHz			*W*-band	
110–300 GHz			–	

2.2 Review of Modulation Schemes

As discussed in Chap. 1, the power amplifier is driven by a signal that is generated by a modulation block. Modulation implies that properties of the carrier signal, that is the signal that can physically be amplified and transmitted, are varied so that the information of interest is superimposed [5, 6]. Modulation is accomplished by means of a modulator. On the receiver side, a demodulator is used to recover the same information.

Modern telecommunication systems are moving from employing analog modulation towards employing digital modulation. Thus, various digital modulation schemes are discussed in this review. In a digital modulation scheme, the carrier signal is modulated by a discrete signal. Digital modulation is therefore a type of analog-to-digital conversion. Some basic digital modulation techniques, already defined in Chap. 1, are redefined here for readers' convenience:

- Phase-shift keying (PSK),
- Frequency-shift keying (FSK),
- Phase-amplitude modulation (PAM),

- Quadrature amplitude modulation (QAM),
- On-off keying (OOK),
- Orthogonal frequency-division multiplexing (OFDM), and
- Various direct-sequence spread spectrum (DSSS) techniques.

2.2.1 *Phase Shift-Keying*

PSK is accomplished by modulating the digital information signal onto the carrier signal by changing its phase. A finite number of phases (*M*) is used, usually two for bits 0 and 1 (binary PSK or BPSK), four for bit combinations 00, 01, 10 and 11 (QPSK) or eight for eight 3-bit combinations (octal PSK). A typical waveform of BPSK is shown in Fig. 2.1, and signal space diagrams (constellations) of BPSK, QPSK and octal PSK are shown in Fig. 2.2.

2.2.2 *Frequency Shift-Keying*

FSK is accomplished by modulating the digital information signal onto the carrier signal by changing its frequency. Usually, two discrete frequencies are used to represent zeros and ones of a binary digital signal. This concept is illustrated in Fig. 2.3.

2.2.3 *Phase-Amplitude Modulation*

In a PAM system, the message is encoded as the amplitude in series of pulses. The number of amplitude levels required is $M = 2^k$ for k-bit blocks of symbols. As in PSK, usually one-bit, two-bit or three-bit symbol combinations are used, corresponding to M = 2, 4 and 8 respectively. This concept is illustrated in Fig. 2.4.

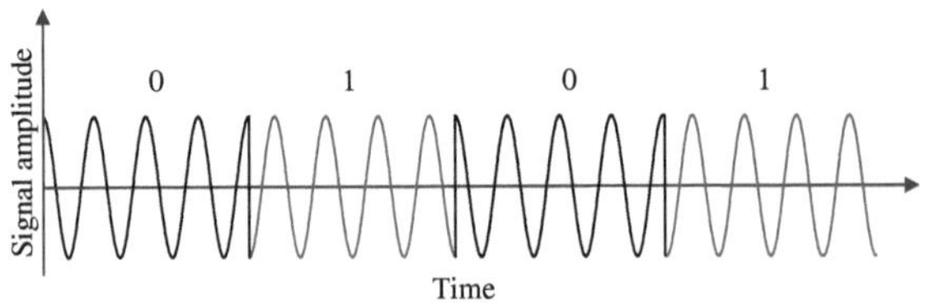

Fig. 2.1 The BPSK waveform

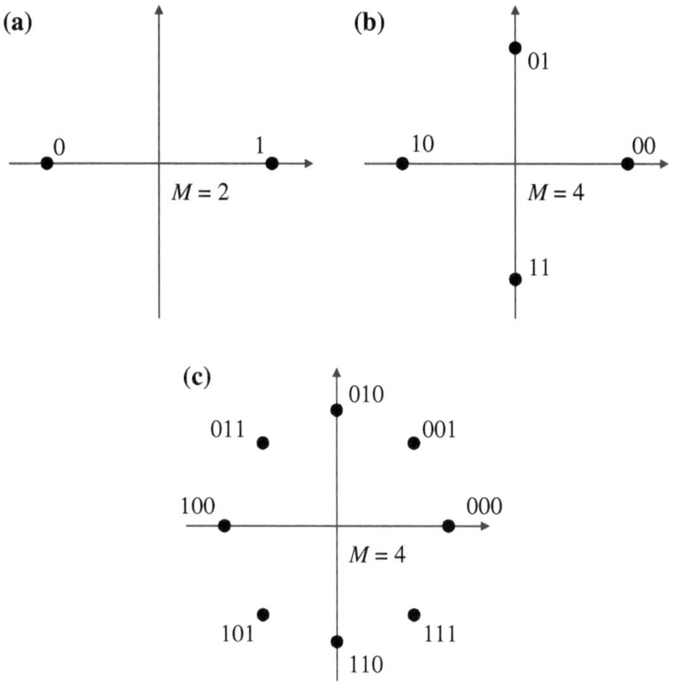

Fig. 2.2 Signal constellations of **a** BPSK, **b** QPSK and **c** octal PSK

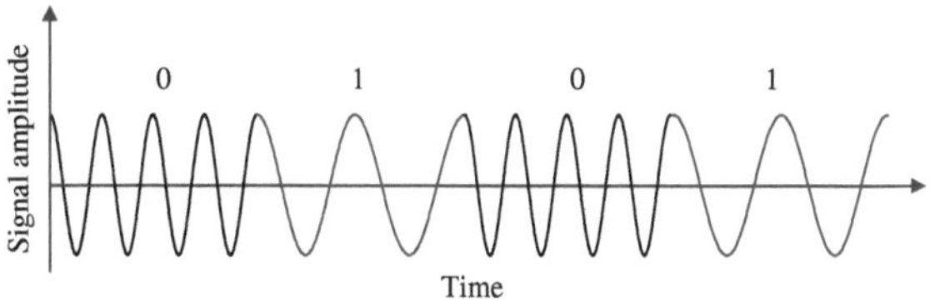

Fig. 2.3 Binary FSK waveform

Fig. 2.4 The PAM signal constellations for **a** $M = 2$, **b** $M = 4$ and **c** $M = 8$

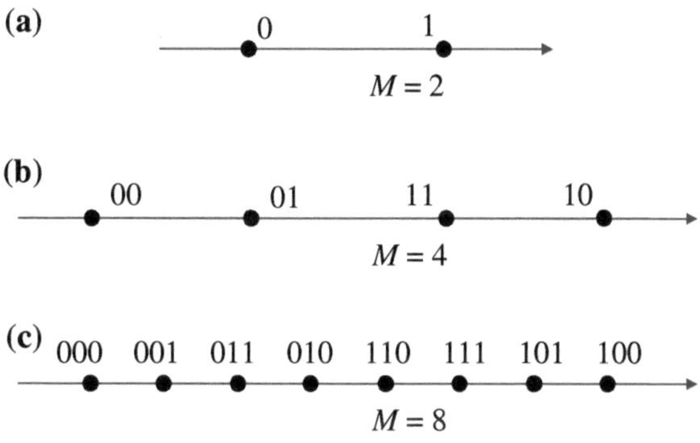

2.2.4 *Quadrature Amplitude Modulation*

QAM is a technique for bandwidth efficiency improvement, where two or more separately modulated signals are combined on the carriers that are out of phase. Thus it is a combination of a PSK scheme with another scheme, usually PAM. When combining M_1-level PAM with M_2-phase PSK, $M = M_1M_2$ combined signal constellation can be created, as illustrated in Fig. 2.5.

2.2.5 *On-Off Keying*

OOK is the simplest modulation technique. In this scheme, if the carrier signal is present, it indicates a digital one, and if the carrier signal is absent, it indicates a digital zero, as illustrated in Fig. 2.6.

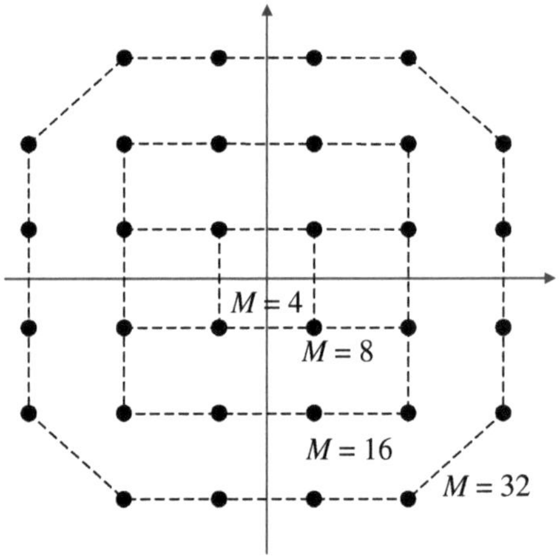

Fig. 2.5 QAM constellations for different values of M

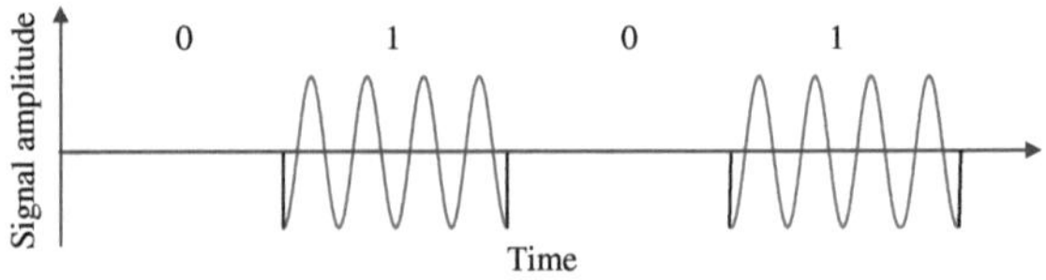

Fig. 2.6 OOK waveform

2.2.6 Orthogonal Frequency-Division Multiplexing

OFDM is a technique used in cases where it is more feasible to transmit data over a large number of carriers simultaneously rather than using a single carrier with a high data rate. This scheme is practical for high data transmission, such as for high-definition television or LTE networks. OFDM combines carriers with the same amplitude and modulation scheme, but separated in frequency so that modulation products arising from one frequency are negligible at the frequencies of the other carriers.

2.3 Antennas and Propagation

After modulation and power amplification, the amplified signal is passed onto an antenna in order to be transmitted. Antennas are therefore an important part of every transmitter and used to radiate the electromagnetic energy into the channel effectively [7]. On the receiving side, antennas are used for receiving the electromagnetic energy from the channel.

Each antenna has characteristic input impedance, which is usually designed to be 50 Ω. For power amplifier design, the antenna efficiency is also important, which is the ratio of radiated power to the power fed to the antenna [2]:

$$\eta_A = \frac{P_{RAD}}{P_{FED}}. \tag{2.2}$$

An antenna also has its radiation characteristics, which are mostly determined by its length and the way in which it is excited. The principle of antenna operation is based on the Ampere-Maxwell's law:

$$\Delta \times \mathbf{H} = \mathbf{J} + \frac{\partial \mathbf{D}}{\partial t}, \tag{2.3}$$

where $\frac{\partial \mathbf{D}}{\partial t}$ is the displacement current, $\mathbf{J}(t)$ is the time varying current density and $\mathbf{H}(t)$ is the time varying magnetic field around the antenna.

The power density at the distance r from the antenna is

$$p(r) = G_T \frac{P_T}{4\pi r^2}, \tag{2.4}$$

where G_T is the gain of the antenna in the particular direction and P_T is the transmitted power. From this equation, it is clear that the power density decreases quadratically with the distance and that high gains are needed to transmit over long distances. The amount of power received by the antenna on the receiver side with gain G_R is given by the Friis formula

$$P_R = \frac{P_T G_T G_R \lambda^2}{(4\pi r)^2}. \tag{2.5}$$

Various types of antennas are commonly used [3]:

- Wire antennas (dipoles, monopoles and others) with low gains, used mostly at HF to UHF,
- Aperture antennas (open-ended waveguides, horns, reflectors) with moderate to high gains, used in microwave or mm-wave bands, and
- Printed antennas on various substrates (slots, dipoles or microstrip) with high gains and also used in microwave or mm-wave bands.

Several antennas can be combined into antenna arrays in order to obtain more directivity and other desirable properties.

The length of an antenna becomes very important when considering packaging, especially when talking about SOP solutions. If an antenna is placed in air, the length of the antenna is proportional to the wavelength. However, if the antenna can be printed in a package, then the length of the antenna scales according to the following equation [8]:

$$l = \frac{\lambda}{\sqrt{\mu_r \varepsilon_r}}, \tag{2.6}$$

where μ_r and ε_r are the relative permeability and permittivity of the substrate on which the antenna is printed. Thus different substrates allow for antennas with different lengths to be placed on SOP which increases the range of options.

For transmission above 30 MHz, propagation is possible only by line-of sight waves. There are two types of these waves: direct waves and ground-reflected waves. Most of the transmissions above HF are accomplished by direct waves. Propagation of waves is influenced by different types of losses, including space loss, atmospheric loss, polarization mismatch loss, impedance mismatch loss and pointing loss.

2.4 The Power Transistor

This section introduces different semiconductor technologies, temperature concerns due to transistor operation and transistor types, and also introduces transistor modeling for power amplification.

2.4.1 Semiconductor Technologies for Transistor Fabrication

Traditionally, several device technologies have been used for fabricating transistors that can be used for power amplification [6, 9, 10]. Initially, gated structures such as MOSFETs, metal-semiconductor field-effect transistors (MESFETs), HEMTs and pseudomorphic HEMTs (pHEMTs) found widespread use. After the introduction of bipolar transistor with a wide-gap emitter, or HBT [11, 12], bipolar transistors emerged as a preferred choice, because of their higher gain and current densities at RF. In the late twentieth century, most of the on-chip power amplifier implementations were deployed with HBTs. Laterally diffused metal oxide semiconductor (LDMOS) transistors are also used for applications below 2 GHz [13]. Figure 2.7 illustrates the relation between different transistors.

In the case of integrated transmitters, this resulted in systems that included at least two ICs in their implementation: a silicon CMOS-based front end and a power amplifier fabricated in another technology. Bipolar complementary metal oxide (BiCMOS) processes are emerging as an alternative to the two-chip integrated solution. This is because it is possible to fabricate both MOSFETs and HBTs in a BiCMOS process, thus allowing BiCMOS technology to bridge the integration gap and reduce the cost of integrated transmitter manufacturing. However, as superior MOSFET technologies emerge, integration in pure CMOS is also becoming practical.

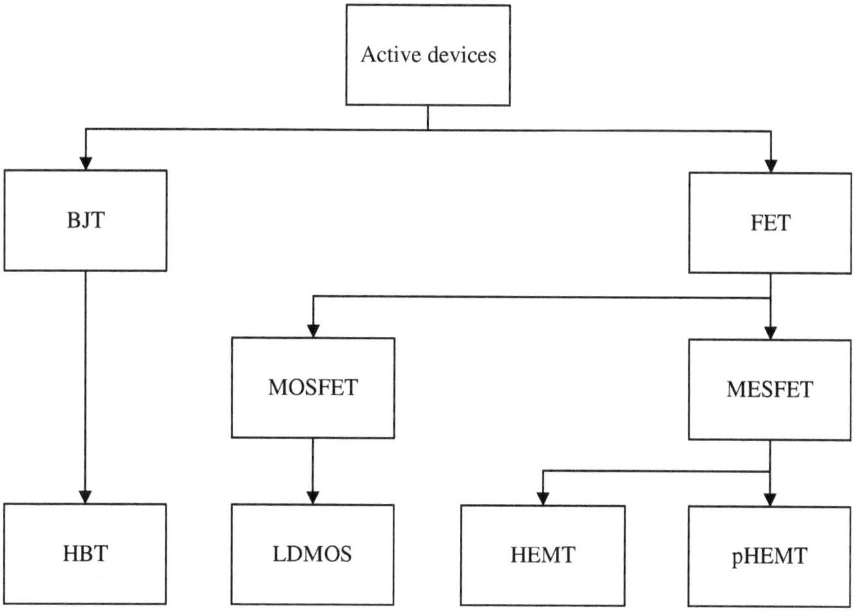

Fig. 2.7 Types of active devices

Table 2.2 Process parameters for Si BJT, SiGe HBT and GaAs HBT technologies [19]

Parameter	Si BJT	SiGe HBT	GaAs HBT
f_T (GHz)	27	44	46
Forward gain, β	100	200	120
Base-emitter voltage, V_{be} (V)	0.8	0.8	1.33
Early voltage, V_A (V)	36	100	1223
Collector-emitter breakdown voltage, V_{ce} (V)	6.2	6	14.3
Collector-base breakdown voltage, V_{cb} (V)	20	12	26
Emitter-base breakdown voltage, V_{eb} (V)	2	5	6.9
Power density (mW/μm$^{2)}$	2	2	0.9
Thermal conductivity (W/cm-°C)	1.5	1.5	0.49
Base-emitter capacitance, C_{be} (fF)	11	10	2.4
Base-collector capacitance, C_{bc} (fF)	3.6	3.3	1
Possibility of NMOS and PMOS integration	Yes	Yes	No

Technologies such as gallium-nitride (GaN), aluminum-gallium-nitride (AlGaN), silicon-carbide (SiC), and indium-phosphate have been struggling for survival in the battle between Si, SiGe and GaAs technologies [9, 14–20]. The AlGaN/GaN is suitable for both high-frequency and high-power designs, but together with the SiC technology it has been sidetracked on the road to microwave applications by technical difficulties and high fabrication costs. The indium-phosphate technology itself does not provide any substantial benefits for the linear power amplifier design.

Table 2.2 shows some typical process parameters for some of the older Si BJT, SiGe and GaAs processes for technology comparison purposes. Silicon technology, as expected, offers the worst process parameters compared to the two HBT technologies. Similar observation is possible by analyzing the PAE, power gain and *S*-parameters obtained from the performance measurement study summarized in Table 2.3.

Emerging technologies can be compared by introducing various figures of merit [2]. Johnson's figure of merit is calculated as

$$JFOM = \frac{E_{BD} v_{sat}}{2\pi}, \tag{2.7}$$

and Baliga's figure of merit is defined as

Table 2.3 Various performance parameters for PAs fabricated in three different technologies ($f = 1.88$ GHz, $V_{CC} = 3.4$ V and $P_{OUT} = 28$ dBm) [19]

Parameter	Si BJT	SiGe HBT	GaAs HBT
PAE (%)	33.1	35	39.3
Gain (dB)	22.1	21.8	27.1
S_{11} (dB)	−13.8	−16.9	−12.6
S_{12} (dB)	20.2	20.5	27.2

Table 2.4 Properties of Si, SiC, GaN and GaAs semiconductors [2]

Property	Unit	Si	SiC	GaN	GaAs
Bandgap energy, E_G	eV	1.12	3.26	3.42	1.42
Electron mobility, μ_n	cm^2/V-s	1360	900	2000	8500
Hole mobility, μ_p	cm^2/V-s	480	120	300	400
Breakdown electric field, E_{BD}	V/cm	2×10^5	2.2×10^6	3.5×10^6	4×10^5
Saturation electron drift velocity, v_{sat}	cm/s	10^7	2.7×10^7	2.5×10^7	1.2×10^7
Relative dielectric constant, ε_r	–	11.7	9.7	9	12.9
Thermal conductivity, k_{th}	W/K-cm	1.5	4.56	1.3	0.55

$$BFOM = \varepsilon_r \varepsilon_0 \mu_n E_G^3, \tag{2.8}$$

where the parameters are defined in Table 2.4 and $\varepsilon_0 = 8.85 \times 10^{-12}$ F/m is the absolute permittivity. Table 2.4 also lists values of these parameters for Si, SiC, GaN and GaAs technologies. These values have been used to form a comparison of JFOM and BFOM of each technology, as shown in Fig. 2.8.

2.4.2 *Temperature Aspects of Transistors*

As mentioned in Chap. 1, excess power dissipated in a transistor due to low efficiency is converted to heat, which decreases the performance of an amplifier [21]. The amount of heat generated and the way in which that heat is dissipated in a power amplifier system depend on the technology in which the IC is fabricated or type of substrate used for discrete implementation, as well as on the type of active device (transistor) used for power amplification.

A MOS transistor is less prone to heating than a BJT (HBT) because of its negative temperature coefficient [6, 21]. This negative coefficient causes the drain current of the MOSFET to decrease with temperature, which allows multiple MOSFETs to be connected in parallel. In HBTs, the situation is reversed. If multiple emitter fingers are used and are not perfectly balanced, the emitter with higher current will tend to take up an even higher proportion of current, which gives rise to local hot spots and possibly permanent device damage. Because current is flowing only in a small hot region, the total gain of the device decreases. Normally, to compensate for this current gain decrease, ballast resistors are added in series with the emitter or base electrodes. This acts as negative feedback, since with the increase in current, the voltage across ballast resistors will also increase,

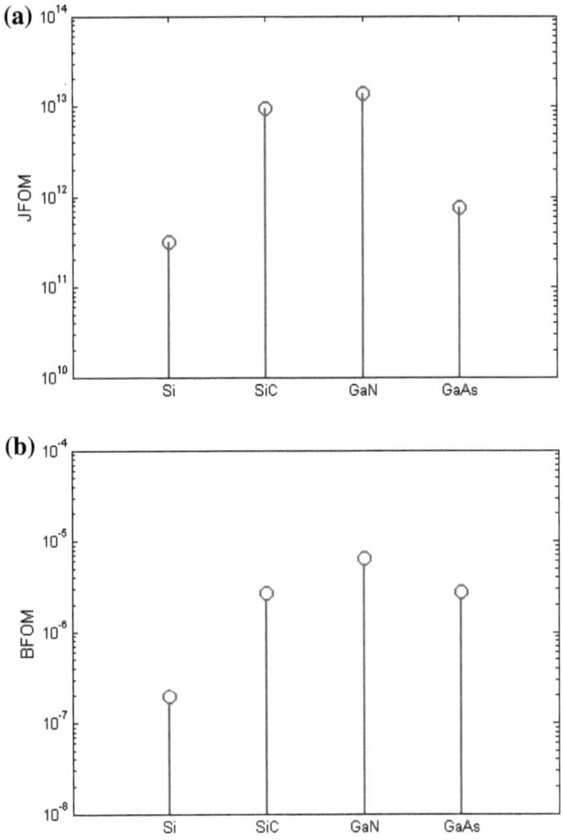

Fig. 2.8 Comparison of **a** JFOM and **b** BFOM of Si, SiC, GaN and GaAs semiconductors [2]

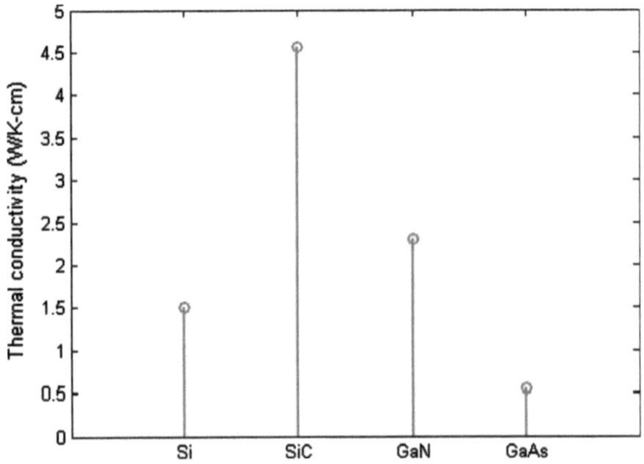

Fig. 2.9 Thermal conductivity of various semiconductor materials

counter-effecting the current increase. Including ballasting resistors, however, decreases the output power, power gain and the efficiency of the amplifier.

Thermal conductivity of various semiconductor materials, listed in Table 2.4, is plotted in Fig. 2.9.

2.4.3 *Transistor Models and Large-Signal Transistor Operation*

Transistor models for a particular technology are usually made available for simulation purposes through a process design kit provided by the foundry. When designing with BJTs, traditionally the Gummel-Poon model was the most popular model for the design of bipolar circuits for a considerable time [22]. Other RF models, such as the Vertical Bipolar Inter-Company (known as VBIC) model for the HBT [23], offer several improvements over the Gummel-Poon model.

For the purpose of this book, however, it is sufficient to use a very simple model of the transistor, with or without specific parasitics as applicable for a particular power amplifier analysis. Throughout this book, in the case of continuous-mode power amplifiers, the transistor is modeled as a dependent current source. In the case of the switch-mode power amplifier, the transistor can be modeled as a simple switch with series resistance (and sufficient gain). When dealing with MOSFETs, the amount of amplification can be set by choosing the transistor width (W) and length (L) parameters. In the case of HBTs, on the other hand, transistor forward gain β is fixed for a chosen transistor, thus an appropriate transistor needs to be selected carefully.

In this section the model and operation of power MOSFET is reviewed, which will be sufficient to analyze any power amplifier stage mentioned later in this book.

2.4.3.1 The Power MOSFET

A high-frequency small signal model of the power MOSFET operating in the pinch-off (saturation) region is shown in Fig. 2.10. For low drain currents, the n-type enhancement-type MOSFET (NMOS) with a long channel and for $v_{GS} \geq V_t$, where v_{GS} is the gate-to-source voltage and V_t is the threshold voltage, is given by a well-known square law:

$$i_D = K(v_{GS} - V_t)^2(1 + \lambda v_{DS}), \tag{2.9}$$

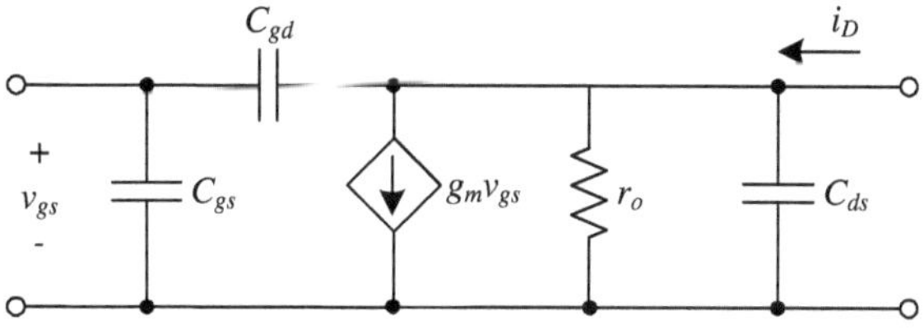

Fig. 2.10 High-frequency model of the high-power MOSFET

where

$$K = \frac{1}{2}\frac{W}{L}\mu_n C_{ox}, \tag{2.10}$$

λ is the channel-length modulation (not to be confused with wavelength) and C_{ox} is the oxide capacitance per unit area of the gate. C_{ox} is defined as

$$C_{ox} = \frac{\varepsilon_{ox}}{t_{ox}}, \tag{2.11}$$

where ε_{ox} is the oxide permittivity and t_{ox} is the oxide thickness.

Channel-length modulation arises because the increase in drain-to-source voltage v_{DS} beyond saturation voltage v_{DSsat} causes the effective channel length to shorten slightly. The relative change is λv_{DS}, as seen in Eq. (2.8). Parameter λ is defined as

$$\lambda = \frac{1}{V_A}, \tag{2.12}$$

where V_A is the MOSFET parameter often referred to as the Early voltage, for the purpose of comparison with BJTs. Channel-length modulation also causes resistance of r_o to appear in the small-signal model shown in Fig. 2.10. If I_D is the DC portion of the drain current, r_o is

$$r_o = \frac{1}{\lambda I_D} = \frac{V_A}{I_D}. \tag{2.13}$$

Transconductance gain g_m, also shown in the small-signal model, is the gain parameter of the transistor applicable to small-signal levels only. It is defined as the derivative of the drain current with respect to gate-to-source voltage, at the point where the transistor is biased (V_{GS}). Thus,

$$g_m = \left.\frac{di_D}{dv_{GS}}\right|_{v_{GS}=V_{GS}} = 2K(V_{GS} - V_t) = 2\sqrt{KI_D}. \tag{2.14}$$

For high drain currents (common in power amplification), NMOS operates according to the linear law, which can be approximated as

$$i_D = K_S(v_{GS} - V_t)(1 + \lambda v_{DS}), \tag{2.15}$$

where

$$K_S = \frac{1}{2}\mu_n C_{ox} W v_{sat}. \tag{2.16}$$

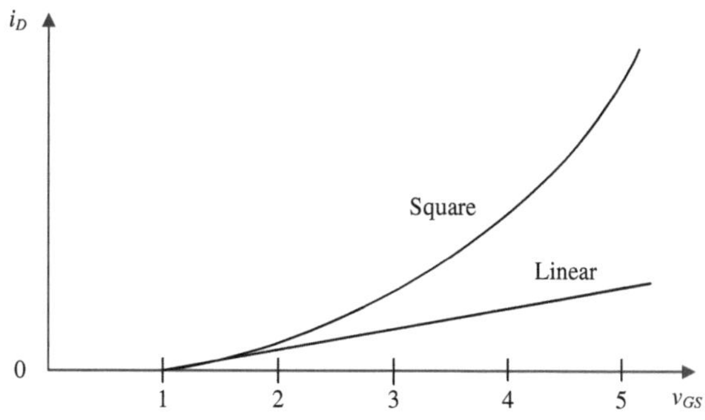

Fig. 2.11 MOSFET linear and square law for V_t = 1 V

The relationship between the square law and the linear is illustrated in Fig. 2.11.

At high frequencies, the effects of the capacitors shown in the model in Fig. 2.10 on the transistor gain become prominent. Current through the gate at high frequencies can be expressed as

$$i_G = j\omega(C_{gs} + C_{gd})v_{GS}, \tag{2.17}$$

where $\omega = 2\pi f$, f is the operating frequency and $j = \sqrt{-1}$. The current gain is then the ratio of the input and output of the current:

$$A(j\omega) = \frac{i_D}{i_G} = \frac{g_m v_{GS}}{j\omega(C_{gs} + C_{gd})v_{GS}} = \frac{g_m}{j\omega(C_{gs} + C_{gd})}. \tag{2.18}$$

An important parameter of any transistor for high frequencies is the unity (0 dB) gain frequency (f_T). It is the frequency at which the magnitude of the short-circuit gain ($|A(j\omega)|$) reaches one (thus making the transistor act as an attenuator instead of an amplifier). For MOSFET, it can be evaluated from

$$\frac{g_m}{2\pi f_T(C_{gs} + C_{gd})} = 1. \tag{2.19}$$

Thus

$$f_T = \frac{g_m}{2\pi(C_{gs} + C_{gd})} = \frac{\sqrt{KI_D}}{\pi(C_{gs} + C_{gd})}. \tag{2.20}$$

The concept of f_T is illustrated in Fig. 2.12.

It is worth noting here that with scaling of technology nodes, transistors with increasing f_T values can be fabricated at the an expense of other parameters, most obviously, the breakdown voltage. This allows for transceiver operation at frequencies much higher than previously possible.

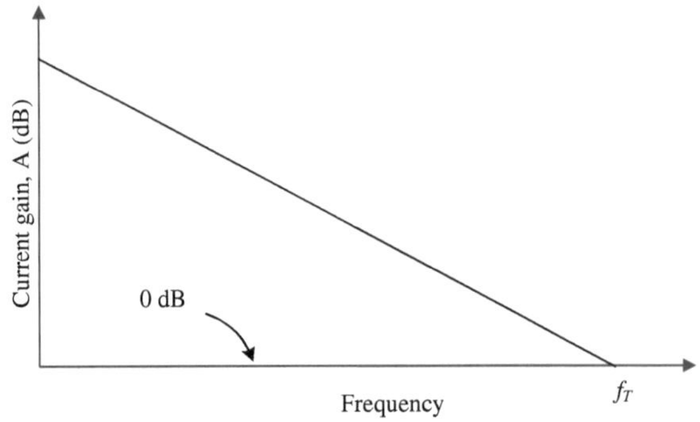

Fig. 2.12 Current gain versus frequency curve of the transistor illustrating the concept of f_T

2.5 Substrates for Discrete Implementations

In integrated designs, substrate refers to a material on which the transistors and other components are deposited, such as Si or SiC, described in the previous section. For discrete power amplifier design, it is important to mention substrates for discrete implementations. When referring to substrates in discrete implementations, one could be referring to [1, 8, 24]:

- Substrates on which discrete components are fabricated (e.g. inductors, capacitors, resistors),
- Substrates for PCBs on which the transmission lines are designed, and
- Substrates for packaged systems and subsystems (SOP).

For discrete components, substrates are chosen so that they exhibit very little parasitic resistance or reactance. Thin-film resistors, for example, are typically produced on alumina or beryllia substrates because they entail a small amount of reactance.

The PCB, which is normally used to support connections of the electronic components mechanically, also features substrates that are non-conductive. Passive components can be laid out on the PCB in the form of microstrip transmission lines. Similar materials are used as the foundation for packaged systems and subsystems. Ceramics and laminates are usually a good option for these PCB-like structures. Different types of ceramic materials, such as high-temperature co-fired alumina or aluminum-nitride with better thermal properties, are available. Hybrid structures, such as laminates that are made of bonded layers, normally with a FR4 material core, may be needed to meet stringent RF requirements.

In SOP, it is also important for the substrate to provide excellent high-frequency electrical properties. It should have good mechanical and chemical resistance and thin-film multilayer capabilities, however, while remaining cost-competitive. Substrate technologies that are able to satisfy these requirements fall under the

Table 2.5 Advantages and disadvantages of using integrated substrates with ICs and discrete substrates with SOP and PCBs in power amplifier design

Type of implementation	Advantages	Disadvantages
IC	• High integration • Small size • Inexpensive	• Low-Q passives • Thermal issues • Lower power
SOP	• Fairly good integration • High density • Moderate- to high-Q passives • High power	• Larger size than that of the IC • Longer time to market than IC • Higher costs
Discrete on PCB	• Readily available • High-Q passives • High power	• Large size • Low density • Component variation

categories of ceramic and organic substrates [8, 24]. Typical ceramic substrates used are low-temperature cofired ceramics and high-temperature cofired ceramics. The organic substrate includes various polymers, including liquid-crystal polymers.

Table 2.5 shows a comparison of advantages and disadvantages between using integrated substrates (ICs), or discrete substrates (on SOP and PCB) in power amplifier design. SOP has more advantages that are not listed in this table, which will be discussed in more detail in Chap. 9 when discussing the packaging of power amplifiers. Kuo et al. also show some performance figures for 5 GHz power amplifiers fabricated on different substrates [25].

2.6 The Smith Chart

The Smith chart is one of the most useful graphical tools available to the RF and microwave designer [1]. It is used to simplify RF and microwave circuit analysis, whereby complex equations can be solved graphically on this chart. An impedance Smith Chart is illustrated in Fig. 2.13. In power amplifier design, the Smith chart can be useful for impedance-to-admittance conversion, as well as for impedance matching, which will be the topic of Chap. 6.

The Smith Chart is basically a combination of a family of constant resistance circles and a family of constant reactance circles, both shown in Fig. 2.14. Each point on the constant resistance or a reactance circle has the same resistance or reactance respectively. A two-port model of a circuit is given in Fig. 1.11 in Chap. 1, where Z_0 is a complex impedance $R = jX$. The reflection coefficient of the load impedance when given a source impedance can be found by formula

$$\rho = \frac{Z_S - Z_L}{Z_S + Z_L} = \frac{Z_0 - 1}{Z_0 + 1}. \tag{2.21}$$

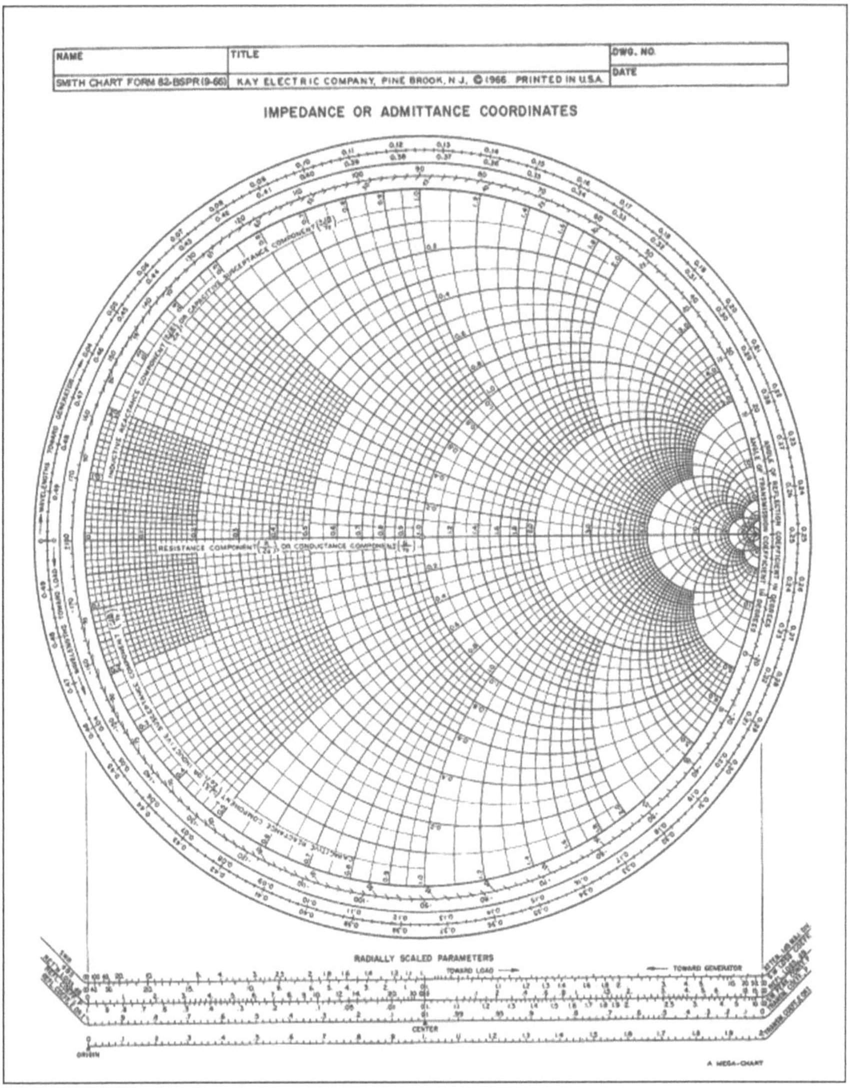

Fig. 2.13 The Smith chart. *Courtesy* Analog Instruments Co.

The polar form of the reflection coefficient is

$$\rho = p + jq = \frac{R + jX - 1}{R + jX + 1}. \tag{2.23}$$

By treating real and imaginary parts separately and removing X from the above equation, the equation for the constant resistance circle is obtained:

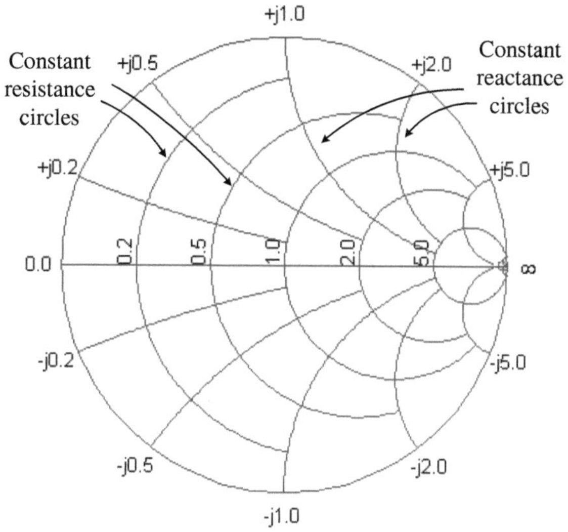

Fig. 2.14 Constant resistance and reactance circles

$$\left(p - \frac{R}{R+1}\right)^2 + q^2 = \left(\frac{1}{R+1}\right)^2, \tag{2.23}$$

with the circuit centers at $p = R/(R + 1)$ and $q = 0$, and radii equal to $1/(R + 1)$. Similarly, by removing R from the same equation, the constant reactance circle is obtained:

$$(p - 1)^2 + \left(q - \frac{1}{X}\right)^2 = \left(\frac{1}{X}\right)^2, \tag{2.24}$$

with the circuit centers at $p = 1$, $q = 1/X$ and radii of $1/X$.

All the circuits have one common intersecting point at coordinate (1, 0). The largest circle is the zero resistance circle and the outer boundary of the Smith chart, and the infinite resistance circle is reduced to a point at (1, 0). To plot the impedances on the Smith chart, the impedances must first be normalized by a certain value. For example, this value could be the characteristic impedance of the transmission line. The concept of characteristic impedance is defined in Chap. 5, and it could be 50 Ω. Thus, if impedance (50 + *j*25) Ω is to be manipulated on the Smith chart, one would typically plot a value of (1 + *j*0.5) Ω. An example of impedance manipulation on the Smith chart will be presented in the next section when showing how to convert impedance into admittance. Several more examples will be presented in Chap. 6 when dealing with impedance matching.

Recently, a 3-D version of the Smith chart was reported [26]. It is based on the extended complex plane (Riemann sphere). The north pole is the perfect matching point, while the south pole is the perfect mismatch point.

2.7 Admittance (Y-) and Scattering (S-) Parameters

2.7.1 *Y-Parameters*

Admittance is defined as the reciprocal of the impedance [1]:

$$Y = \frac{1}{Z} = G \pm jB, \tag{2.25}$$

where G is the conductance and B is the susceptance, both expressed in siemens (S). On the Smith chart, any impedance can be converted to admittance by plotting the impedance on the chart and drawing a straight line from the plotted point through the origin of the graph. The admittance is then at the point on the same line, and the same distance from the origin but on the other side. This essentially means that the Smith chart can simply be flipped over. Let us illustrate this by means of an example, where we want to convert an impedance of $(50 + j25)\ \Omega$ into admittance. First we start by normalizing the value by a factor of 50, thus we need to plot $(1 + j0.5)\ \Omega$. Then we draw a line through the center of the graph, and extend it to the other side of the graph. The value that is read off at that point is $(0.8 - j0.4)$ S. If we normalize it back with the factor of 50 we obtain $(0.016 - j0.008)$ S (note that for the admittance we have to divide by 50). This process is illustrated in Fig. 2.15. To verify the procedure, we can use Eq. (2.25):

$$Y = \frac{1}{Z} = \frac{1}{50+j25} = \frac{50-j25}{(50+j25)(50-j25)} = \frac{50-j25}{50^2+25^2} = \frac{50-j25}{3125}$$
$$= (0.016 - j0.008)\ \text{S}.$$

As mentioned, the Smith chart shown in Fig. 2.13 is the impedance Smith chart, but from previous calculations it became evident that if we rotate the chart by 180°, the chart becomes an impedance chart. If we now overlay the impedance and admittance Smith charts, we get the impedance-admittance Smith chart, which is extremely useful for impedance matching, as will be seen later in Chap. 6.

The admittance parameters (Y-parameters) of a two-port network were introduced as a tool to present the characteristics of an amplifier unambiguously at a certain frequency. An amplifier as a two-port black box with Y-parameters is shown

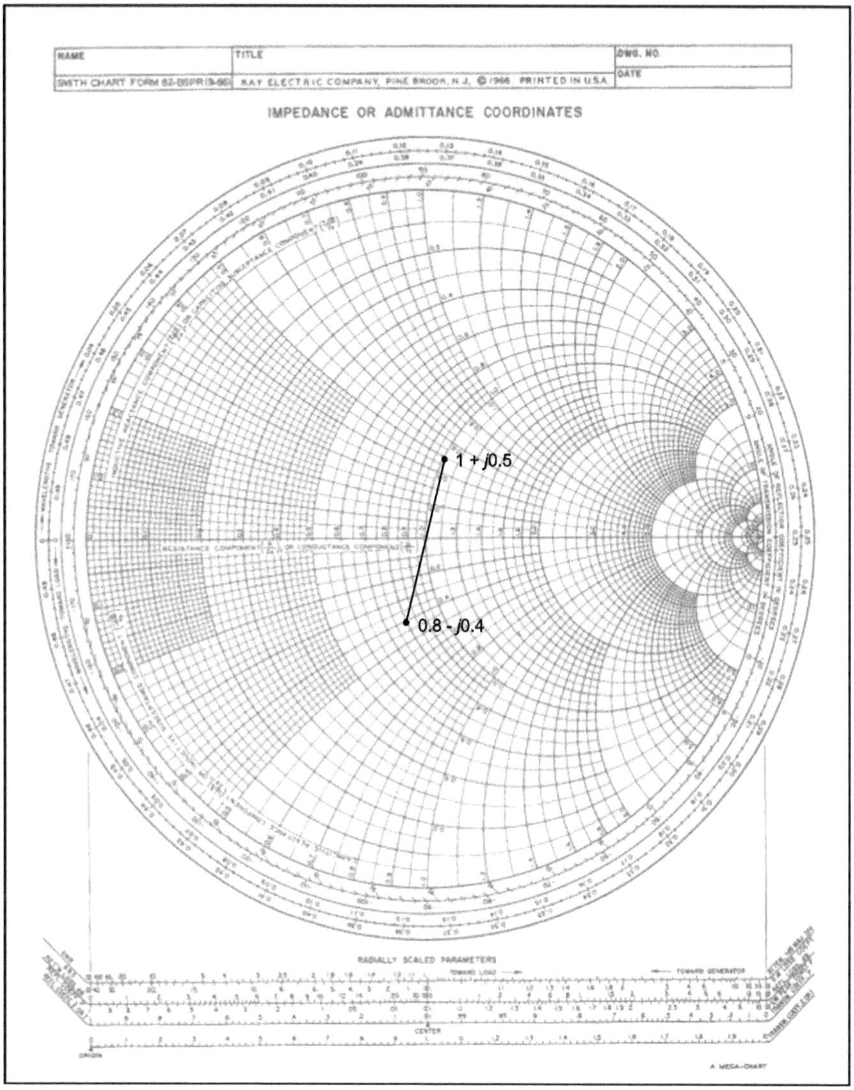

Fig. 2.15 Illustration of impedance to admittance conversion using the Smith chart

in Fig. 2.16, where I_1 and V_1 are the input current and voltage and I_2 and V_2 are the output current and voltage. The short-circuit Y-parameters are then by definition:

$$y_i = \left.\frac{I_1}{V_1}\right|_{V_2=0}, \tag{2.26}$$

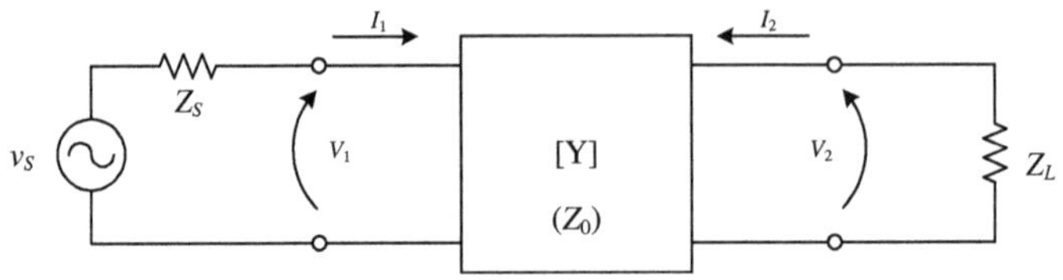

Fig. 2.16 Two port Y-parameter network

$$y_r = \left.\frac{I_1}{V_2}\right|_{V_1=0}, \tag{2.27}$$

$$y_f = \left.\frac{I_2}{V_1}\right|_{V_2=0}, \tag{2.28}$$

and

$$y_o = \left.\frac{I_2}{V_2}\right|_{V_1=0}, \tag{2.29}$$

where y_i is the short-circuit input admittance, y_r is the short-circuit reverse transfer admittance, y_f is the short-circuit forward transfer admittance and y_o is the short-circuit output admittance. Thus,

$$I_1 = y_i V_1 + y_r V_2, \tag{2.30}$$

and

$$I_2 = y_f V_1 + y_o V_2. \tag{2.31}$$

2.7.2 S-Parameters

Scattering parameters (S-parameters) are much easier to measure and work with than Y-parameters. S-parameters are also more intuitive than Y-parameters, since they are the measure of the reflection and gain, as opposed to being a measure of a just an abstract quantity such as admittance. Scattering parameters can be defined with the aid of Fig. 2.17, which is identical to Fig. 1.11, but with incident and reflected waves shown instead of source and load reflection coefficients.

A travelling wave has the following characteristics [1]:

1. A part of the traveling wave originating from the source and incident upon the two-port device (a_1) will be reflected as b_1 and another part will be transmitted through the two-port device;

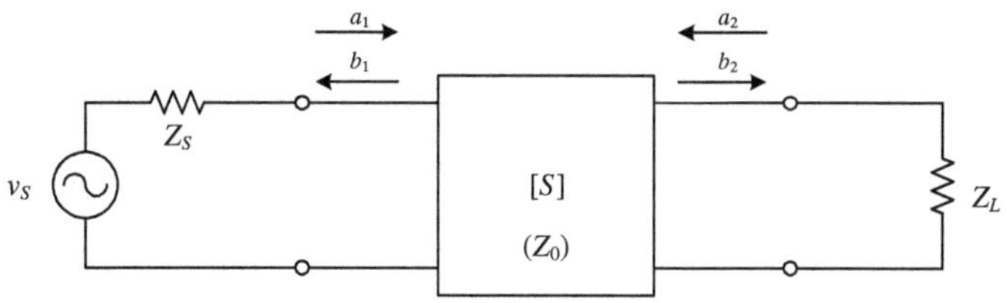

Fig. 2.17 *S*-parameter two-port model with incident and reflected waves shown

2. A part of the transmitted signal is reflected from the load and becomes incident upon the output of the two-port device (a_2); and
3. A part of the signal (a_2) is reflected from the output port back toward the load as b_2 and another part is transmitted through the two-port device back to the source.

This indicates a need for two reflection coefficients (S_{11} and S_{22}) and two gain coefficients. Thus, the input port reflection coefficient S_{11} is defined as

$$S_{11} = \left.\frac{b_1}{a_1}\right|_{a_2=0}, \tag{2.32}$$

the output port reflection coefficient S_{22} is

$$S_{22} = \left.\frac{b_2}{a_2}\right|_{a_1=0}, \tag{2.33}$$

the forward gain coefficient is

$$S_{21} = \left.\frac{b_2}{a_1}\right|_{a_2=0}, \tag{2.34}$$

and the reverse gain coefficient is

$$S_{12} = \left.\frac{b_1}{a_2}\right|_{a_1=0}. \tag{2.35}$$

Finally, the following two equations can be set

$$b_1 = S_{11}a_1 + S_{12}a_2, \tag{2.36}$$

and

$$b_2 = S_{21}a_1 + S_{22}a_2. \tag{2.37}$$

2.7.3 Conversion Between Y-Parameters and S-Parameters

Conversion between Y-parameters and S-parameters is fairly simple when conversion formulas are known. To obtain S-parameters from Y-parameters, the following equations are used:

$$S_{11} = \frac{(1 - y_i)(1 + y_o) + y_r y_f}{(1 + y_i)(1 + y_o) - y_r y_f}, \tag{2.38}$$

$$S_{12} = \frac{-2y_r}{(1 + y_i)(1 + y_o) - y_r y_f}, \tag{2.39}$$

$$S_{21} = \frac{-2y_f}{(1 + y_i)(1 + y_o) - y_r y_f}, \tag{2.40}$$

and

$$S_{22} = \frac{(1 + y_i)(1 - y_o) + y_r y_f}{(1 + y_i)(1 + y_o) - y_r y_f}. \tag{2.41}$$

To obtain Y-parameters from S-parameters, the following equations are used:

$$y_i = \frac{(1 + S_{22})(1 - S_{11}) + S_{12}S_{21}}{(1 + S_{11})(1 + S_{22}) - S_{12}S_{21}} \cdot \frac{1}{Z_0}, \tag{2.42}$$

$$y_r = \frac{-2S_{12}}{(1 + S_{11})(1 + S_{22}) - S_{12}S_{21}} \cdot \frac{1}{Z_0}, \tag{2.43}$$

$$y_f = \frac{-2S_{21}}{(1 + S_{11})(1 + S_{22}) - S_{12}S_{21}} \cdot \frac{1}{Z_0}, \tag{2.44}$$

and

$$y_o = \frac{(1 + S_{11})(1 - S_{22}) + S_{12}S_{21}}{(1 + S_{11})(1 + S_{22}) - S_{12}S_{21}} \cdot \frac{1}{Z_0}. \tag{2.45}$$

2.8 Resonant Circuits

The resonant circuit is commonly used in power amplifiers to selectively pass a frequency or a group of frequencies while attenuating all other frequencies [1, 2, 27]. It is essentially a band-pass filter (BPF) and is characterized by bandwidth. A typical BPF response is shown in Fig. 2.18. The resonant circuit can be

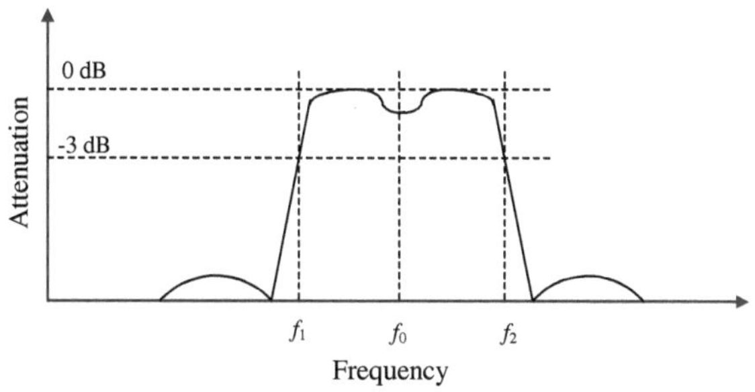

Fig. 2.18 A typical BPF response

implemented by either lumped elements or transmission lines. In this book we stick to lumped elements and parallel and series resonant circuits.

2.8.1 *Bandwidth*

The bandwidth of the resonant circuit is the difference between the upper frequency f_2 and lower frequency f_1, which mark the point where the magnitude of signal passing through the resonator is 3 dB below (a half of) the maximum signal magnitude:

$$BW = f_2 - f_1. \tag{2.46}$$

2.8.2 *Resonant Frequency*

A circuit diagram of a parallel resonant circuit is shown in Fig. 2.19.

The reactance of capacitor C at any frequency is

$$X_C = \frac{1}{\omega C}, \tag{2.47}$$

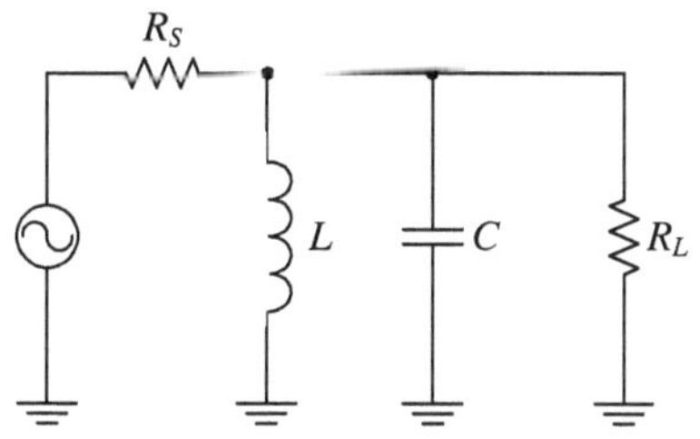

Fig. 2.19 Schematic of a parallel resonant circuit

where $\omega = 2\pi f$ is the angular frequency. The reactance of the inductor is

$$X_L = \omega L. \tag{2.48}$$

Resonance is achieved at the frequency where both reactances are equal:

$$X_C = X_L = \frac{1}{\omega C} = \omega L. \tag{2.49}$$

Resonant frequency is then

$$f_o = \frac{1}{2\pi\sqrt{LC}}. \tag{2.50}$$

Assuming that the combined reactance of the capacitor and inductor close to resonance is sufficiently small so that load resistance in parallel can be ignored, the magnitude of the output voltage of the resonant circuit, V_{out}, in terms of the magnitude input voltage V_{in} is

$$V_{out} = \frac{Z_C||Z_L}{R_S + Z_C||Z_L} V_{in}, \tag{2.51}$$

where $Z = jX$ is the impedance of a capacitor or inductor. Thus, the attenuation at any frequency ω is

$$\frac{V_{out}}{V_{in}} = \left|\frac{j\omega L}{R_S - \omega^2 R_S LC + j\omega L}\right|. \tag{2.52}$$

At ω_o,

$$\frac{V_{out}}{V_{in}} = 1 \tag{2.53}$$

which was expected from the definition of the resonant circuit.

In practice, the load and/or source resistances are not much larger than the reactances of the capacitor and inductor, thus increasing the difficulty of the resonant circuit design. One solution is to use impedance transformation in order to increase the apparent value of resistance. Various power amplifier stages use this method by design.

2.8.3 *Quality Factor*

The quality factor or Q-factor of the resonant circuit is the ratio of the center frequency of the resonant circuit to its 3-dB bandwidth:

$$Q = \frac{f_o}{BW}. \tag{2.54}$$

The quality factor of the circuit is often called a loaded Q-factor (Q_L) because it describes the passband characteristics of the resonant circuit under loaded conditions. The loaded Q-factor is influenced by both the resistance of the previous circuit stage (source) and the resistance of the following circuit stage (load), and either the inductive or capacitive reactance:

$$Q = \frac{R_p}{X_p}. \tag{2.55}$$

In the case of a power amplifier, the resonant circuit is usually followed by the antenna with resistance of $R = 50\ \Omega$, and preceded by a driving transistor with output resistance r_o, which is much larger, thus

$$R_p = R||r_o \approx R. \tag{2.56}$$

2.8.4 Component Quality Factor

The loaded Q-factor should not be confused with the Q-factor of an inductor or a capacitor, which will be discussed later in Chap. 5. However, the Q-factor of each passive influences the loaded quality factor. If a lossy passive with serial reactance X_s and serial resistance R_s is used in a resonant circuit, the additional parallel resistance seen by the circuit will be

$$R_p = (Q_{comp}^2 + 1)R_s \tag{2.57}$$

and the equivalent reactance of that component will be

$$X_p = \frac{R_p}{Q_{comp}}. \tag{2.58}$$

For $Q_{comp} >> 1$, $X_p \approx X_s$. Thus, in order for the resistance of components to be as large as possible and reactance to remain unchanged from specified values, components should have high Q values.

2.8.5 Insertion Loss

Insertion loss is the amount of attenuation due to a component or a group of components inserted between a source and a load. In the resonant circuit, it arises from resistive losses of the inductor and capacitor. It is closely related to the component Q-factor, discussed earlier.

2.8.6 Coupling of Resonant Circuits

If a single resonant circuit is not sufficient for a particular application, resonant circuits can be coupled together to achieve steeper passband skirts. The choice among capacitive coupling (Fig. 2.20a), inductive coupling (Fig. 2.20b), transformer coupling (Fig. 2.20c) or active coupling (Fig. 2.20d) will depend on the type of application.

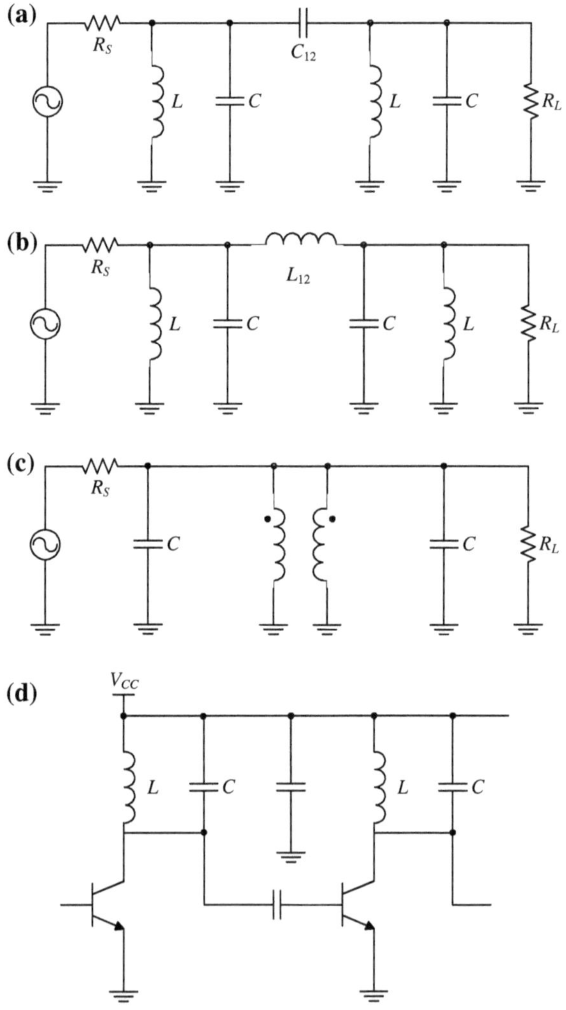

Fig. 2.20 Different ways of resonant circuit coupling: **a** capacitive coupling, **b** inductor coupling, **c** transformer coupling and **d** transistor coupling

For capacitive coupling, the value of capacitance to couple two identical resonant circuits is

$$C_{12} = \frac{C}{Q_L}, \tag{2.59}$$

where C is the resonant circuit capacitance and Q_L is the loaded Q factor of a single resonator. Similarly, for inductive coupling, the combined inductance is

$$L_{12} = Q_L L, \tag{2.60}$$

where L is the resonant circuit inductance.

Active coupling is accomplished by using a transistor. The combined Q is

$$Q_{total} = \frac{Q_L}{\left(2^{1/n} - 1\right)^{1/2}}, \tag{2.61}$$

where n is the number of resonant circuits.

2.9 Fourier Analysis of Periodic Signals

Periodic signals are used to analyze power amplifier output stages. A function is periodic if the following is satisfied [2]:

$$f(\omega t) = f(\omega t \pm 2\pi n). \tag{2.62}$$

Any periodic function can be expressed as an infinite sum of sinusoidal signals, called Fourier series. Fourier analysis is thus used to calculate the DC and harmonic components of periodic signals throughout this book and is discussed here.

Function $f(\omega t)$ can be expressed as

$$\begin{aligned} f(\omega t) = a_0 + \sum_{n=1}^{\infty} (a_n \cos n\omega t + b_n \sin n\omega t) &= a_0 + a_1 \cos \omega t + b_1 \sin \omega t \\ + a_2 \cos 2\omega t + b_2 \sin 2\omega t + a_3 \cos 3\omega t + b_3 \sin 3\omega t \\ + a_n \cos n\omega t + b_n \sin n\omega t, \end{aligned} \tag{2.63}$$

where a_n and b_n are called Fourier coefficients and can be calculated for one cycle from

$$a_0 = \frac{1}{T} \int_0^T f(t)dt = \frac{1}{2\pi} \int_0^{2\pi} f(\omega t)d(\omega t), \tag{2.64}$$

$$a_n = \frac{2}{T}\int_0^T f(t)\cos ntd(t) = \frac{1}{\pi}\int_0^{2\pi} f(\omega t)\cos n\omega td(\omega t) \tag{2.65}$$

and

$$b_n = \frac{2}{T}\int_0^T f(t)\sin ntdt = \frac{1}{\pi}\int_0^{2\pi} f(\omega t)\sin n\omega\, td(\omega t). \tag{2.66}$$

Two special cases of function in Eq. (2.63) will be used often. If the function $f(\omega t)$ is odd, meaning that $f(\omega t) = -f(\omega t)$, then the calculation of coefficients simplifies to

$$a_0 = a_n = 0, \tag{2.67}$$

and

$$b_n = \frac{4}{T}\int_0^{T/2} f(t)\sin n\omega tdt = \frac{2}{\pi}\int_0^{\pi} f(\omega t)\sin n\omega td(\omega t). \tag{2.68}$$

Sine function is an example of an odd function. If the same function is even, meaning that $f(\omega t) = f(-\omega t)$, then

$$a_0 = \frac{2}{T}\int_0^T f(t)dt = \frac{1}{\pi}\int_0^{2\pi} f(\omega t)d(\omega t), \tag{2.69}$$

$$a_n = \frac{4}{T}\int_0^{T/2} f(t)\cos ntdt = \frac{2}{\pi}\int_0^{\pi} f(\omega t)\cos n\omega td(\omega t) \tag{2.70}$$

and

$$b_n = 0. \tag{2.71}$$

An example of an even function is the cosine function. Since both sine and cosine functions are defined with an arbitrary origin, it can be shifted on the time axis to simplify the analysis where necessary.

2.10 Summary

In this chapter we reviewed the frequency spectrum and identified that the frequency bands mentioned in the title of the book occupy frequencies between 2 and 18 GHz, where typically the power amplifiers can be implemented with lumped passives, and that transmission lines, if needed, can be implemented in discrete implementations. We reviewed the modulation schemes, where we presented constellations and/or waveforms for PSK, FSK, QAM, OOK and OFDM schemes. Power amplifiers are typically used to drive antennas, thus we discussed some basic antenna concepts. Technologies and substrates for systems-on-chip, SOP and discrete implementations were also discussed, where we concluded that each implementation has its advantages and disadvantages. Some practical aspects of various implementations and packaging will form part of more discussions in Chap. 9. We also presented the theory behind the active device, but we leave the discussion of passives until Chap. 5. The Smith chart was presented as an excellent graphical tool for various RF and microwave calculations, and we demonstrated its power in impedance-to-admittance conversions. However, we will use the Smith chart again in Chap. 6 with impedance matching. Defining equations of *S*- and *Y*-parameters were also given here. Various aspects of resonant circuits were discussed, as these form the basic building blocks of any power amplifier. Fourier analysis, necessary for the analysis of the power amplifier waveforms, was also discussed.

Theory presented in this chapter laid the groundwork for Chaps. 3 and 4, where various aspects will be used when defining and describing different power amplifier output stages. Although some of the concepts presented in this chapter are not paramount to understanding Part 2 of this book, they add to the better understanding of power amplifier application.

References

1. Bowick C, Blyler J, Ajluni C. RF circuit design. 2nd ed. Burlington: Elsevier; 2008.
2. Kazimierczuk MK. RF power amplifiers. 2nd ed. Chiechester: Wiley; 2015.
3. Pozar M. Microwave engineering. 4th ed. Hoboken: Wiley; 2012.
4. International Telecommunication Union. Nomenclature of the frequency and wavelength bands used in telecommunications. ITU-R Recommendation V.431 [Internet]. 2000 May [cited 2015 May 19]. Available from: http://www.itu.int/rec/R-REC-V.431/en.
5. Proakis JG, Salehi M. Digital communications. 5th ed. New York: McGraw-Hill; 2008.
6. Raab FH, Asbeck P, Kenington PB, Cripps S, Popovic ZB, Pothecary N, Sevic JF, Sokal NO. RF and microwave power amplifier and transmitter technologies—part 1. High Freq Electron. 2003;2:22–36.
7. Cheng DK. Fundamentals of engineering electromagnetics. 1st ed. Reading: Addison-Wesley Publishing Company; 1993.
8. Tummala RR, Swaminathan M. System-on-package: miniaturization of the entire system. 1st ed. New York: McGraw-Hill Professional; 2008.
9. Poulin D. The III-V vs. silicon battle. Microw J. 2009;52(4):22–38.

10. Johnson JB, Joseph AJ, Sheridan DC, Maladi RM, Brandt PO, Persson J, Andersson J, Bjomeklett A, Persson U, Abasi F, et al. Silicon-Germanium BiCMOS HBT technology for wireless power amplifier applications. IEEE J Solid-State Circuits. 2004;39(10):1605–14.
11. Kroemer H. Heterostructure bipolar transistors and integrated circuits. Proc IEEE. 1982;70 (1):13–25.
12. Chiou HK, Liao HY, Chen CC, Wang SM, Chen CC. A, 2.6-GHz fully integrated CMOS power amplifier using power combining transformer. Microw Opt Technol Lett. 2010;52 (2):299–302.
13. Cripps SC. RF power amplifiers for wireless communications. Norwood: Artech House; 2006.
14. Avenier G, Diop M, Chevalier P, Troillard G, Loubet N, Bouvier J, Depoyan L. 0.13 m SiGe BiCMOS technology fully dedicated to mm-wave applications. IEEE J Solid-State Circuits. 2009;44(9):2312–21.
15. Nikandish G, Medi A. A design procedure for high-efficiency and compact-size 5–10-W MMIC power amplifiers in GaAs pHEMT technology. IEEE Trans Microw Theory Tech. 2013;61(8):2922–33.
16. Sheppard ST, Doverspike K, Pribble WL, Allen ST, Palmour JW, Kehias LT, Jenkins TJ. High-power microwave GaN/AlGaN HEMTs on semi-insulating silicon carbide substrates. IEEE Electron Device Lett. 1999;20(4):161–3.
17. Nguyen C, Micovic M. The state-of-the-art of GaAs and InP power devices and amplifiers. IEEE Trans Electron Devices. 2001;48(3):472–8.
18. Pribble WL, Palmour W, Sheppard ST, Smith RP, Allen ST, Smith TJ, Ring Z, Sumakeris JJ, Saxler AW, Milligan JW. Applications of SiC MESFETs and GaN HEMTs in power amplifier design. In: 2002 IEEE MTT-S International Microwave Symposium Digest; 2002; Seattle. p. 1819–22.
19. Nellis K, Zampardi P. A comparison of linear handset power amplifier in different bipolar technologies. IEEE J Solid-State Circuits. 2004;39(10):1746–54.
20. Vitusevich SA, Kurakin AM, Klein N, Petrychuk MV, Naumov AV, Belyaev AE. AlGaN/GaN high electron mobility transistor structures: self-heating effect and performance degradation. IEEE Trans Device Mater Reliab. 2008;8(3):543–8.
21. Dongyue J, Wanrong Z, Pei S, Hongyun X, Yang W, Wei Z, Lijian H, Yongping S, Jia L, Junning G. Multi-finger power SiGe HBTs for thermal stability enhancement over a wide biasing range. Solid-State Electron. 2008;52(6):937–40.
22. Degachi L, Ghannouchi F. An augmented small-signal HBT model with its analytical based parameter extraction technique. IEEE Trans Electron Devices. 2008;55(4):968–72.
23. Corporation IBM. BiCMOS7WL design manual. Armonk: IBM Corporation; 2008.
24. Greig WJ. Integrated circuit packaging, assembly and interconnections. 1st ed. New York: Springer; 2007.
25. Kuo CC, Hsu YW, Huang WC, Wang H, Lu H. Performance comparison of flip-chip-assembled 5-GHz 0.18-CMOS power amplifiers on different packaging substrates. IEEE. Trans Compon Packag Manuf Technol. 2013;3(12):2014–21.
26. Muller AA, Soto P, Dascalu D, Neculoiu D, Boria VE. A 3-D smith chart based on the riemann sphere for active and passive microwave circuits. IEEE Microw Wirel Compon Lett. 2011;21(6):286–8.
27. Grebennikov A, Sokal NO, Franco MJ. Switchmode RF and microwave power amplifiers. 2nd ed. Burlington: Elsevier; 2012.

Chapter 3
Continuous-Mode Power Amplifiers

In this chapter, continuous-mode power amplifiers, or Class-A, B, AB and C amplifiers, are analyzed. They are usually treated together because of the minimal differences in their basic circuit diagrams and their waveforms [1–18].

A Class-A power amplifier is very similar to small-signal amplifiers. It typically consists of a transistor, a parallel-resonant circuit *LC* and an RFC. The driving transistor operates in the active region at all times and acts as a dependent current source, determined by the gate or base drive and bias. The drain or collector voltage and current waveforms are amplified reproductions of the input signal. Both the envelope and phase of the driving signal are retained. The conduction angle (2θ) of Class-A stage is 360°. The DC-input power is constant and the instantaneous efficiency is proportional to the output power. A Class A amplifier exhibits high linearity, high gain and operation close to the maximum operating frequency of the transistor, but in turn suffers low drain efficiency (maximum of 50 % in theory) and dissipates the maximum power at zero output power. Total harmonic distortion and intermodulation distortion are determined only by transistor non-linearities.

A Class-B power amplifier circuit is similar to that of the Class A, but the resonant circuit role as a bandpass filter becomes prominent in its function to isolate the fundamental harmonic and aid linearity, because the gate bias is adjusted so that the transistor is active only during half of the drive cycle and the resulting drain current is a half-sinusoid. The drain current is proportional to the drive amplitude. For the part of the cycle during which it conducts, it acts as a dependent current source. The conduction angle of Class-B stage is 180°. The instantaneous efficiency varies linearly with the output voltage and reaches 78.5 % in theory, which makes it much more efficient than Class A, especially for low-level signals. A Class-B power amplifier is often used in a push-pull configuration.

A Class-AB power amplifier is the same as a Class-B amplifier, with the exception that the transistor is biased so that it conducts current for more than half of the drive cycle. The resulting conduction angle is between 180° and 360° and the efficiency is between 50 and 78.5 %. The distortion of Class AB is less than that of the Class B but higher than that of Class A, thus making the Class AB power a compromise between Class-A and Class-B.

A Class-C power amplifier is the continuous stage in which the gate bias of the driving transistor is adjusted so that it conducts for less than half of the drive cycle.

M. Božanić and S. Sinha, *Power Amplifiers for the S-, C-, X- and Ku-bands*, Signals and Communication Technology, DOI 10.1007/978-3-319-28376-0_3

The resulting conduction angle is between 0° and 180°, allowing the efficiency to get close to 100 %. As a result, linearity decreases. It is only effective when the transistor is driven close to saturation, which makes it suitable for peaking amplifiers.

A Class-J power amplifier is similar to a "deep" Class-AB power amplifier, but it has been found to be more realizable, with a better compromise between power and efficiency and a substantial bandwidth. Its theoretical efficiency can reach that of a Class-B amplifier (78.5 %).

Doherty amplifiers are also discussed in this chapter. A Doherty power amplifier combines two different output stages in order to boost efficiency.

3.1 Class-A Power Amplifier

Figure 3.1 shows a circuit operating as a Class-A power amplifier [1]. In this circuit, the transistor is shown as an n-type MOSFET. The MOSFET is chosen specifically instead of a BJT to simplify circuit analysis. L and C form a resonant circuit at the fundamental (operating) frequency f_0, thus

$$f_0 = \frac{1}{2\pi\sqrt{LC}}. \tag{3.1}$$

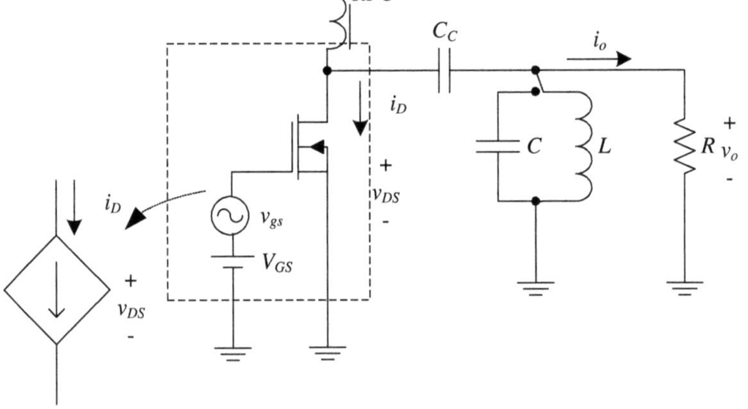

Fig. 3.1 Circuit schematic of a single-ended Class-A power amplifier used for theoretical analysis. This figure also shows how the transistor can be replaced with an ideal current source

Furthermore, R is the required load resistance (matched to the antenna via an output-matching network), and RFC is the RF choke. V_{DD} denotes positive supply voltage.

To be able to analyze the Class-A power amplifier for voltage and current waveforms, power dissipation, drain efficiency and output power capability, some assumptions need to be made:

1. **Waveforms**:
    - The gate of the transistor is driven by a sinusoidal waveform, given by

$$v_{GS} = V_{GS} + v_{gs} = V_{gsm} \cos \omega t + V_{GS}, \tag{3.2}$$

    where v_{GS} is total gate-to-source voltage, V_{GS} is its DC component, v_{gs} is its AC component, m denotes the amplitude, and ω is the angular frequency $2\pi f$, where f denotes the frequency.
    - At any moment, $v_{GS} > V_t$, where V_t is transistor threshold voltage. This ensures that the conduction angle remains 360°.
    - Drain-to-source voltage v_{DS} can swing from 0 to $2V_{DD}$ and drain current i_D can swing from 0 to $2I_D$, where I_D is the DC portion of the transistor drain current (quiescent current).
2. **Transistor**:
    - The output capacitance of the transistor can be absorbed into resonant capacitance, i.e. $C = C_{OUT} + \Delta C$.
    - Drain current is sufficiently high so that a linear relationship exists between i_D and the difference between the gate-source voltage and threshold voltage, i.e.

$$i_D = K_S(v_{GS} - V_t), \tag{3.3}$$

    where K_S is the proportionality constant specific to each MOSFET and defined in Chap. 2.
    - Transistor channel length-modulation is zero; therefore the output resistance of the transistor is infinite.
    - Transistor drain-source saturation voltage $v_{DS(sat)}$ is zero, so that the maximum amplitude of the signal V_m can reach V_{DD}.
3. **Passives**:
    - RFC is ideal (zero resistance).
    - All other passive components are ideal.

3.1.1 Current and Voltage Waveform Analysis

If all assumptions listed in the previous section hold, then from Eqs. (3.2) and (3.3), the drain current of the transistor is:

$$i_D = K_S[V_{gsm} \cos \omega t + V_{GS} - V_t]. \tag{3.4}$$

If I_m is defined as $K_S V_{gsm}$, the DC and AC portions of the current are inherently

$$I_D = K_S(V_{GS} - V_t) \tag{3.5}$$

and

$$i_d = I_m \cos \omega t. \tag{3.6}$$

At the resonant frequency, the parallel *LC* circuit will act as an open circuit (owing to its infinite reactance), and all AC current will be going through the load resistor:

$$i_o = I_D - (I_D + I_m \cos \omega t) = -I_m \cos \omega t. \tag{3.7}$$

By multiplying both sides of Eq. (3.7) with *R* and noting that

$$V_m = RI_m, \tag{3.8}$$

the output voltage of the Class-A stage over the load resistor can be derived:

$$v_o = Ri_o = -V_m \cos \omega t. \tag{3.9}$$

Again because of resonance, the AC component of the drain-source voltage v_{ds} equates to v_o. If V_{DS} is the DC component of the drain source voltage and because the RFC is present it assumes V_{DD} level, then $v_{DS} = V_{DS} + v_{ds}$ can be expressed as

$$v_{DS} = V_{DD} - V_m \cos \omega t. \tag{3.10}$$

To avoid distortion, no current or voltage can be negative at any time, i.e. amplitudes of both sinusoidal signals must be less than or equal to I_D and V_{DD} respectively. Thus,

$$I_{m(\max)} = I_D \tag{3.11}$$

and

$$V_{m(\max)} = V_{DD}. \tag{3.12}$$

Hence the maximum drain current for Class-A stage is

$$I_{DM(\max)} = I_D + I_{m(\max)} = 2I_D \tag{3.13}$$

and likewise, the maximum drain-to-source voltage is

$$V_{DS(\max)} = V_{DD} + V_{m(\max)} = 2V_{DD}. \tag{3.14}$$

In a practical system, because of the transistor saturation voltage, a slightly smaller maximum drain current and voltage are allowed. The saturation voltage is

$$V_{DSsat(\max)} = v_{GS(\max)} - V_t = V_{GS} + V_{gsm} - V_t. \tag{3.15}$$

Current and voltage waveforms resulting from this analysis are shown in Fig. 3.2.

3.1.2 Power and Efficiency

The instantaneous power delivered to the load at the operating frequency is

$$p_o(\omega t) = i_o(\omega t)v_o(\omega t) = (-I_m \cos\omega t)(-V_m \cos\omega t) = I_m V_m \cos^2\omega t. \tag{3.16}$$

The average power is then the integral over the complete cycle, i.e.

$$\begin{aligned} P_O &= \frac{1}{2\pi}\int_0^{2\pi} p_o d(\omega t) = \frac{1}{2\pi}\int_0^{2\pi} I_m V_m \cos^2\omega t \\ &= \frac{1}{2\pi}\int_0^{2\pi} \frac{V_m I_m}{2}(1+\cos 2\omega t)d(\omega t) \\ &= \frac{1}{2}V_m I_m = \frac{1}{2}\frac{V_m^2}{R} \end{aligned} \tag{3.17}$$

The instantaneous power dissipated in the transistor is

$$\begin{aligned} p_D(\omega t) &= i_D(\omega t)v_{DS}(\omega t) = (I_D + I_m\cos\omega t)(V_{DD} - V_m\cos\omega t) \\ &= I_D\left(1+\frac{I_m}{I_D}\cos\omega t\right)V_{DD}\left(1-\frac{V_m}{V_{DD}}\cos\omega t\right) \\ &= P_{DD}(1+\gamma_1\cos\omega t)(1-\zeta_1\cos\omega t), \end{aligned} \tag{3.18}$$

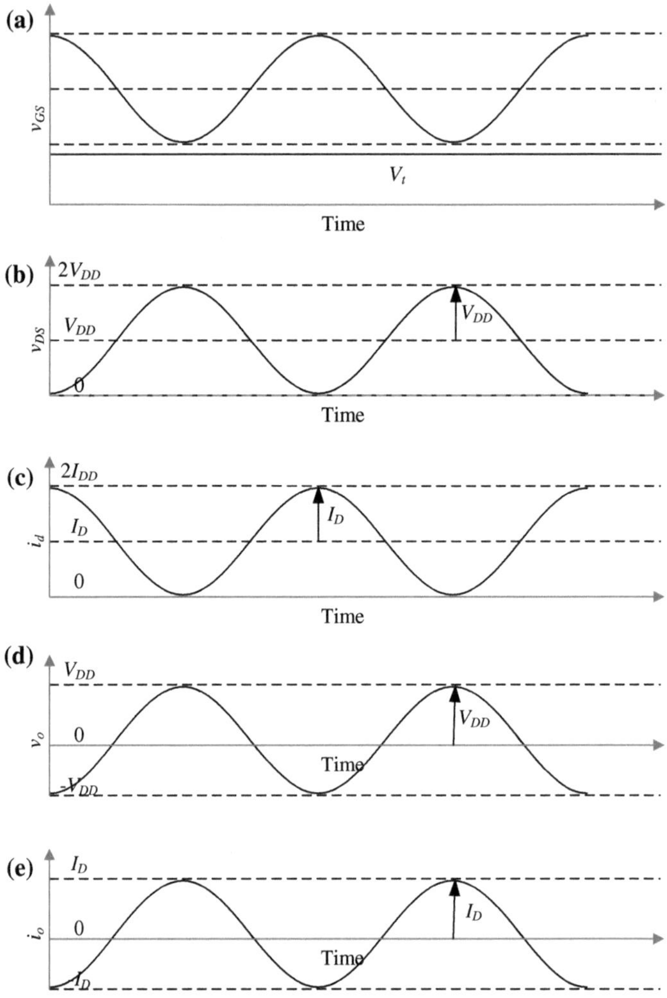

Fig. 3.2 Waveforms of the Class-A power amplifier output stage: **a** gate-source voltage, **b** drain-source voltage, **c** drain current, **d** output voltage and **e** output current

where $\gamma_1 = I_m/I_D$ and $\zeta_1 = V_m/V_{DD}$ are normalized amplitudes of drain current and drain-to-source voltage respectively. For maximum swing, recall that $I_m = I_D$ and $V_m = V_{DD}$, thus

$$\begin{aligned} p_D(\omega t) &= P_{DD}(1 + \cos \omega t)(1 - \cos \omega t) \\ &= P_{DD}(1 - \cos^2 \omega t) \\ &= P_{DD}(1 - \cos 2\omega t). \end{aligned} \tag{3.19}$$

The average power dissipated over a cycle is

$$
\begin{aligned}
P_D &= \frac{1}{2\pi}\int_0^{2\pi} [I_D V_{DD} - I_m V_m \cos^2 \omega t + (V_{DD} I_m - I_D V_m)\cos \omega t] d(\omega t) \\
&= \frac{1}{2\pi}\int_0^{2\pi} \left[I_D V_{DD} - \frac{I_m V_m}{2}(1 + \cos 2\omega t) + (V_{DD} I_m - I_D V_m)\cos \omega t\right] d(\omega t) \\
&= I_D V_{DD} - \frac{I_m V_m}{2} = P_{DD} - P_O.
\end{aligned}
\tag{3.20}
$$

For zero swing, $I_m = 0$ and $V_m = 0$, hence $P_O = 0$, thus

$$P_D = P_{DD}, \tag{3.21}$$

which means that for idle operation of the Class-A power amplifier, the transistor will continue to draw all available power from the supply. The instantaneous power dissipated in the transistor for different values of $\zeta_1 = \gamma_1$ in terms of supply power P_{DD} is illustrated in Fig. 3.3.

Drain efficiency, defined in Chap. 1, is a ratio of output power and power provided by the supply:

$$\eta = \frac{P_{out}}{P_{dc}} \tag{3.22}$$

In the case of Class A,

$$\eta = \frac{P_O}{P_{DD}} = \frac{\frac{1}{2} I_m V_m}{I_D V_{DD}} = \frac{1}{2}\left(\frac{I_m}{I_D}\right)\left(\frac{V_m}{V_D}\right) = \frac{1}{2}\gamma_1 \zeta_1. \tag{3.23}$$

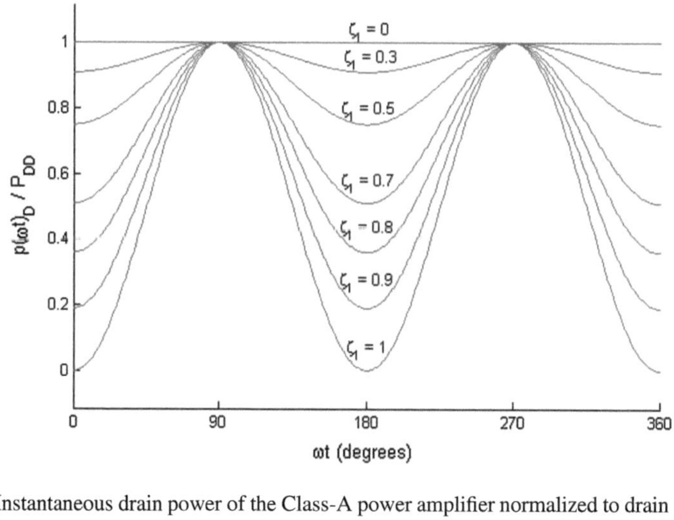

Fig. 3.3 Instantaneous drain power of the Class-A power amplifier normalized to drain power P_{DD}

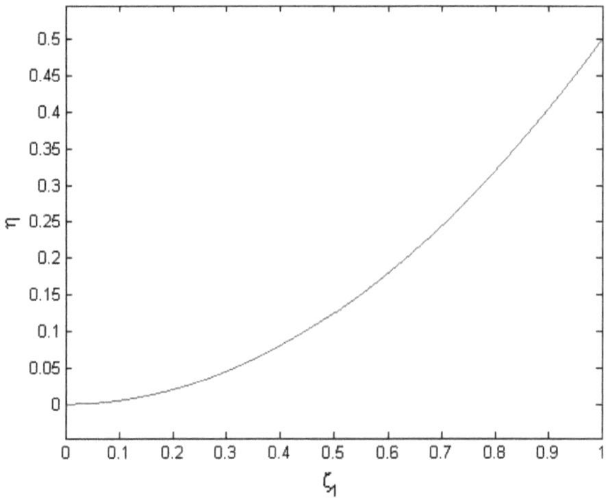

Fig. 3.4 The efficiency of a Class-A power amplifier as a function of ζ_1

The maximum reachable efficiency is for maximum swing, i.e.

$$\eta = \frac{1}{2}\left(\frac{I_D}{I_D}\right)\left(\frac{V_{DD}}{V_{DD}}\right) = 50\,\%. \tag{3.24}$$

The efficiency as a function of $\zeta_1 = \gamma_1$ is shown in Fig. 3.4.
The output power capability is

$$c_p = \frac{P_{out(\max)}}{I_{D\max} V_{DS(\max)}}. \tag{3.25}$$

For Class A, using Eqs. (3.11)–(3.14) and (3.17),

$$c_p = \frac{\frac{1}{2} I_{m(\max)} V_{m(\max)}}{I_{D\max} V_{DS(\max)}} = \frac{1}{2} \cdot \frac{I_D V_{DD}}{(2I_D)(2V_{DD})} = \frac{1}{8}. \tag{3.26}$$

3.1.3 Bandwidth

The bandwidth of the Class-A power amplifier is determined by the parallel resonant circuit, which acts as a bandpass filter. In Chap. 2, the 3-dB bandwidth was defined as

$$BW = \frac{f_0}{Q_L}, \tag{3.27}$$

where Q_L is the loaded quality factor of the parallel resonant circuit inserted between the source and load resistance of total resistance R, given by

$$Q_L = \frac{2\pi f_0 L}{R}. \tag{3.28}$$

3.1.4 *RFC and Coupling Capacitor*

The role of the RFC is to serve as an inductor used to block AC signals while passing DC signals. In order to keep the current ripple to more than ten times lower than I_D, the value of the reactance of the RFC, denoted X_{RFC} should be set so that

$$X_{RFC} \geq 10R. \tag{3.29}$$

Thus, the inductance of the RFC should be

$$L_{RFC} = \frac{X_{RFC}}{\omega_0} \geq \frac{10R}{\omega_0}. \tag{3.30}$$

Using a similar approach, for the acceptable ripple over the coupling capacitor, its reactance should be at least ten times less than that of the resistor:

$$X_{C_C} \geq \frac{R}{10}. \tag{3.31}$$

Thus, the capacitance C_C should be

$$C_C = \frac{1}{\omega_0 X_{C_C}} \geq \frac{R}{10\omega_0}. \tag{3.32}$$

3.2 Class-B Power Amplifier

3.2.1 *Current and Voltage Waveform Analysis*

Structurally, the Class-B power amplifier is identical to the Class-A amplifier and the circuit diagram shown in Fig. 3.1 can be used for its analysis. Most assumptions listed for Class-A amplifiers still hold and L, C, RFC and C_C are all chosen in the same way as for the Class-A amplifier. However, the driving transistor of a Class-B amplifier is biased so that its conduction angle is 180°. This means that the DC portion of the v_{GS} signal must be set to be equal to the threshold voltage, i.e. $V_{GS} = V_t$. Thus, Eq. (3.2) becomes

$$v_{GS} = V_{GS} + v_{gs} = V_{gsm} \cos \omega t + V_t. \tag{3.33}$$

The saturation voltage in Eq. (3.15) is

$$V_{DSsat} = V_{GS} + V_{gsm} - V_t = V_{gsm}. \tag{3.34}$$

The magnitude of the drain-to-source voltage V_m is then

$$V_m = V_{DD} - V_{DSsat} = V_{DD} - V_{gsm}. \tag{3.35}$$

For the part during which the transistor conducts ($-\pi/2 < \omega t < \pi/2$), drain current in Eq. (3.4) is

$$i_D = K_S V_{gsm} \cos \omega t = I_{DM} \cos \omega t, \tag{3.36}$$

where I_{DM} is defined as $i_D(0) = K_S V_{gsm}$. The amplitude of the fundamental component of drain current and the DC component of the drain current is determined using Fourier analysis:

$$I_m = \frac{1}{\pi} \int_{-\frac{\pi}{2}}^{\frac{\pi}{2}} i_D \cos \omega t d(\omega t) = \frac{1}{\pi} \int_{-\frac{\pi}{2}}^{\frac{\pi}{2}} I_{DM} \cos^2 \omega t d(\omega t) = \frac{I_{DM}}{2} \tag{3.37}$$

and

$$I_D = \frac{1}{2\pi} \int_{-\frac{\pi}{2}}^{\frac{\pi}{2}} i_D d(\omega t) = \frac{1}{\pi} \int_{0}^{\frac{\pi}{2}} I_{DM} \cos \omega t d(\omega t) = \frac{I_{DM}}{\pi} = \frac{2I_m}{\pi}. \tag{3.38}$$

The resonant circuit filters out the harmonics of this current, hence the output current is still given by Eq. (3.7), drain-to-source voltage is given by Eq. (3.10) and its maximum value is given by Eq. (3.14). Output voltage v_o is given by Eq. (3.9). Current and voltage waveforms resulting from this analysis are shown in Fig. 3.5.

3.2.2 *Power and Efficiency*

The instantaneous power delivered to the load at the operating frequency is the same as for the Class-A amplifier and hence it is given by Eq. (3.16). The average power over the complete cycle is thus given by Eq. (3.17).

The instantaneous power dissipated in the transistor at resonant frequency is zero for the part of the cycle when the transistor does not conduct and for the part when it conducts ($-\pi/2 < \omega t < \pi/2$) it is

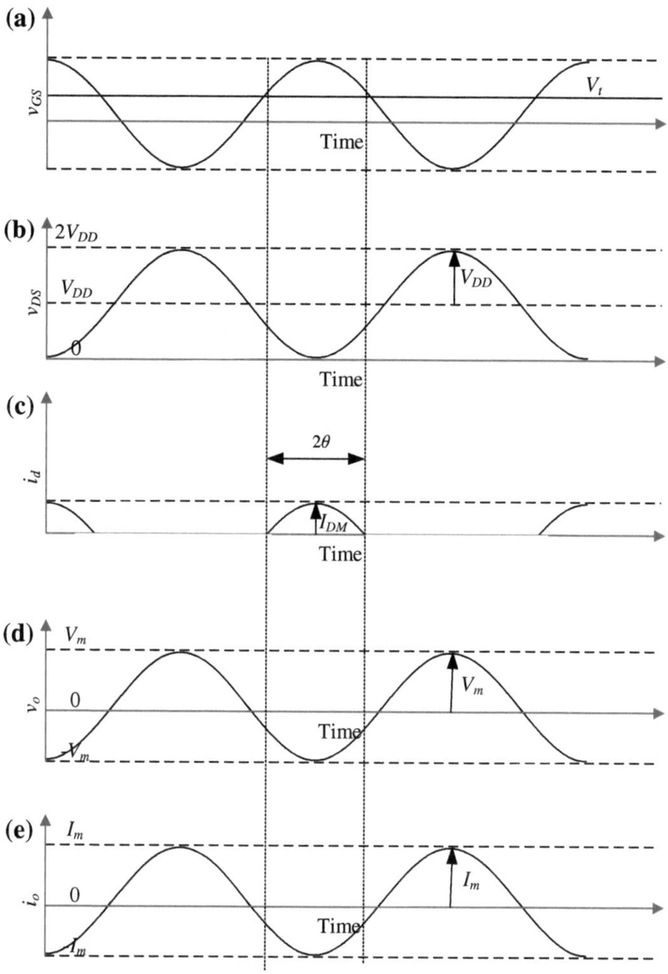

Fig. 3.5 Waveforms of the Class-B power amplifier output stage: **a** gate-source voltage, **b** drain-source voltage, **c** drain current **d** output voltage and **e** output current

$$\begin{aligned} p_D(\omega t) &= i_D(\omega t) v_{DS}(\omega t) = (I_{DM} \cos \omega t)(V_{DD} - V_m \cos \omega t) \\ &= I_{DM} V_{DD} \cos \omega t \left(1 - \frac{V_m}{V_{DD}} \cos \omega t\right) \\ &= \pi I_D V_{DD} \cos \omega t \left(1 - \frac{V_m}{V_{DD}} \cos \omega t\right) \\ &= \pi P_{DD} \cos \omega t (1 - \zeta_1 \cos \omega t). \end{aligned} \tag{3.39}$$

The DC power from the supply is

$$P_{DD} = I_D V_{DD} = \frac{2I_m}{\pi} V_{DD} = \frac{2}{\pi} \frac{V_{DD} V_m}{R}. \tag{3.40}$$

Unlike for the Class-A power amplifier, the DC supply power is a function of V_m.

The average power dissipated in the transistor can be calculated by subtracting the output power from the DC supply power:

$$P_D = P_{DD} - P_O = \frac{2}{\pi} \frac{V_{DD} V_m}{R} - \frac{V_m^2}{2R}. \tag{3.41}$$

An interesting result for the Class-B amplifier is that for zero swing ($V_m = 0$) and thus zero output power, both the supply power and power dissipated in the transistor are zero, i.e. the Class-B power amplifier does not waste any power when it is idle. The instantaneous power dissipated in the transistor for different values of ζ_1 in terms of supply power P_{DD} is illustrated in Fig. 3.6.

The drain efficiency of Class-B is

$$\eta = \frac{P_O}{P_{DD}} = \frac{\frac{1}{2} I_m V_m}{\frac{2}{\pi} \frac{V_{DD} V_m}{R}} = \frac{\pi}{4} \left(\frac{V_m}{V_{DD}} \right) \left(\frac{I_m R}{V_m} \right) = \frac{\pi}{4} \zeta_1. \tag{3.42}$$

The maximum reachable efficiency is for maximum swing, i.e.

$$\eta = \frac{\pi}{4} \left(\frac{V_{DD}}{V_{DD}} \right) \approx 78.5\,\%. \tag{3.43}$$

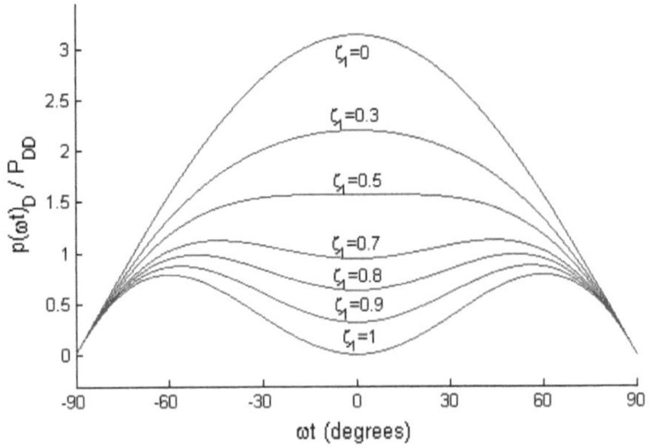

Fig. 3.6 Instantaneous drain power of the Class-B power amplifier normalized to drain power P_{DD}

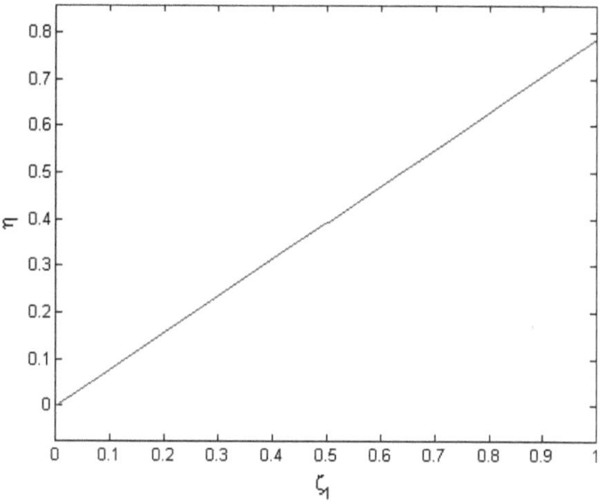

Fig. 3.7 Efficiency of Class-B power amplifier as a function of ζ_1

The efficiency as a function of ζ_1 is shown in Fig. 3.7.

The output power capability of a Class-B power amplifier is

$$c_p = \frac{P_{out(\max)}}{I_{D\max} V_{DS(\max)}} = \frac{\frac{1}{2} I_{m(\max)} V_{m(\max)}}{I_{D\max} V_{DS\max}} = \frac{1}{2} \cdot \frac{\frac{I_{DM}}{2} V_{DD}}{I_{DM}(2V_{DD})} = \frac{1}{8}, \tag{3.44}$$

which is the same as for a Class-A amplifier.

3.3 Class-AB and Class-C Power Amplifiers

3.3.1 Current and Voltage Waveform Analysis

The circuit of the Class-AB and Class-C power amplifier is the same as that of the Class-A and Class-B. For Class-AB operation, the conduction angle is set between 180° and 360°. For Class-C operation, it is set at less than 180°. The Class-B amplifier is therefore a special case of both Class-AB and Class-C power amplifiers, and the Class-A amplifier is a special case of the Class-AB amplifier.

The analysis of the Class-AB and Class-C (and Class B) amplifiers can begin by noting that the gate-to-source voltage is expressed by the Class-A (Eq. 3.2). If the conduction angle is defined as 2θ, the circuit conducts if

$$-\theta \leq \omega t \leq \theta. \tag{3.45}$$

If $\omega t = \theta$, voltage v_{GS} is at the threshold level, thus

$$V_{gsm}\cos\theta + V_{GS} = V_t. \tag{3.46}$$

Half of the conduction angle is then

$$\theta = \arccos\frac{V_t - V_{GS}}{V_{gsm}} \tag{3.47}$$

for Class-C amplifiers and

$$\theta = \pi - \arccos\frac{V_t - V_{GS}}{V_{gsm}} \tag{3.48}$$

for Class-AB amplifiers.

Drain current is specified by Eq. (3.4) for the portion of the cycle during which the transistors conducts ($-\theta \le \omega t \le \theta$) and it is zero for the rest of the cycle. The maximum value of the drain current is

$$\begin{aligned} I_{DM} &= i_D(0) = K_S(V_{gsm}\cos 0 + V_{GS} - V_t) \\ &= K_S V_{gsm}\left(1 - \frac{V_t - V_{GS}}{V_{gsm}}\right) = K_S V_{gsm}(1 - \cos\theta). \end{aligned} \tag{3.49}$$

Drain current in terms of half of conduction angle is then

$$i_D = K_S V_{gsm}\left(-\frac{V_t - V_{GS}}{V_{gsm}} + \cos\omega t\right) = \frac{I_{DM}}{1 - \cos\theta}(\cos\omega t - \cos\theta). \tag{3.50}$$

The amplitude of the fundamental component of the drain current and the DC component of the drain current is determined using Fourier analysis and shown in Eqs. (3.51) and (3.52) respectively:

$$\begin{aligned} I_m &= \frac{1}{\pi}\int_{-\theta}^{\theta} i_D\cos\omega t d(\omega t) = \frac{2}{\pi}\int_{0}^{\theta} i_D\cos\omega t d(\omega t) \\ &= \frac{2I_{DM}}{\pi}\int_{0}^{\theta}\frac{\cos\omega t - \cos\theta}{1 - \cos\theta}\cos\omega t d(\omega t) = \frac{I_{DM}(\theta - \sin\theta\cos\theta)}{\pi(1 - \cos\theta)} \end{aligned} \tag{3.51}$$

and

$$I_D = \frac{1}{2\pi}\int_{-\theta}^{\theta} i_D d(\omega t) = \frac{1}{\pi}\int_{0}^{\theta} i_D(\omega t) = \frac{I_{DM}}{\pi}\int_{0}^{\theta} \frac{\cos\omega t - \cos\theta}{1-\cos\theta} d(\omega t)$$
$$= \frac{I_{DM}(\sin\theta - \theta\cos\theta)}{\pi(1-\cos\theta)}. \tag{3.52}$$

As in the case of the Class-B amplifier, the resonant circuit filters out the harmonics of this current, hence Eqs. (3.7), (3.9)–(3.10) and (3.14) apply. Current and voltage waveforms resulting from this analysis for Class-C and Class-AB amplifiers are shown in Figs. 3.8 and 3.9 respectively.

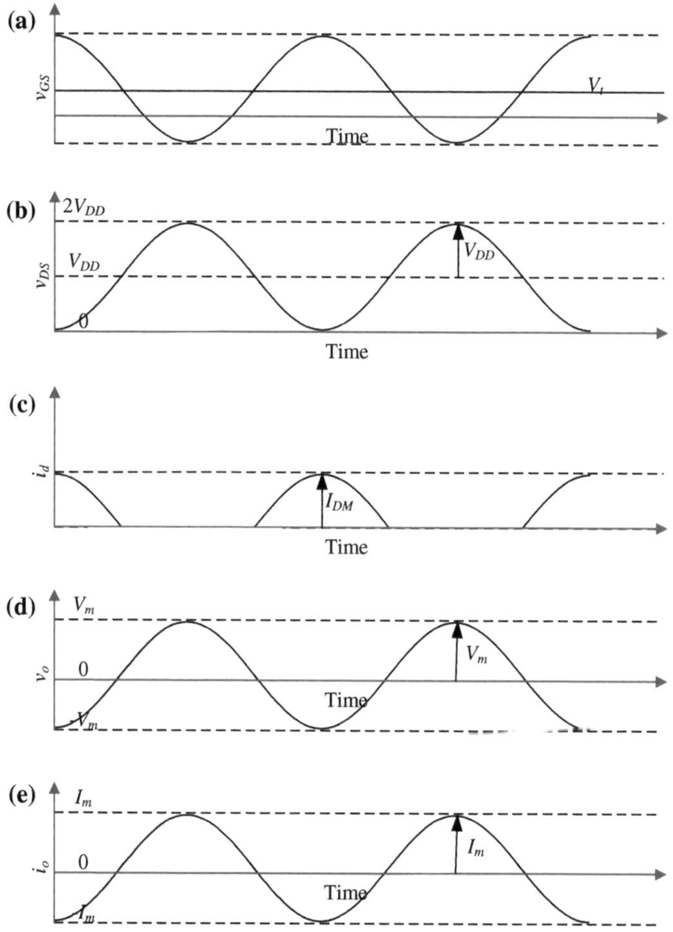

Fig. 3.8 Waveforms of the Class-C power amplifier output stage: **a** gate-source voltage, **b** drain-source voltage, **c** drain current **d** output voltage and **e** output current

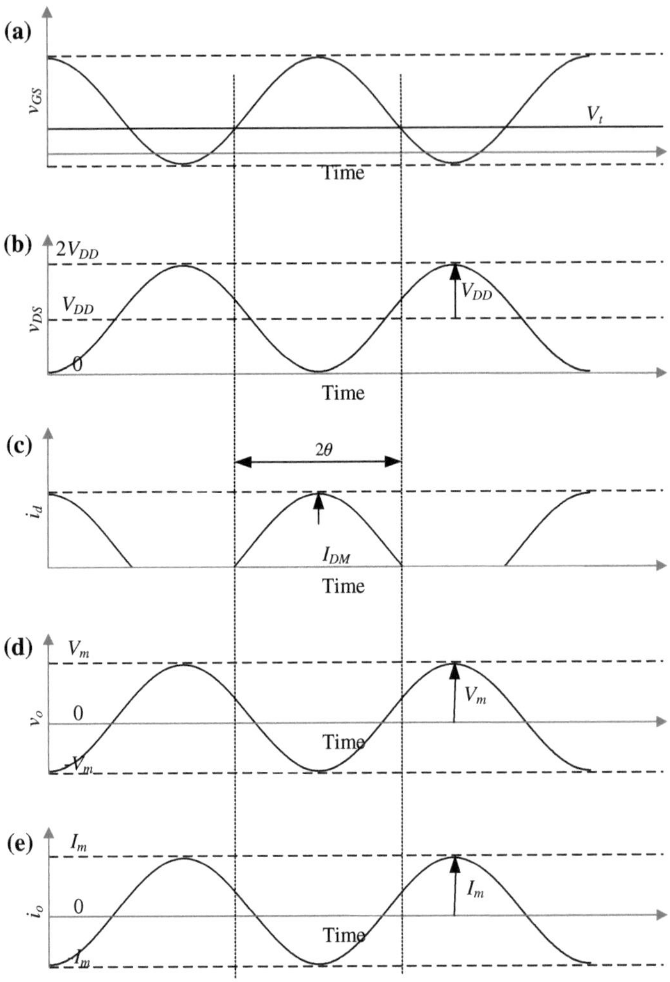

Fig. 3.9 Waveforms of the Class-AB power amplifier output stage: **a** gate-source voltage, **b** drain-source voltage, **c** drain current **d** output voltage and **e** output current

3.3.2 *Power and Efficiency*

The instantaneous and average powers delivered to the load at the operating frequency are the same as for the special case of Class-A and Class-B amplifiers and hence they are given by Eqs. (3.16)–(3.17).

The instantaneous power dissipated in the transistor at resonant frequency is zero for the part of the cycle when the transistor does not conduct and for the part when it conducts ($-\theta \leq \omega t \leq \theta$) it is

$$p_D(\omega t) = i_D(\omega t)v_{DS}(\omega t) = \frac{I_{DM}}{1-\cos\theta}(\cos\omega t - \cos\theta)(V_{DD} - V_m\cos\omega t)$$
$$= \frac{I_{DM}V_{DD}}{1-\cos\theta}(\cos\omega t - \cos\theta)\left(1 - \frac{V_m}{V_{DD}}\cos\omega t\right). \quad (3.53)$$

Noting that

$$I_{DM} = \frac{\pi I_D(1-\cos\theta)}{\sin\theta - \theta\cos\theta} \quad (3.54)$$

instantaneous power becomes

$$p_D(\omega t) = \frac{\pi I_D V_{DD}(\cos\omega t - \cos\theta)}{\sin\theta - \cos\theta}\left(1 - \frac{V_m}{V_{DD}}\cos\omega t\right)$$
$$= \frac{\pi P_{DD}(\cos\omega t - \cos\theta)}{\sin\theta - \theta\cos\theta}(1 - \zeta_1\cos\omega t). \quad (3.55)$$

The DC power from the supply is

$$P_{DD} = I_D V_{DD} = \frac{I_{DM}V_{DD}}{\pi(1-\cos\theta)}(\sin\theta - \theta\cos\theta). \quad (3.56)$$

The average power dissipated in the transistor can be calculated by subtracting the output power from the DC supply power:

$$P_D = P_{DD} - P_O = I_D V_{DD} - \frac{1}{2}I_m V_m$$
$$= \frac{I_{DM}V_{DD}}{\pi(1-\cos\theta)}(\sin\theta - \theta\cos\theta) - \frac{I_{DM}V_m}{2\pi(1-\cos\theta)}(\theta - \sin\theta\cos\theta). \quad (3.57)$$

The instantaneous power dissipated in the transistor for different values of ζ_1 in terms of supply power P_{DD} is illustrated in Fig. 3.10 for the Class-C power amplifier ($\theta = 45°$) and in Fig. 3.11 for the Class-AB power amplifier ($\theta = 120°$).

Drain efficiency of Class-A/AB/B/C is

$$\eta = \frac{P_O}{P_{DD}} = \frac{\frac{1}{2}I_m V_m}{I_D V_{DD}} = \frac{1}{2}\left(\frac{V_m}{V_{DD}}\right)\cdot\frac{\theta - \sin\theta\cos\theta}{\sin\theta - \theta\cos\theta} = \frac{1}{2}\zeta_1\cdot\frac{\theta - \sin\theta\cos\theta}{\sin\theta - \theta\cos\theta}. \quad (3.58)$$

The maximum efficiency occurs when maximum swing is reached, i.e.

$$\eta = \frac{1}{2}\cdot\frac{\theta - \sin\theta\cos\theta}{\sin\theta - \theta\cos\theta}. \quad (3.59)$$

The efficiency as a function of θ for different values of ζ_1 is shown in Fig. 3.12. The output power capability of the Class-A/AB/B/C power amplifier is

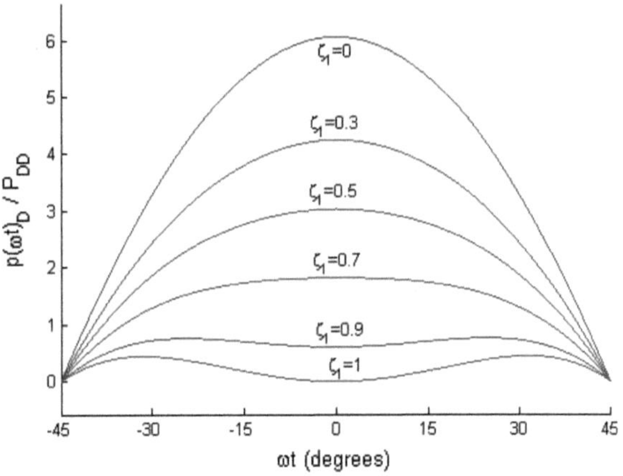

Fig. 3.10 Instantaneous drain power of the Class-C power amplifier normalized to drain power P_{DD} ($\theta = 45°$)

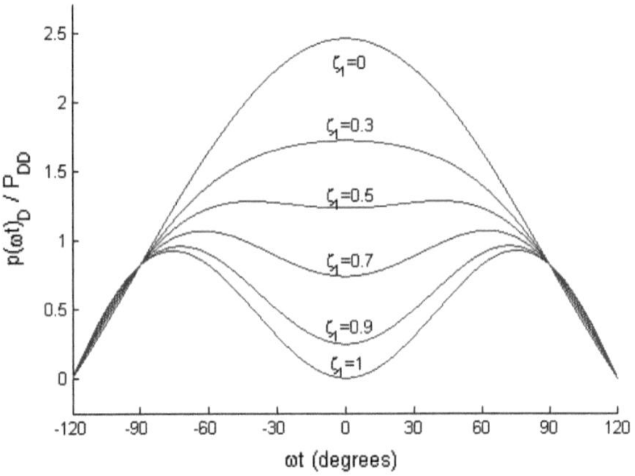

Fig. 3.11 Instantaneous drain power of the Class-AB power amplifier normalized to drain power P_{DD} ($\theta = 120°$)

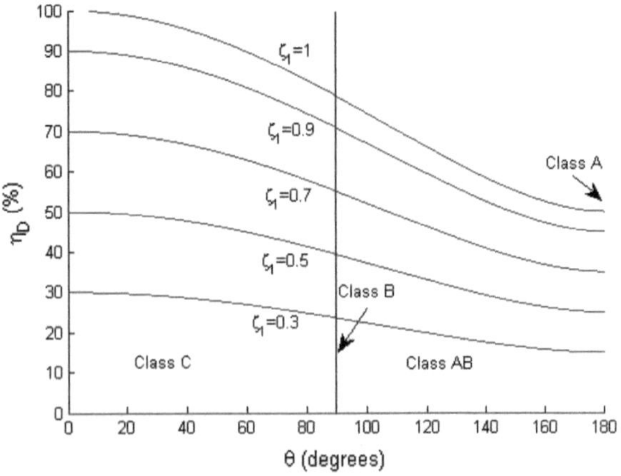

Fig. 3.12 Efficiency of Class-A/AB/B/C power amplifier as a function of θ for different values of ζ_1

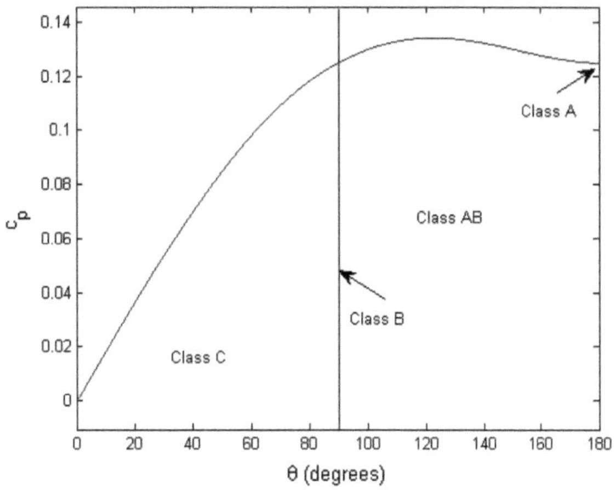

Fig. 3.13 Output power capability as a function of θ for Class-A/AB/B/C power amplifier

$$c_p = \frac{P_{out(\max)}}{I_{D\max} V_{DS(\max)}} = \frac{\frac{1}{2} I_{m(\max)} V_{m(\max)}}{I_{D\max} V_{DS\max}} = \frac{1}{2} \cdot \frac{I_{m(\max)} V_{m(\max)}}{I_{DM}(2V_{DD})} = \frac{1}{4} \cdot \frac{\theta - \sin\theta\cos\theta}{\pi(1-\cos\theta)}. \quad (3.60)$$

The output power capability as a function of θ is illustrated in Fig. 3.13. The highest output power capability is reached for Class-AB operation.

The efficiency is highest for the lowest value of the conduction angle (0 degrees), but this amplifier has zero output power compatibility. A good compromise for both efficiency and output power capability is when θ is between 45° and 60°.

3.4 Class-A/AB/B/C Power Amplifier

In Sect. 3.3, it became apparent that Class-A, Class, AB, Class-B and Class-C power amplifiers are special cases of a common Class-A/AB/B/C power amplifier. Equations applicable to Class-AB and Class-C power amplifiers are thus generic for all classes mentioned, if the correct conduction angle is used. Table 3.1 lists some

Table 3.1 Important quantities for Class-A/AB/B/C power amplifier design

θ (°)	Class	$\frac{\theta-\sin\theta\cos\theta}{\pi(1-\cos\theta)}$	$\frac{\sin\theta-\theta\cos\theta}{\pi(1-\cos\theta)}$	$\frac{\theta-\sin\theta\cos\theta}{\sin\theta-\theta\cos\theta}$	η	c_p
30	C	0.2152	0.1106	1.9460	0.9730	0.0538
45	C	0.3102	0.1649	1.8808	0.9404	0.0775
60	C	0.3910	0.2180	1.7936	0.8968	0.0978
90	B	0.5000	0.3183	1.5708	0.7854	0.1250
120	AB	0.5363	0.4060	1.3210	0.6605	0.1341
150	AB	0.5204	0.4720	1.1025	0.5513	0.1301
180	A	0.5000	0.5000	1.0000	0.5000	0.1250

θ-dependent quantities used in calculations when designing a Class/A/AB/B/C power amplifier.

3.5 Push-Pull Class-A/AB/B/C Power Amplifier

Another implementation of the Class-A/AB/B/C power amplifier is using two complementary driving transistors. Such a circuit requires a p-type and an n-type transistor, a coupling (DC blocking) capacitor and resonant circuit connected to the load, as illustrated in Fig. 3.14, using n-type and p-type MOSFET transistors (NMOS and PMOS). Such an arrangement is most practical for Class-B mode operation, where for half of the waveform cycle, the NMOS transistor is conducting (pushing current), and for the other half of the cycle, the PMOS transistor is conducting (pulling current). Voltage at the node connecting the two emitters settles at $V_{DD}/2$.

3.5.1 *Current and Voltage Waveform Analysis*

The waveforms for this configuration are shown in Fig. 3.15.

The main reason why this complementary stage found such widespread use is that THD lower can be accomplished if two matched transistors are used instead of a single driving transistor. This phenomenon can be analyzed by looking at the total drain current, which is the difference in currents i_{Dn} and I_{Dp} originating from the

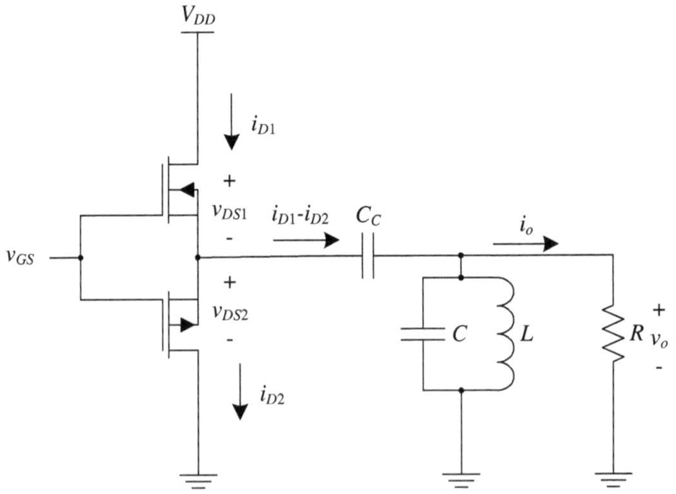

Fig. 3.14 Class-A/AB/B/C power amplifier push-pull configuration

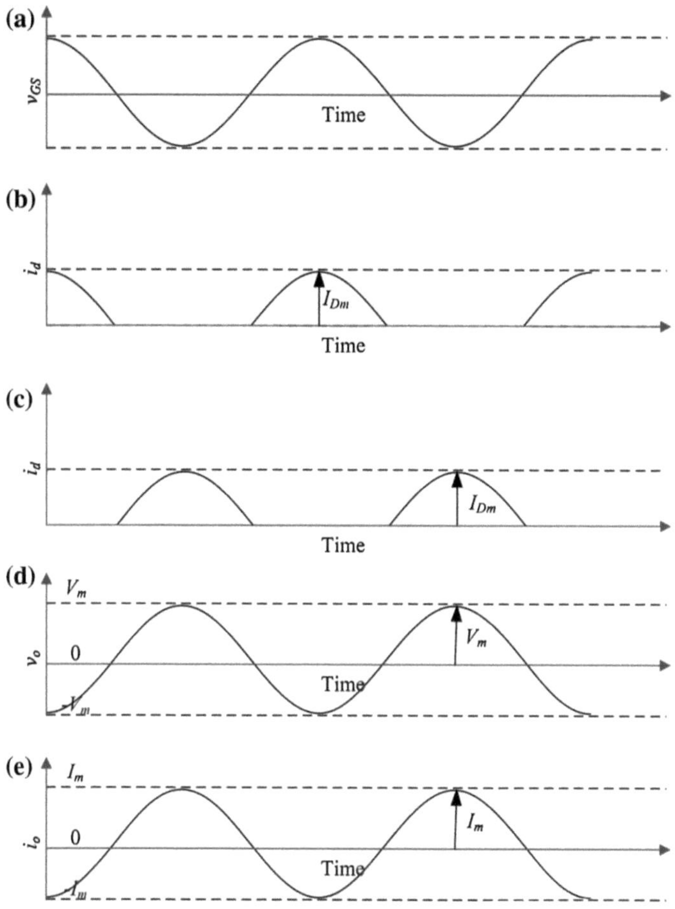

Fig. 3.15 Waveforms of the Class-B push-pull power amplifier: **a** driving voltage, **b** NMOS current, **c** PMOS current, **d** output voltage and **e** output current

NMOS and PMOS respectively. Using Fourier series expansion, the half-sine waveform of i_{Dn} can be expressed as

$$\begin{aligned} i_{Dn} &= I_D + i_{d1} + i_{d2} + i_{d3} + \cdots \\ &= I_D + I_{dm1}\cos\omega t + I_{dm2}\cos 2\omega t + I_{dm3}\cos 3\omega t + \cdots. \end{aligned} \tag{3.61}$$

The drain current of i_{Dp}, however, is shifted in phase by 180°, thus

$$\begin{aligned} i_{Dp} &= I_D + I_{dm1}\cos(\omega t - \pi) + I_{dm2}\cos 2(\omega t - \pi) + I_{dm3}\cos 3(\omega t - \pi) + \cdots \\ &= I_D - I_{dm1}\cos\omega t + I_{dm2}\cos 2\omega t - I_{dm3}\cos 3\omega t + \cdots. \end{aligned} \tag{3.62}$$

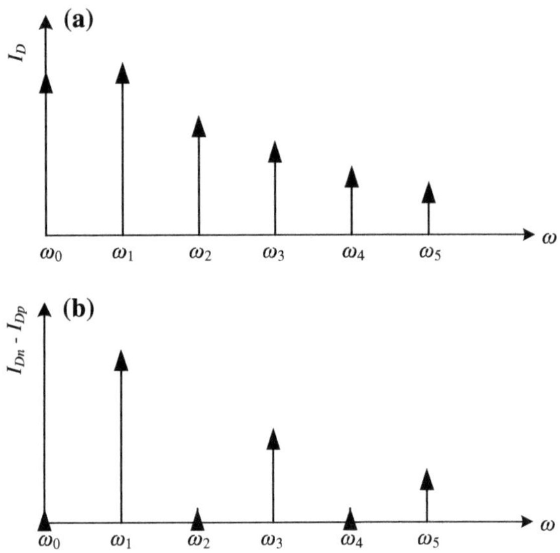

Fig. 3.16 Spectrum of the current waveform of **a** single-transistor and **b** complementary push-pull Class-B power amplifier

The difference in current is then

$$i_{Dn} - i_{Dp} = 2(I_{dm1}\cos\omega t + I_{dm3}\cos 3\omega t + I_{dm5}\cos 5\omega t + \cdots) \tag{3.63}$$

and it contains no even harmonics. The bandpass filter formed by the resonant circuit only has to remove odd harmonics. The spectrum of the current waveforms up to the fifth harmonic for a single-transistor Class-B power amplifier and for the complementary push-pull configuration is shown in Fig. 3.16.

When designing this stage, care needs to be taken not to introduce another type of distortion. A prominent type of distortion in push-pull configurations is cross-over distortion. This distortion occurs because the driving transistors do not switch on exactly when $v_{GS} = 0$ V, but rather when $v_{GS} = V_t$. Cross-over distortion can be reduced by introducing separate biasing circuits for NMOS and PMOS transistors, or by incorporating feedback.

3.5.2 *Power and Efficiency*

The DC supply power of this configuration is

$$P_{DD} = \frac{2V_{DD}V_m}{\pi R}, \tag{3.64}$$

the drain power dissipated in both transistors is

$$P_D = \frac{2V_{DD}V_m}{\pi R} - \frac{V_m}{2R} \tag{3.65}$$

and the drain efficiency is

$$\eta = \frac{\pi}{4}\left(\frac{V_m}{V_{DD}}\right). \tag{3.66}$$

Thus, the maximum theoretical efficiency of the push-pull Class-B power amplifier remains unchanged at 78.5 %. The output power capability is also unchanged and it is 1/8.

3.6 Class-J Power Amplifiers

A Class-J power amplifier is similar to a Class-AB/B power amplifier, the main difference being that the resonant circuit is omitted to allow for wideband operation. The result is that both drain voltage and drain current are half-sinusoidal waveforms, i.e. they are both composed of even harmonics. The bias is that of the conventional Class-AB/B amplifier and the Class-B efficiency of 78.5 % can still be reached. A simplified Class-J power amplifier circuit, shown in Fig. 3.17 [2, 3], consists of a transistor, an *RFC* and a capacitor in parallel to the load Z_L. A parallel capacitor C may consist entirely of the transistor parasitic capacitance C_{OUT}. The load is shown such that it includes a matching network.

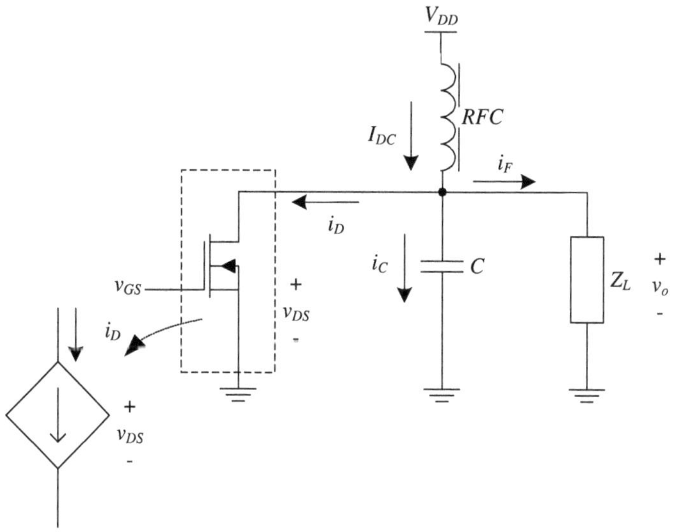

Fig. 3.17 Circuit of the Class-J power amplifier

3.6.1 Current and Voltage Waveform Analysis

Under the assumption that the Class-J amplifier circuit is biased at the bias of Class-B, i.e. the conduction angle is 180°, and if the part of the cycle during which the transistor conducts is $0 \leq \omega t \leq \pi$, then the current can be expressed in terms of $\sin \omega t$ as:

$$i_D = \begin{cases} I_{DM} \sin \omega t, & \text{when } 0 \leq \omega t \leq \pi \\ 0, & \text{when } \pi \leq \omega t \leq 2\pi \end{cases}. \tag{3.67}$$

Furthermore, we define the current flowing through the complex load with matching network as

$$i_F = I_1 \sin(\omega t + \phi), \tag{3.68}$$

where I_1 and ϕ are independent Class-J design parameters and ϕ denotes the initial phase shift. Current through the capacitor is

$$i_C = I_{DC} - i_D - i_F. \tag{3.69}$$

I_{DC} flows entirely into the transistor, thus using Fourier expansion in a similar manner as for Class-B amplifier, it can be calculated as

$$I_{DC} = \frac{I_{DM}}{\pi}. \tag{3.70}$$

Output voltage over load Z_L is the same as the voltage over the shunt capacitor C. Thus it can be calculated by integrating the current through the capacitor. For $0 \leq \omega t \leq \pi$, and using substitution $\theta = \omega t$, it is

$$\begin{aligned} v_o &= \frac{1}{\omega C} \int_0^\theta \left[I_{DM} \sin \theta - \frac{I_{DM}}{\pi} - I_1 \sin(\theta + \phi) d\theta \right] + V_{OFF} \\ &= \frac{1}{\omega C} \left[I_{DM} \left(1 - \frac{\theta}{\pi} - \cos \theta \right) + I_1(\cos(\theta + \phi) - \cos \phi) \right] + V_{OFF} \end{aligned} \tag{3.71}$$

and for $\pi \leq \omega t \leq 2\pi$ it is

$$v_o = \frac{1}{\omega C}\left[\int_0^{\pi}\left(I_{DM}\sin\theta - \frac{I_{DM}}{\pi} - I_1\sin(\theta+\phi)\right)d\theta + \int_{\pi}^{\theta}\left(-\frac{I_{DM}}{\pi} - I_1\sin(\theta+\phi)d\theta\right)\right] + V_{OFF}$$
$$= \frac{1}{\omega C}\left[I_{DM}\left(2-\frac{\theta}{\pi}\right) + I_1(\cos(\theta+\phi) - \cos\phi)\right] + V_{OFF}, \tag{3.72}$$

where V_{OFF} is an arbitrary DC offset chosen such that v_o remains positive for any value of θ. Value of C then has to be chosen so that the mean value of the voltage function is unity.

3.6.2 Power and Efficiency

DC power delivered from the supplies is

$$P_{DD} = V_{DD}I_{DC}. \tag{3.73}$$

The output power is

$$P_O = \mathrm{Re}\left(\frac{I_1^2 Z_L}{2}\right). \tag{3.74}$$

The resulting efficiency is

$$\eta = \frac{\mathrm{Re}\left(I_1^2 Z_L\right)}{2V_{DD}I_{DC}}. \tag{3.75}$$

Maximum efficiency is obtained sweeping through all possible combinations of I_1 and ϕ and X_C/R_L, ratio of reactance of the capacitor C and real part of the load, Re $(Z_L) = R_L$. Then, with C calculated from the voltage equation, R_L can be obtained with ease. The imaginary part of the load, $X_L = \mathrm{Im}(Z_L)$, is then adjusted accordingly, as discussed in the next section [3].

As already discussed, all the theory in this section has been reviewed assuming that the transistor is biased at Class B level. By biasing the Class-J amplifier deeper in the Class-AB bias region, the second harmonic components are reduced and the usable range of X_C/R_L increases. Figure 3.18 shows the efficiency plots for Class-B and Class-AB bias. Thus, efficiency of 78.5 % is possible for different design parameter choices. The parameter choices, two optimal and one suboptimal for Class-B bias, have been listed in Table 3.2 and are taken from [2].

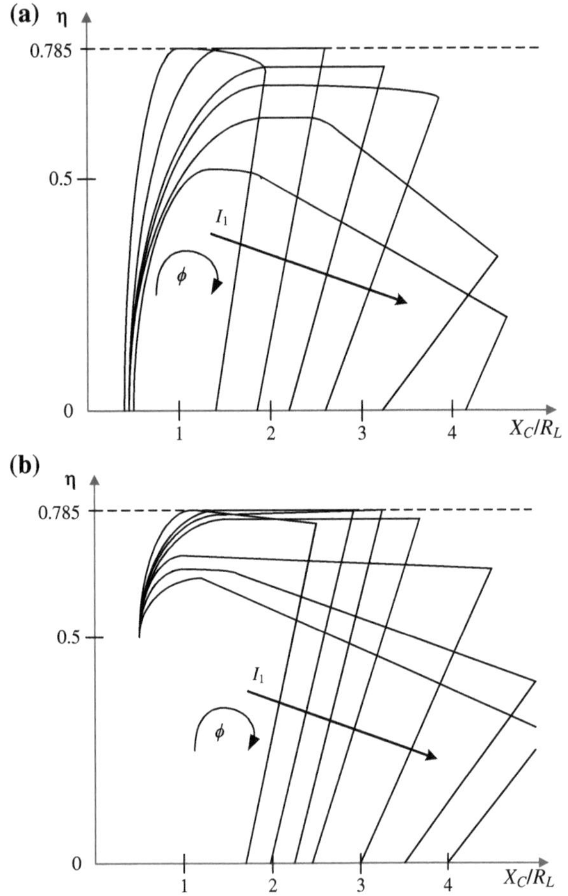

Fig. 3.18 Efficiency of Class-J power amplifier for **a** Class-B and **b** deep Class-AB bias

Table 3.2 Efficiency for different Class-J parameter choices

I_1	ϕ (°)	X_C/R_L	η (%)
2	68	0.27	78.5
0.7	−15.6	2.65	78.5
0.6	−10.2	3.82	63.9

3.6.3 *Calculating X_L*

The choice of initial phase shift ϕ requires the correct load impedance. Thus, the correct value of X_L needs to be calculated. The impedance seen by the source is

$$Z_i = \frac{Z_L}{j\omega C Z_L + 1}. \tag{3.76}$$

Now, because the voltage over the transistor and over the load is the same, we can write

$$I_F Z_L = I_D Z_i, \tag{3.77}$$

where I_D is the fundamental component of transistor current, $I_D = I_{DM}/2$. Thus,

$$I_F = \frac{I_{DM}}{2}\frac{Z_i}{Z_L} = \frac{I_{DM}}{2}\frac{1}{j\omega C Z_L + 1} = \frac{I_{DM}}{2}\frac{1 - \omega C X_L - j\omega C R_L}{(\omega C R_L)^2 + (1 - \omega C X_L)^2}. \tag{3.78}$$

Finally, X_L is calculated by solving

$$\phi = \arg(I_F), \tag{3.79}$$

where $\arg(x + jy)$ is the complex number argument function, defined as

$$\arg(x + jy) = \arctan\left(\frac{y}{x}\right) + \frac{\pi}{2}\operatorname{sign}(y)(1 - \operatorname{sign}(x)). \tag{3.80}$$

and $\operatorname{sign}(x)$ is the signum function. In Eq. (3.80), $\operatorname{sign}(y) = -1$, $\operatorname{sign}(x)$ is sign $(1 - \omega C X_L)$, and $\arctan(y/x)$ is

$$\arctan\left(\frac{y}{x}\right) = -\arctan\left(\frac{\omega C R_L}{1 - \omega C X_L}\right). \tag{3.81}$$

Thus we need to solve

$$\phi = -\arctan\left(\frac{\omega C R_L}{1 - \omega C X_L}\right), \quad \text{if } X_L < \frac{1}{\omega C}, \tag{3.82}$$

and

$$\phi = -\pi - \arctan\left(\frac{\omega C R_L}{1 - \omega C X_L}\right), \quad \text{if } X_L > \frac{1}{\omega C}. \tag{3.83}$$

Because of the periodicity of the tangent function, there is only one solution, namely:

$$X_L = \frac{\tan\phi + \omega C R_L}{\omega C \tan\phi}. \tag{3.84}$$

3.7 Doherty Power Amplifiers

It was demonstrated earlier in this text that the efficiency of any continuous-mode power amplifier decreases with a decrease in the swing of the output waveform. As a result, a power amplifier at low signal drives is not practical from a power consumption perspective.

There are, however, several techniques that have found widespread use in practical designs for efficiency improvement at different voltage levels. In this chapter we discuss the Doherty technique.

A typical Doherty architecture combines a carrier (main) power amplifier of moderate efficiency at lower drive voltages, and an auxiliary (peaking) power amplifier of high efficiency at high drive voltages, through two quarter-wavelength transmission lines or networks, as shown in Fig. 3.19. The quarter-wavelength transmission line, discussed in Chap. 5, acts as an impedance converter. Traditionally, the main power amplifier operates in Class-A or Class-B configuration, while the auxiliary power amplifier operates in Class-C configuration. Because the Class-F amplifier has efficiency comparable to that of Class C, Doherty amplifiers combining a pair of Class-F amplifiers, a Class-F amplifier and an inverse Class-F amplifier and a Class-F amplifier and a Class-C amplifier have been demonstrated recently. One recent implementation is shown in Fig. 3.20.

Only the main power amplifier is active when the signal amplitude is half or less of the full drive swing amplitude and both power amplifiers contribute to output power when the signal amplitude is larger than half of the full swing. This operation can be understood by looking at a Class-B main/Class-C peaking Doherty system and dividing it into low-power, medium-power and peak-power regions. In the low-power region, the auxiliary power amplifier remains cut off and appears as an open circuit and the main power amplifier operates as an ordinary Class-B amplifier, with the efficiency reaching the theoretical 78.5 % of the ideal Class-B circuit

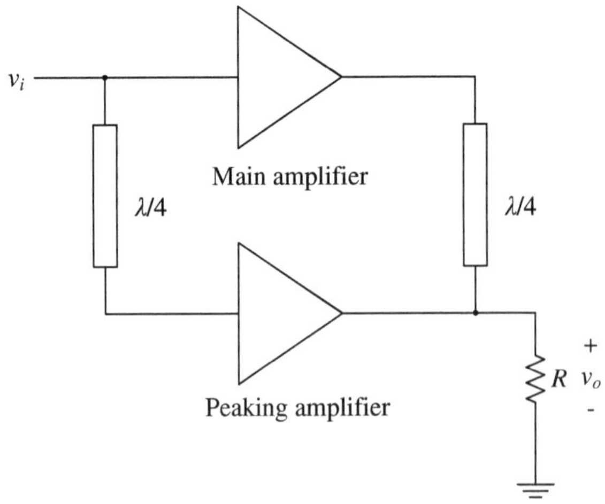

Fig. 3.19 A Class-B-main/Class-C-peaking Doherty power amplifier

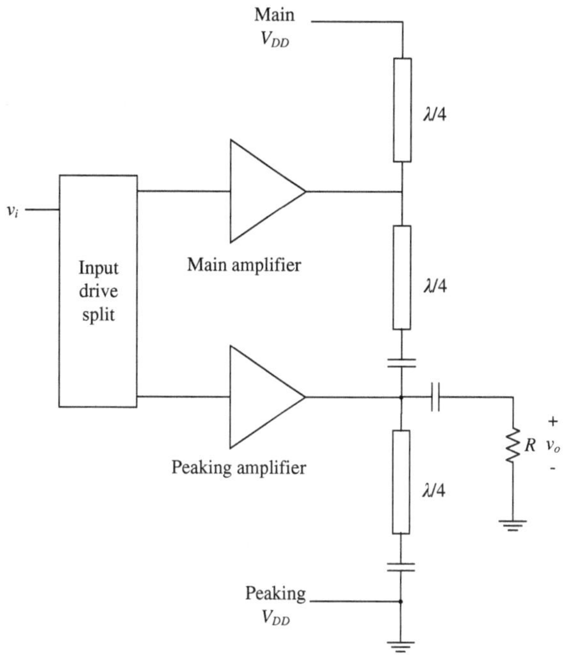

Fig. 3.20 A Class-F-main/Class-C-peaking Doherty power amplifier

at half of the maximum drive. As the signal amplitude increases into the medium-power region, the auxiliary power amplifier becomes active and with a further power increase, the main amplifier remains in saturation and the auxiliary amplifier delivers an increased portion of power. At peak drive, both power amplifiers deliver half of the total system power and ideally, the efficiency for the ideal system remains at maximum of Class-B configuration: 78.5 %.

The operation of the Doherty power amplifier is based on the fact that for maximum efficiency, the ratio $\zeta_1 = V_m/V_{DD}$ should be kept relatively constant. Because $V_m = RI_m$ and I_m varies with the signal, load resistance R must also be varied to counter the effect. This is accomplished by the quarter-wavelength transformers in Fig. 3.19.

The efficiency of the Doherty stage using quarter-wavelength transformers can be expressed in terms of the drive voltage V_i and maximum drive voltage V_{imax} as

$$\eta = \begin{cases} \frac{\pi}{2}\left(\frac{V_i}{V_{imax}}\right)^2, & \text{when } 0 \leq V_i \leq \frac{V_{imax}}{2} \\ \frac{\pi}{2}\frac{\left(\frac{V_i}{V_{imax}}\right)}{3\left(\frac{V_i}{V_{imax}}\right)-1}, & \text{when } \frac{V_{imax}}{2} \leq V_i \leq V_i \end{cases} \tag{3.85}$$

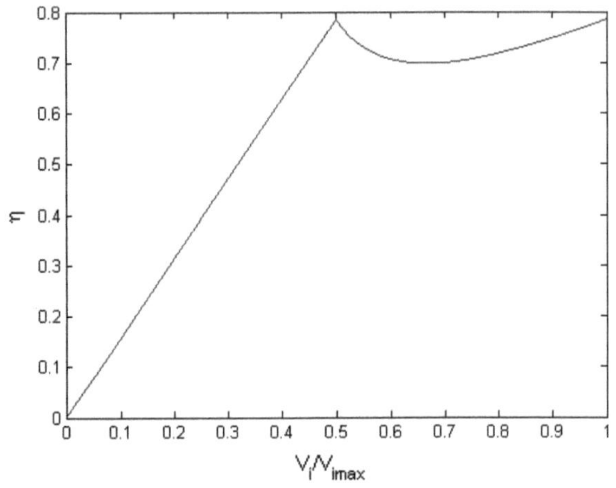

Fig. 3.21 Efficiency of Class-B main/Class-C peaking Doherty power amplifier

The efficiency is illustrated in Fig. 3.21, where it is apparent that the efficiency varies by around 10 % for drives higher than about 40 % of the maximum driving voltage.

3.8 Recent State-of-the-Art Examples

In this section, we list several state-of-the-art examples of continuous-mode power amplifiers published in recent years.

Kaynak et al. [4] demonstrated a fully-integrated Class-A power amplifier for biomedical applications delivering 3.75 mW of power from a supply voltage of 2 V and 1.25 W from a 1-V supply. The amplifier operates at 2.4 GHz.

Luo et al. [5] reported a 24-W GaN Class-AB power amplifier operating in *Ku* band. The amplifier has the largest gain at 14 GHz.

Vatankhahghadim and Boumaiza [6] demonstrated two broadband Class-AB power amplifiers for operation around 2 GHz. The amplifier reached 70 % efficiency for 400 MHz bandwidth and 60 % for 900 MHz bandwidth for 45.5 dBm outputs.

Yang et al. presented a Class-AB CMOS power amplifier operating from 1.2-V supply [7]. The amplifier incorporates an L-shaped concentric vertical transformer and achieves a power level of 22 dBm with efficiency of 47 % at 2.4 GHz.

Zihir et al. [8] demonstrated an *X*-band SiGe power amplifier for phased radar applications. The amplifier delivers 23.2 dBm output power at 9 GHz, and consists of two stages, one biased in Class-A configuration, and an output stage biased in Class-AB configuration.

Razaei et al. [9] presented a design of a wideband Class-J power amplifier fabricated in GaN technology. The power amplifier exhibits greater than 50 % drain efficiency in a wide band from 2.25 to 3.075 GHz.

Camarchia et al. [10] demonstrated two high-efficiency 7-GHz Doherty GaN monolitic power amplifiers for microwave backhaul radio links. Both amplifiers are capable of delivering 5 W output power.

3.9 Summary

In this chapter, the theory of operations of continuous power amplifiers was presented, and specifically, that of Classes A, B, AB, C and J. Although some of these power amplifier classes are not necessarily widely used at RF and microwave frequencies, the theory behind the operation of each class is important for thorough understanding of the aspects of power amplification. We noticed that important design parameters of each power amplifier are its efficiency, output power, output power capability and linearity. A well-designed power amplifier will try to strike a balance between the output power and the efficiency at which that power is reached, but the overall performance will affect the linearity as well. Typically, in a continuous-mode power amplifier, the choice of a quantity called the conduction angle determines the overall performance of the amplifier. We noticed that Classes A, B, AB and C are actually variations of the same amplifier, which we call the Class-A/AB/B/C power amplifier. Furthermore, the circuit of the Class-J power amplifier is simpler than that of Class-A/AB/B/C, but its design is tedious and requires computer-aided calculations, but if performed properly, it can lead to very efficient implementation. Finally, we discussed that efficiency and output power can be increased with power combining, and we gave an example of a Doherty power amplifier.

When discussing the Class C power amplifier, we realized that it has by far the highest efficiency if Classes A, AB, B and C are considered. We can note that the operation of Class C, however, tends to resemble that of the switch, which leads to the conclusion that switch-mode power amplifiers are expected to have higher efficiency than continuous-mode power amplifiers. Thus, the switch-mode power amplifier variations are going to form the topic of the next chapter.

References

1. Kazimierczuk MK. RF power amplifiers. 2nd ed. Chiechester: Wiley; 2015.
2. Cripps SC. RF power amplifiers for wireless communications. Norwood: Artech House; 2006.
3. Andersson CM, Gustafsson D, Yamanaka K, Kuwata E, Otsuka H, Nakayama M, Hirano Y, Angelov I, Fager C, Rorsman N. Theory and design of class-J power amplifiers with dynamic load modulation. IEEE Trans Microw Theor Tech. 2012;60(12):3778–86.
4. Kaynak M, Tekin I, Gurbuz Y. Fully integrated low-power SiGe power amplifier for biomedical applications. IET Microw Antennas Propag. 2011;5(2):214–9.

5. Luo WJ, Chen XJ, Pang L, Yuan TT, Liu XY. A, 24 W Ku Band GaN based power amplifier with 9.1 dB linear gain. Microelectron J. 2012;43(8):569–72.
6. Vatankhahghadim A, Boumaiza S. Design of high-efficiency and broadband-tuned class AB power amplifier. Microw Opt Technol Lett. 2011;53(2):395–8.
7. Yang HS, Chen YH, Chen YJE. A 1.2-V 90-nm fully integrated compact CMOS linear power amplifier using the coupled L-shape concentric vortical transformer. IEEE Trans Microw Theory Tech. 2014;62(11):2689–99.
8. Zihir S, Dinc T, Gurbuz Y. Compact X-band SiGe power amplifier for single-chip phased array radar applications. IET Microw Antennas Propag. 2012;6(8):956–61.
9. Rezaei S, Belostotski L, Ghannouchi FM, Aflaki P. Integrated design of a class-J power amplifier. IEEE Trans Microw Theor Tech. 2013;61(4):1639–48.
10. Camarchia V, Rubio JJM, Pirola M, Quaglia P, Colantonio P, Giannini F, Giofre P, Piazzon L, Emanuelsson T, Wegeland T. High-efficiency 7 GHz Doherty GaN MMIC power amplifiers for microwave backhaul radio links. IEEE Trans Electron Devices. 2013;60(10):3592–5.
11. Holmes D. Asymetric Doherty power amplifier design. Microw J. 2014;57(4):32–40.
12. Hayati M, Roshani S. A novel miniaturized power amplifier with nth harmonic suppression. AEU-Int J Electron Comm. 2014;68(10):1016–21.
13. Corrado F, Cignani R, Santarelli A, Filicori F. Design of 40-W AlGaN/GaN MMIC high power amplifiers for C-band SAR applications. IEEE Trans Microw Theor Tech. 2013;61 (12):4492–504.
14. Grebennikov A, Sokal NO. Switchmode RF power amplifiers. 1st ed. Burlington: Elsevier; 2007.
15. Eccleston KW, Smith KJI, Gough PT. A compact class-F/class-C Doherty amplifier. Microw Opt Technol Lett. 2011;53(7):1606–10.
16. Rogers J, Plett C. Radio frequency integrated circuit design. 2nd ed. Boston: Artech House; 2010.
17. Wright P, Lees J, Benedikt J, Tasker PJ, Cripps SC. A methodology for realizing high efficiency class-J in a linear and broadband PA. IEEE Trans Microw Theor Tech. 2009;57 (12):3196–204.
18. Raab FH, Asbeck P, Cripps S, Kenington PB, Popovic ZB, Pothecary N, Sevic JF, Sokal NO. Power amplifiers and transmitters for RF and microwave. IEEE Trans Microw Theor Tech. 2002;50(3):814–26.

Chapter 4
Switch-Mode Power Amplifiers

In the previous chapter it became obvious that the efficiency of a power amplifier increases if the transistor is configured to operate under non-linear conditions instead as a voltage-dependent current source. If the transistor is operated under saturation, the operation will resemble that of the switch, and as a result, it will have higher efficiency. This led to several classes of switch-mode power amplifiers finding widespread use [1–20]. Class D, Class, E, Class F and combinations thereof all use a different method for waveform shaping and/or harmonic termination in order to accomplish efficiency improvement and thus form the topic of this chapter.

Class-D power amplifiers use two or more transistors to generate square drain (collector) voltage waveforms. This class of amplifier has a series resonant filter that passes only the fundamental harmonic to the load. Current is only drawn through the transistor that is on. This results in the maximum power output (in transformer coupled configuration) of just under V_{DD}^2/R and the theoretical efficiency of 100 % with the output power capability of 0.159. Output power capability is the largest attained by any class. Practical implementations of Class-D amplifiers, however, suffer from losses due to saturation, switching speed and drain capacitance. Current device technology limitations do not allow Class-D amplifiers to reach the higher end of the UHF band. Recently, research efforts have been directed towards current-switching Class-D configurations that are capable of operating up to the *S*-band.

The Class-E power amplifier, on the other hand, only requires one transistor and an additional shunt capacitance, a series resonant filter, and an RFC or a finite DC feed inductor. Other configurations are also possible, as long as the basic idea that the current and voltage waveforms are never simultaneously on is followed. This implies that switching losses are reduced and efficiency is once more 100 % in theory. However, the output power is just over half of V_{DD}^2/R and the output power capability is 0.098. The output power capability is much less than that of Class D, but is comparable to practical Class-A/AB/B/C power amplifier implementations ($\sim$0.125). Inverse Class-E power amplifiers (Class-E^{-1}), where the load network inductor and capacitor are swapped around, are also discussed. Class-E power amplifiers have been successfully demonstrated up to mm-wave frequencies.

Class-F power amplifiers use multiple harmonic resonators or transmission lines to shape the drain or collector waveforms. An infinite number of odd harmonic

M. Božanić and S. Sinha, *Power Amplifiers for the S-, C-, X- and Ku-bands*,
Signals and Communication Technology, DOI 10.1007/978-3-319-28376-0_4

resonators results in a voltage that is shaped as a square wave and in a current that is shaped as half-sine wave. The efficiency is directly influenced by the number of resonators used and increases from 50 % for only one harmonic trap to 100 % when an infinite number of resonators (or quarter-wave transmission line) are used. The output power capability is better than that of Class-A/AB/B/C and Class-E power amplifiers and increases from 0.125 to 0.159, similarly depending on the number of resonators. Inverse Class-F (Class F^{-1}) is essentially identical to Class-F, but even-harmonic resonators are used to shape voltage as half-sine wave and current as square wave.

Numerous hybrid classes are also possible. In this chapter, Class-DE, Class-FE, Class-E/F and Class-E_M will also be mentioned.

4.1 Class-D Power Amplifier

Class-D power amplifiers can further be classified into Class-D voltage-switching amplifiers, fed by a DC voltage source, and Class-D current-switching amplifiers, fed by a DC voltage source through an RFC, thus forming the current source. The Class-D voltage-switching amplifier requires a variation of a series resonant circuit and the Class-D current-switching amplifier requires a variation of a shunt resonant circuit.

Numerous topologies of both voltage-switching and current-switching variations involving different combinations of n-type and p-type transistors with different types of resonant circuits are possible. For additional details regarding these topologies, the reader is referred to [1], but in this book, only one complementary voltage-switching and one complementary current-switching variation will be discussed.

4.1.1 Class-D Complementary Push-Pull Voltage-Switching Power Amplifier

The complementary Class-D voltage-switching power amplifier employs two MOSFETs, or any two matched transistors, one pushing the current into and the other one pulling the current out of the series resonant circuit, as shown in Fig. 4.1. In this figure, a three-tap transformer is used at the input side to place a non-inverting and an inverting drive waveform at the top and bottom transistor. This causes one transistor to turn on as the other one turns off, which defines a square voltage waveform over the drain. The resonant *LC* circuit is tuned to the operating frequency so that a sinusoidal waveform appears across the load resistor. A large blocking capacitor C_b is needed to isolate the DC power supply by bypassing the RF current to ground. This capacitor also sets the apparent value of

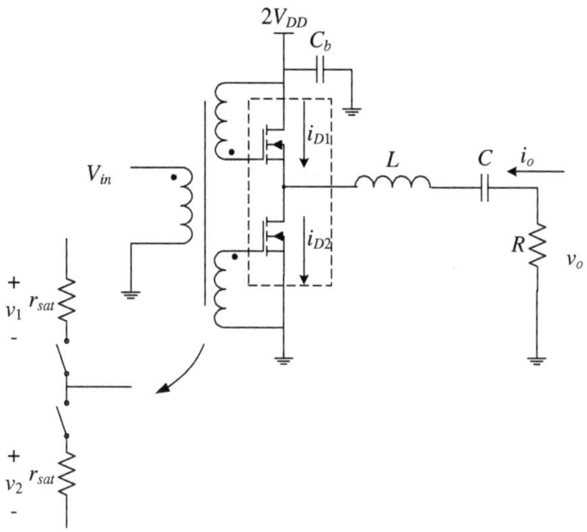

Fig. 4.1 Complementary Class-D voltage-switching push-pull configuration

resistance to each transistor to $2R$. The circuit is powered from dual rail supplies V_{DD} and $-V_{DD}$, or alternatively, $2V_{DD}$ and 0 V.

4.1.1.1 Current and Voltage Waveform Analysis

To begin the analysis of the Class-D voltage-switching output stage, it is assumed that power losses due to transistor leakage currents, transistor-switching and non-ideal tuning are negligible. The two transistors are perfectly matched and conduct at a conduction angle of 180°. Furthermore, the transistor can be modeled as an ideal switch with the series resistance r_{sat} and the driving signal is a square wave with a duty cycle of 50 %. The resonant circuit is tuned to the fundamental harmonic with frequency

$$f_0 = \frac{1}{2\pi\sqrt{LC}} \tag{4.1}$$

and the loaded quality factor Q_L

$$Q_L = \frac{2\pi f_0 L}{R}, \tag{4.2}$$

which have all been discussed in previous chapters. The voltage through each switch will be

$$v_{sat}(\omega t) = r_{sat} i_D(\omega t), \tag{4.3}$$

where ω is the angular frequency ($2\pi f$) and $i_D(\omega t) = i_{D1}(\omega t) = i_{D2}(\omega t)$. The voltage at the input of the LC circuit is

$$v(\omega t) = v_{D1}(\omega t) - V_{DD}. \tag{4.4}$$

During the part of the cycle when the switch is on ($-\pi/2 \leq \omega t \leq \pi/2$), this voltage becomes

$$v(\omega t) = -V_{DD} + r_{sat} I_{DM} \cos \omega t \tag{4.5}$$

and the drain current is

$$i_{D1}(\omega t) = I_{DM} \cos \omega t, \tag{4.6}$$

where $I_{DM} = i_D(0)$. During the part of the cycle when the switch is off ($\pi/2 \leq \omega t \leq 3\pi/2$), the voltage and the current are

$$v(\omega t) = V_{DD} + r_{sat} I_{DM} \cos \omega t \tag{4.7}$$

and

$$i_{D1}(\omega t) = 0. \tag{4.8}$$

The fundamental frequency voltage V_m of voltage $v(\omega t)$ can be calculated using Fourier analysis:

$$V_m = -\frac{1}{\pi} \int_{-\frac{\pi}{2}}^{\frac{3\pi}{2}} v \cos \omega t d(\omega t) = \frac{4}{\pi} V_{DD} - r_{sat} I_{DM}. \tag{4.9}$$

The DC current I_{DC} can be obtained as

$$I_{DC} = \frac{1}{2\pi} \int_{-\frac{\pi}{2}}^{\frac{\pi}{2}} i_{D1}(\omega t) d(\omega t) = \frac{I_{DM}}{\pi}. \tag{4.10}$$

Note that one of the two transistors will always be on and that the filter will remove all components other than the fundamental. Thus, the current through the load will have a magnitude of $I_m = V_m/R$, which will also be equal to I_{DM}:

$$I_m = \frac{V_m}{R} = \frac{4}{\pi} \frac{V_{DD} - r_{sat} I_m}{R}, \tag{4.11}$$

which can be rearranged as

$$I_m = \frac{4}{\pi}\frac{V_{DD}}{R}\frac{1}{1+\frac{r_{sat}}{R}}. \tag{4.12}$$

Class-D voltage-switching waveforms are shown in Fig. 4.2.

4.1.1.2 Power and Efficiency

The DC power from the supply can be obtained as

$$P_{DD} = 2V_{DD}I_{DM} = \frac{8}{\pi^2}\frac{V_{DD}^2}{R}\frac{1}{1+\frac{r_{sat}}{R}}. \tag{4.13}$$

The power through the load is

$$P_O = \frac{1}{2}I_m V_m = \frac{1}{2}I_m^2 R = \frac{8}{\pi^2}\frac{V_{DD}^2}{R}\frac{1}{\left(1+\frac{r_{sat}}{R}\right)^2}. \tag{4.14}$$

The efficiency is the ratio of the DC power from the supply to the power delivered to the load:

$$\eta = \frac{P_O}{P_{DD}} = \frac{\frac{8}{\pi^2}\frac{V_{DD}^2}{R}\frac{1}{\left(1+\frac{r_{sat}}{R}\right)^2}}{\frac{8}{\pi^2}\frac{V_{DD}^2}{R}\frac{1}{1+\frac{r_{sat}}{R}}} = \frac{1}{1+\frac{r_{sat}}{R}}. \tag{4.15}$$

The efficiency is thus only dependent on the load resistance and switch saturation resistance and it can reach 100 % for the idealized case where $r_{sat} = 0$. In the ideal case, the drain voltage waveform is a square wave. Figure 4.3 shows the efficiency as the function of r_{sat}/R.

In this analysis, we modeled a switch by a constant resistance. In fact, the total impedance contributing to the transistor loss increases with frequency. As a result, the losses make Class-D voltage-switching configurations already impractical at UHF.

The output power capability is

$$c_p = \frac{1}{N}\frac{P_{out(\max)}}{I_{D\max}V_{DS(\max)}}, \tag{4.16}$$

where N was introduced to cater for the number of transistors greater than one and $V_{DS(\max)}$ is the total drain-source voltage for both transistors. Thus, for this Class-D configuration with two transistors and $r_{sat} = 0$,

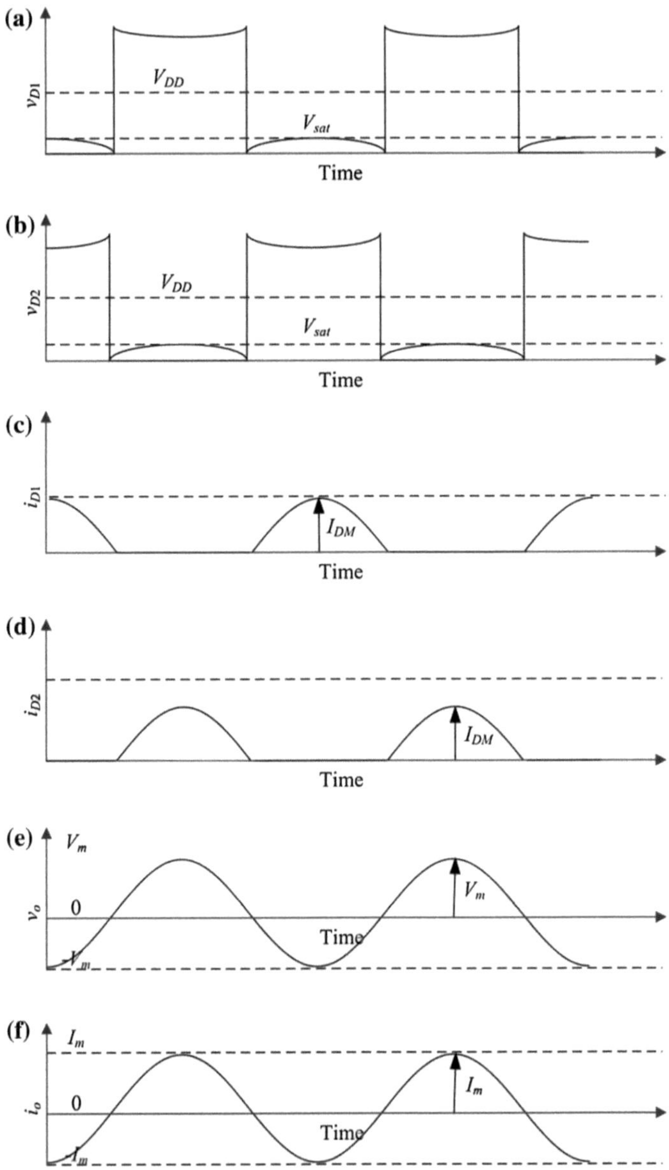

Fig. 4.2 Waveforms of the Class-D voltage-switching power amplifier: **a** switch voltage v_{D1}, **b** switch voltage v_{D2}, **c** drain current i_{D1}, **d** drain current i_{D2}, **e** output voltage, **f** output current

$$c_p = \frac{1}{2}\frac{\frac{1}{2}V_m I_m}{I_{DM}(2V_{DD})} = \frac{1}{8}\frac{\frac{4}{\pi}V_{DD}I_m}{V_{DD}I_m} = \frac{1}{2\pi} \approx 0.159. \tag{4.17}$$

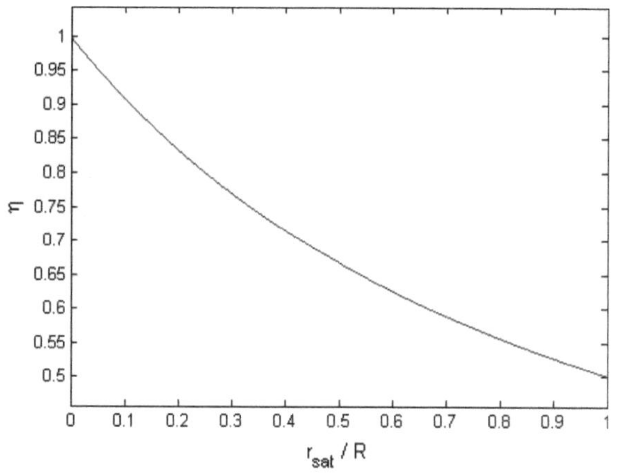

Fig. 4.3 Efficiency of Class-D voltage-switching power amplifier versus normalized saturation resistance

4.1.2 *Class-D Complementary Push-Pull Current-Switching Power Amplifier*

The schematic of the Class-D complementary current-switching power amplifier is similar to the schematic of its voltage-switching counterpart, with the change that the V_{DD} is supplied via an RFC to the center tap of inductor L forming a part of the parallel LC resonant circuit, as depicted in Fig. 4.4.

4.1.2.1 Current and Voltage Waveform Analysis

Once again, the transistors can be modeled as ideal switches with resistance r_{sat}, and are driven with the square wave of the duty cycle of 50 % through a

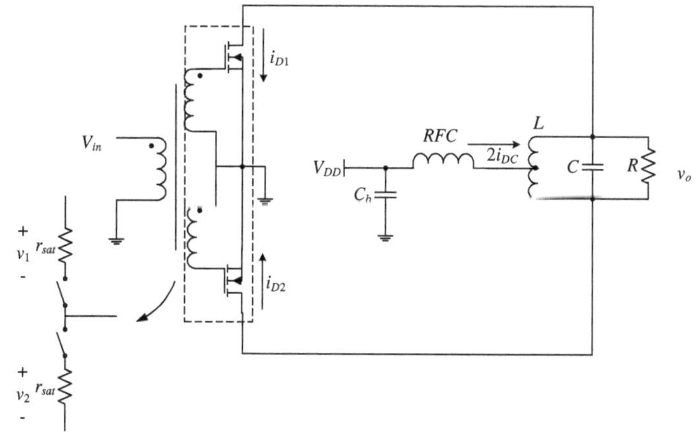

Fig. 4.4 Class-D current-switching power amplifier circuit

transformer, so that the transistors are driven in a complementary fashion. The resonant circuit is still tuned to f_0 according to Eq. (4.1), with quality factor given in Eq. (4.2). Because only DC current can flow through the RFC, a rectangular current waveform of amplitude $2I_{DC}$ is formed. The resonant circuit results in a sinusoidal voltage with amplitude V_m across the load R. The half wave of the load voltage is established over the drain of the transistor that is off, and the only voltage over the transistor that is on is due to the current $2I_{DC}$ flowing through the switch resistance r_{sat}. The load resistance is in series to each transistor, which means that each device sees a resistance of $R/2$.

If the current through the resonant circuit is

$$i(\omega t) = i_{D1}(\omega t) - I_{DC}, \tag{4.18}$$

then for the part of the cycle during which the switch in on ($-\pi/2 \leq \omega t \leq \pi/2$), $i_{D1} = 2I_{DC}$, thus

$$i(\omega t) = I_{DC}, \tag{4.19}$$

and the voltage is

$$v_{D1}(\omega t) = r_{sat}(2I_{DC}) = 2r_{sat}I_{DC}. \tag{4.20}$$

When the other switch is on ($\pi/2 \leq \omega t \leq 3\pi/2$), current is

$$i(\omega t) = -I_{DC}, \tag{4.21}$$

and the voltage is

$$v_{D1}(\omega t) = -(V_{\max} - 2r_{sat}I_{DC}) \cos \omega t + 2r_{sat}I_{DC}, \tag{4.22}$$

where $V_{\max}$ has been introduced as the maximum drain voltage.Using Fourier analysis, the magnitude of the fundamental current component can be determined as

$$I_m = \frac{1}{\pi} \int_{-\frac{\pi}{2}}^{\frac{3\pi}{2}} i \cos \omega t d(\omega t) = \frac{4}{\pi} I_{DC}. \tag{4.23}$$

Supply voltage V_{DD} will be the DC component of the voltage waveform Fourier expansion:

$$\begin{aligned}V_{DD} &= \frac{1}{2\pi}\int_{-\frac{\pi}{2}}^{\frac{3\pi}{2}} v_{D1}(\omega t)d(\omega t)\\ &= \frac{1}{2\pi}\int_{-\frac{\pi}{2}}^{\frac{\pi}{2}} (V_{\max} - 2r_{sat}I_{DC})\cos\omega t d(\omega t) + \frac{1}{2\pi}\int_{-\frac{\pi}{2}}^{\frac{3\pi}{2}} 2r_{sat}I_{DC}d(\omega t)\\ &= \frac{1}{\pi}(V_{\max} - 2r_{sat}I_{DC} + 2\pi r_{sat}I_{DC}) = \frac{1}{\pi}(I_m R + 2\pi r_{sat}I_{DC}),\end{aligned} \tag{4.24}$$

where we note that $V_m = I_m R = V_{\max} - 2r_{sat}I_{DC}$. Substituting Eq. (4.23) into Eq. (4.24), we get

$$V_{DD} = \frac{1}{\pi}\left(4\frac{I_{DC}}{\pi}R + 2\pi r_{sat}I_{DC}\right) = \frac{I_{DC}R}{\left(\frac{\pi}{2}\right)^2}\left(1 + \frac{\pi^2}{2}\frac{r_{sat}}{R}\right). \tag{4.25}$$

Thus, I_{DC} in terms of V_{DD} and resistances is

$$I_{DC} = \frac{\left(\frac{\pi}{2}\right)^2}{1 + \frac{\pi^2}{2}\frac{r_{sat}}{R}}\frac{V_{DD}}{R}. \tag{4.26}$$

V_m is then

$$V_m = \frac{\pi}{1 + \frac{\pi^2}{2}\frac{r_{sat}}{R}}V_{DD}. \tag{4.27}$$

Class-D current-switching power amplifier waveforms are shown in Fig. 4.5.

4.1.2.2 Power and Efficiency

The DC power from the supply can be obtained as

$$P_{DD} = 2V_{DD}I_{DC} = \frac{\pi^2}{2}\frac{V_{DD}^2}{R}\frac{1}{1 + \frac{\pi^2}{2}\frac{r_{sat}}{R}}. \tag{4.28}$$

The power through the load is

$$P_O = \frac{1}{2}I_m V_m = \frac{1}{2}\frac{V_m^2}{R} = \frac{\pi^2}{2}\frac{V_{DD}^2}{R}\frac{1}{\left(1 + \frac{\pi^2}{2}\frac{r_{sat}}{R}\right)^2}. \tag{4.29}$$

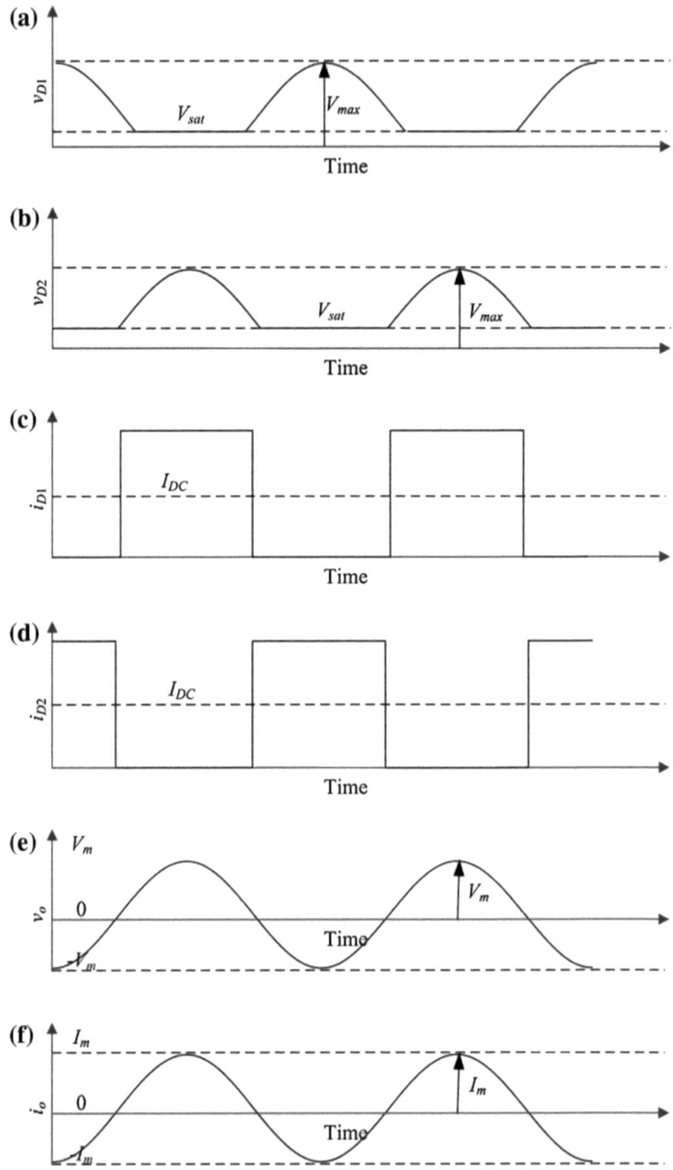

Fig. 4.5 Waveforms of the Class-D voltage-switching power amplifier: **a** switch voltage v_{D1}, **b** switch voltage v_{D2}, **c** drain current i_{D1}, **d** drain current i_{D2}, **e** output voltage, **f** output current

The efficiency is then

$$\eta = \frac{P_O}{P_{DD}} = \frac{1}{1 + \frac{\pi^2}{2} \frac{r_{sat}}{R}}. \tag{4.30}$$

Fig. 4.6 Efficiency of Class-D voltage-switching power amplifier versus normalized saturation resistance

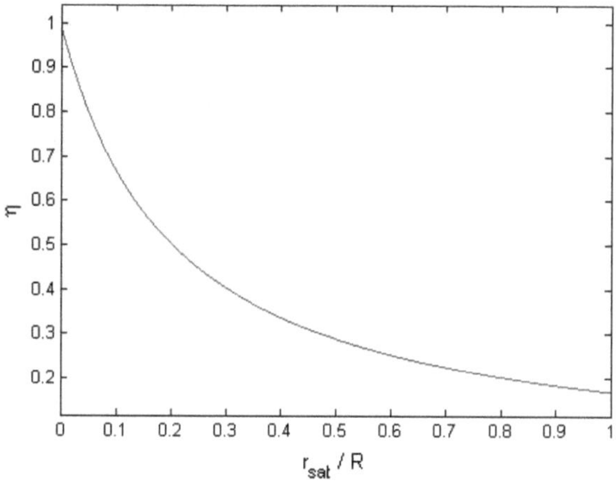

The efficiency of the current-switching variant of Class-D can still reach 100 % for $r_{sat} = 0$, but it depends more strongly on r_{sat} for $r_{sat} \neq 0$, as shown in Fig. 4.6. However, for practical implementations where r_{sat} represent losses due to transistor capacitances, r_{sat} can be partially absorbed into the output tank capacitor C. Thus, current-switching Class-D power amplifiers have been demonstrated at gigahertz frequencies, as discussed at the end of this chapter.The output power capability is

$$c_p = \frac{1}{2}\frac{\frac{1}{2}\frac{V_m}{R}V_m}{I_{DC}(2V_m)} = \frac{1}{2}\frac{\frac{1}{2}\frac{\pi V_{DD}}{R}}{\left(\frac{\pi}{2}\right)^2\frac{V_{DD}}{R}\cdot 2} = \frac{1}{2\pi} \approx 0.159, \tag{4.31}$$

which is, as expected by definition, the same as for the voltage-switching configuration.

4.2 Class-E Power Amplifier

Class E power amplifiers can further be classified into Class-E zero-voltage-switching (ZVS) power amplifiers (Class-E proper power amplifiers) and Class-E zero-current-switching (ZCS) power amplifiers (Class E^{-1} power amplifiers). Class-E power amplifiers are most efficient so far in practice, because voltage and current waveforms are displaced in time so that very low power dissipation in the transistor occurs.

4.2.1 Class-E Zero-Voltage-Switching Power Amplifier

The circuit of the Class-E power amplifier consists of a driving transistor, powered through an *RFC*, a shunt capacitor, a series inductor $L + L_0$ and a series capacitor C_0, as shown in Fig. 4.7. L_0 and C_0 form a resonant circuit with

$$f_0 = \frac{1}{2\pi\sqrt{L_0 C_0}}. \tag{4.32}$$

The loaded Q-factor is dependent on total inductance $L + L_0$:

$$Q_L = \frac{2\pi f_0 (L + L_0)}{R}. \tag{4.33}$$

The transistor can be modeled as a switch, thus the drain voltage is determined by the switch when the transistor is turned on and by the response of the resonant circuit when the switch is turned off. A Class-E ZVS circuit can be analyzed if some assumptions are introduced:

1. **Waveforms**:
   - The transistor is driven by a square wave voltage with duty cycle of 50 %.
2. **Transistor**:
   - The transistor is ideally switched with zero saturation voltage and zero saturation resistance.
   - It switches on and off instantaneously.
   - The output capacitance of the transistor can be absorbed into resonant capacitance, i.e. $C = C_{OUT} + \Delta C$.

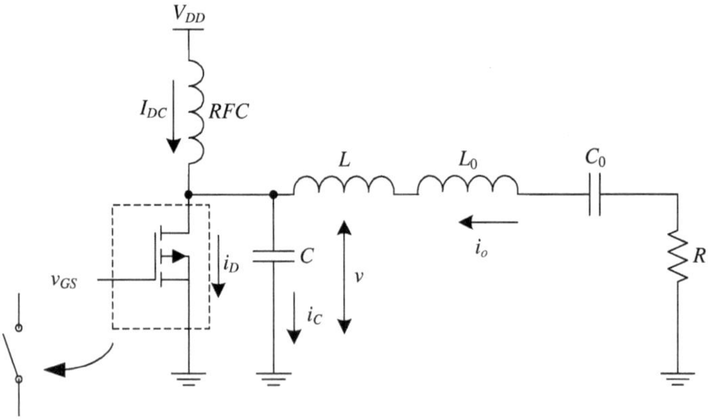

Fig. 4.7 Class-E ZVS circuit

3. **Passives**:

- *RFC* is ideal (zero resistance).
- All other passive components are ideal.
- The quality factor of the resonant tank is sufficiently high so that the output current is sinusoidal at the switching frequency.

4.2.1.1 Current and Voltage Waveform Analysis

For lossless operation of the transistor, the following conditions need to be imposed for the voltage over the switch when the transistor is saturated:

$$v(\omega t)|_{\omega t=2\pi} = 0 \tag{4.34}$$

and

$$\left.\frac{dv(\omega t)}{d(\omega t)}\right|_{\omega t=2\pi} = 0. \tag{4.35}$$

Because of the high quality of the resonant circuit, the current through the load can be modeled as

$$i_o = I_m \sin(\omega t + \phi), \tag{4.36}$$

where ϕ was introduced as the initial phase shift. For the part of the cycle during which the transistor is on ($0 \leq \omega t \leq \pi$), the current over the capacitor C will be

$$i_C(\omega t) = \omega C \frac{dv(\omega t)}{d(\omega t)} = 0. \tag{4.37}$$

Thus the drain current will be

$$i_D = I_{DC} + I_m \sin(\omega t + \phi), \tag{4.38}$$

where I_{DC} is the current through the RFC and is defined as

$$I_{DC} = -I_m \sin \phi. \tag{4.39}$$

Thus the drain current is

$$i_D = I_m[\sin(\omega t + \phi) - \sin \phi]. \tag{4.40}$$

For the part of the cycle during which the transistor is off ($\pi \leq \omega t \leq 2\pi$), no current flows through the drain, thus all current flows through the shunt capacitor:

$$i_C = I_{DC} + I_m \sin(\omega t + \phi). \tag{4.41}$$

Voltage drop over the transistor is equal to the voltage of the shunt capacitor, and it can thus be obtained by integration of the capacitor current:

$$v(\omega t) = \frac{1}{\omega C}\int_{\pi}^{\omega t} i_C(\omega t)d(\omega t) = -\frac{I_m}{\omega C}[\cos(\omega t + \phi) + \cos\phi + (\omega t - \pi)\sin\phi]. \tag{4.42}$$

Phase angle can then calculated by applying the zero condition in Eq. (4.34) to Eq. (4.42):

$$\phi = \arctan\left(-\frac{2}{\pi}\right) = 0.5569 \text{ rad}. \tag{4.43}$$

and sine and cosine of the phase angle can be evaluated to

$$\sin\phi = -\frac{2}{\sqrt{\pi^2 + 4}} \tag{4.44}$$

and

$$\cos\phi = \frac{\pi}{\sqrt{\pi^2 + 4}}. \tag{4.45}$$

With phase angle known, I_{DC} is

$$I_{DC} = \frac{2I_m}{\sqrt{\pi^2 + 4}}, \tag{4.46}$$

and the voltage across the capacitor in terms of I_{DC} ($\pi \le \omega t \le 2\pi$) is

$$v(\omega t) = \frac{I_{DC}}{\omega C}\left(\omega t - \frac{3\pi}{2} - \frac{\pi}{2}\cos\omega t - \sin\omega t\right). \tag{4.47}$$

Supply voltage V_{DD} is the DC harmonic of the voltage waveform, thus it can be evaluated by Fourier analysis

$$V_{DD} = \frac{1}{2\pi}\int_{0}^{2\pi} v(\omega t)d(\omega t) = \frac{I_{DC}}{\pi\omega C}. \tag{4.48}$$

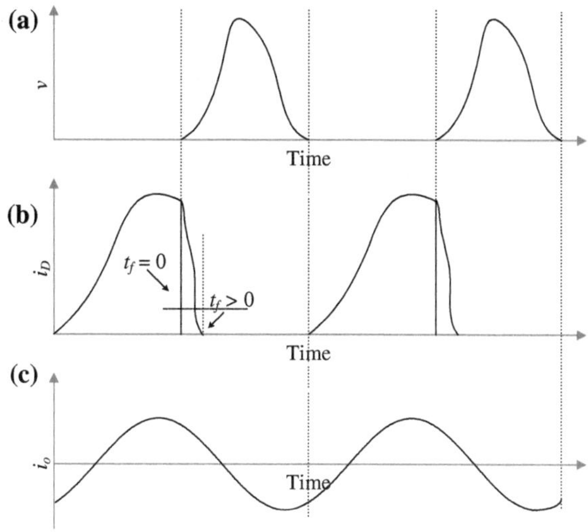

Fig. 4.8 Class-E ZVS power amplifier waveforms: **a** drain voltage, **b** drain current and **c** load current

The current waveform for period $0 \leq \omega t \leq 2\pi$ is

$$i(\omega t) = I_{DC}\left(\frac{\pi}{2}\sin\omega t - \cos\omega t + 1\right). \tag{4.49}$$

Class-E ZVS waveforms are illustrated in Fig. 4.8.

The maximum values of drain voltage and current can be found by differentiating the waveform equations, setting the derivative to zero, and evaluating for ωt, and thereafter substituting the value back into the applicable equation. Thus

$$V_{\max} = 3.562V_{DD} \tag{4.50}$$

and

$$I_{\max} = 2.8621I_{DC}. \tag{4.51}$$

Because the voltage and current waveforms are never simultaneously on, the power drawn from the supply will always be equal to the power dissipated in the load:

$$P_O = P_{DD}. \tag{4.52}$$

Thus

$$I_{DC}V_{DD} = \frac{1}{2}I_m^2 R. \tag{4.53}$$

Substituting Eq. (4.46) and solving for I_{DC}, results in

$$I_{DC} = \frac{8}{\pi^2 + 4}\frac{V_{DD}}{R} = 0.577\frac{V_{DD}}{R}. \tag{4.54}$$

The amplitude of the output voltage $V_m = I_m R$ is

$$V_m = \frac{4V_{DD}}{\sqrt{\pi^2 + 4}} = 1.074V_{DD}. \tag{4.55}$$

The optimum load resistance R for specified output power can be calculated by multiplying Eq. (4.54) by V_{DD}:

$$R = \frac{8}{\pi^2 + 4}\frac{V_{DD}^2}{P_O} = 0.577\frac{V_{DD}^2}{P_O}. \tag{4.56}$$

By manipulating Eqs. (4.46) and (4.48), I_m can be expressed in terms of C and V_{DD}:

$$I_m = \frac{\sqrt{\pi^2 + 4}}{2}\pi\omega C V_{DD}, \tag{4.57}$$

and load resistance is then

$$R = \frac{V_m}{I_m} = \frac{\frac{4}{\sqrt{\pi^2+4}}V_{DD}}{\frac{\sqrt{\pi^2+4}}{2}\pi\omega C V_{DD}} = \frac{8}{\pi(\pi^2 + 4)\omega C} = \frac{0.1836}{\omega C}. \tag{4.58}$$

Shunt capacitance C is thus

$$C = \frac{1}{5.447\omega R}. \tag{4.59}$$

The magnitude of voltage over inductor L is

$$V_L = -\frac{1}{\pi}\int_0^{2\pi} v\cos(\omega t + \phi)d(\omega t) = \frac{\pi(\pi^2 - 4)}{4\sqrt{\pi^2 + 4}}V_{DD}. \tag{4.60}$$

The reactance of inductor L is the ratio of voltage over the inductor to the current through the inductor:

$$\omega L = \frac{V_L}{I_m} = \frac{\frac{\pi(\pi^2-4)}{4\sqrt{\pi^2+4}}V_{DD}}{\frac{\sqrt{\pi^2+4}}{2}\pi\omega C V_{DD}} = \frac{\pi^2 - 4}{2(\pi^2 + 4)\omega C} \tag{4.61}$$

and inductance is

$$L = \frac{\pi^2 - 4}{2(\pi^2 + 4)\omega^2 C} = \frac{0.2116}{\omega^2 C} = \frac{1}{4.726\omega^2 C}. \tag{4.62}$$

The reactance of the resonant inductor and resonant capacitor in terms of loaded Q-factor are

$$\omega L_0 = \frac{1}{\omega C_0} = Q_L R - \omega L = R\left(Q_L - \frac{\omega L}{R}\right). \tag{4.63}$$

By noting that

$$\frac{\omega L}{R} = \frac{\pi(\pi^2 - 4)}{16} = 1.1525, \tag{4.64}$$

$$\omega L_0 = \frac{1}{\omega C_0} = R(Q_L - 1.1525). \tag{4.65}$$

The minimum value of RFC inductance can be calculated for the chosen ripple. As before, a ripple of 10 % is acceptable. The voltage over the RFC inductor is

$$v_{L_f} = V_{DD}, \tag{4.66}$$

and the current is the integral of the voltage

$$i_{L_f} = \frac{1}{L_f}\int_0^t V_{DD} dt + i_{L_f}(0) = \frac{V_{DD}}{L_f} t + i_{L_f}(0). \tag{4.67}$$

At $T/2$, i.e. $1/(2f)$, the instantaneous current is

$$i_{L_f}\left(\frac{T}{2}\right) = \frac{V_{DD}}{2fL_f} + i_{L_f}(0). \tag{4.68}$$

The peak-to-peak current ripple is

$$\Delta i_{L_f} = i_{L_f}\left(\frac{T}{2}\right) - i_{L_f}(0) = \frac{V_{DD}}{2fL_f}. \tag{4.69}$$

The minimum choke inductance is then in terms of resistance

$$L_f = \frac{V_{DD}}{2f\Delta i_{L_f}} = \frac{V_{DD}}{2fI_{DC}\left(\frac{\Delta i_{L_f}}{I_{DC}}\right)} = \frac{\frac{I_{DC}}{\pi\omega C}}{2fI_{DC}(0.1)} = \frac{\frac{5.447\omega R}{\pi\omega}}{0.2f} = 8.669\frac{R}{f}. \tag{4.70}$$

4.2.1.2 Power and Efficiency

Power and efficiency have already been discussed when making assumptions for derivation of Class-E equation. For the sake of completeness, the output power in terms of resistance, capacitance and inductance is listed:

$$P_O = 0.577\frac{V_{DD}^2}{R}, \tag{4.71}$$

$$P_O = \pi\omega C V_{DD}^2 = 2\pi^2 f C V_{DD}^2 = 19.739 f C V_{DD}^2, \tag{4.72}$$

$$P_O = \frac{\pi(\pi^2 - 4)V_{DD}^2}{2(\pi^2 + 4)\omega L} = 0.1058\frac{V_{DD}^2}{fL}. \tag{4.73}$$

Efficiency of 100 % was already established based on the non-overlapping waveform characteristics. However, the current cannot reach zero the moment the transistor turns off, as illustrated in Fig. 4.8 by introducing t_f. Also, the saturation voltage of the transistor, V_{sat}, can affect the efficiency. It can be shown that the efficiency can be expressed as

$$\eta = \frac{1 - \frac{(2\pi A)^2}{6} - \frac{V_{sat}}{V_{DD}}\left[1 + A - \frac{(2\pi A)^2}{6}\right]}{1 - \frac{(2\pi A)^2}{12}}, \tag{4.74}$$

where

$$A = \left(1 + \frac{0.82}{Q_L}\right) f t_f. \tag{4.75}$$

For the ideal power amplifier, both t_f and V_{sat} are zero, and efficiency of 100 % is restored. The output power capability is

$$c_p = \frac{P_{out(\max)}}{I_{D\max} V_{DS(\max)}} = \frac{I_{DC} I_{DD}}{2.862 I_{DC} 3.562 V_{DD}} = 0.0981. \tag{4.76}$$

An interesting result is that the output power capability of the Class-E power amplifier is exactly 78.5 % of the output power capability of the Class-B power amplifier (0.125). Thus, the gain in efficiency is traded for the same percentage of gain in output power capability.Another important consideration is the maximum operating frequency. The maximum operating frequency will be reached when the output capacitance of the transistor cannot be absorbed into the shunt capacitance C anymore. From Eq. (4.58):

$$2\pi f_{\max} C_{OUT} R = \frac{8}{\pi(\pi^2+4)}. \tag{4.77}$$

Removing R in this equation and solving for $f_{\max}$ will yield

$$f_{\max} = \frac{P_O}{2\pi^2 C_{OUT} V_{DD}^2} = 0.0507 \frac{P_O}{C_{OUT} V_{DD}}. \tag{4.78}$$

4.2.2 Class-E Zero-Current-Switching Power Amplifier

High efficiency of the Class-E power amplifier output stage can also achieved if the capacitors and inductors in the circuit of Fig. 4.7 are swapped with each other (inverted). This leads to the introduction of the inverse Class-E (Class-E^{-1}) power amplifier, or alternatively, Class-E ZCS power amplifier. The modified circuit is shown in Fig. 4.9. Note that this circuit does not contain an RFC, which is replaced with finite shunt inductance L so that the current through the transistor can be controlled. L_0 and C_0 form a series resonant circuit with frequency given in Eq. (4.32). The loaded Q-factor is dependent on total capacitance $C + C_0$:

$$Q_L = \frac{1}{2\pi f_0 R(C||C_0)} = \frac{(C+C_0)}{2\pi f_0 R(CC_0)}. \tag{4.79}$$

The analysis of the Class-E ZCS power amplifier is based on similar assumptions stated for the Class-E ZVS analysis, with an additional condition that the output capacitance of the power amplifier needs to be assumed to be zero.

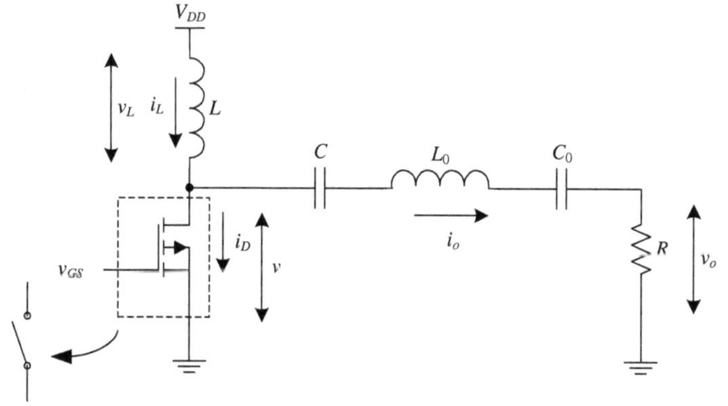

Fig. 4.9 Class-E ZCS power amplifier circuit diagram

4.2.2.1 Current and Voltage Waveform Analysis

For lossless operation of the transistor, the following conditions need to be imposed for the drain current flowing into the transistor operating as a switch:

$$i_D(\omega t)\big|_{\omega t=2\pi} = 0 \tag{4.80}$$

and

$$\left.\frac{di_D(\omega t)}{d(\omega t)}\right|_{\omega t=2\pi} = 0. \tag{4.81}$$

Current in the load is sinusoidal and modeled by Eq. (4.36). Currents and voltages in the circuit can be written as

$$i_D(\omega t) = i_L(\omega t) - i_o(\omega t) \tag{4.82}$$

and

$$v(\omega t) = V_{DD} - v_L(\omega t). \tag{4.83}$$

During the part of the cycle during which the transistor is off ($0 \le \omega t \le \pi$), $i_D(\omega t) = 0$. Thus,

$$i_L(\omega t) = i_o(\omega t) = I_m \sin(\omega t + \phi). \tag{4.84}$$

Voltage over the inductor can be found by differentiation:

$$v_L(\omega t) = \omega L \frac{di_L(\omega t)}{d(\omega t)} = \omega L I_m \cos(\omega t + \phi). \tag{4.85}$$

and voltage $v(\omega t)$ is then

$$v(\omega t) = V_{DD} - \omega L I_m \cos(\omega t + \phi). \tag{4.86}$$

When the transistor is on ($\pi \le \omega t \le 2\pi$), $v(\omega t)$ is zero and

$$v_L(\omega t) = \omega L I_m \cos(\omega t + \phi) = V_{DD} \tag{4.87}$$

and the current through the inductor is

$$i_L(\omega t) = \frac{V_{DD}}{\omega L}(\omega t - \pi) - I_m \sin\phi. \tag{4.88}$$

Device current is then obtained from Eqs. (4.82) and (4.84):

$$i_D(\omega t) = \frac{V_{DD}}{\omega L}(\omega t - \pi) - I_m[\sin(\omega t + \phi) + \sin \phi]. \tag{4.89}$$

The zero condition in Eq. (4.80) can now be applied to Eq. (4.89) to get

$$I_m = \frac{\pi V_{DD}}{2\omega L \sin \phi}, \tag{4.90}$$

where $0 < \phi < \pi$ so that $I_m > 0$. Current through the inductor thus simplifies to

$$i_D(\omega t) = \frac{V_{DD}}{\omega L}\left[\omega t - \frac{3\pi}{2} - \frac{\pi}{2 \sin \phi} \sin(\omega t + \phi)\right]. \tag{4.91}$$

The zero condition in Eq. (4.81) can be applied to Eq. (4.91) to solve for phase angle

$$\phi = \arctan\left(\frac{\pi}{2}\right) = 1.0039 \text{ rad}, \tag{4.92}$$

yielding

$$\sin \phi = \frac{\pi}{\sqrt{\pi^2 + 4}} \tag{4.93}$$

and

$$\cos \phi = \frac{2}{\sqrt{\pi^2 + 4}}. \tag{4.94}$$

Drain current is thus

$$i_D(\omega t) = \frac{V_{DD}}{\omega L}\left[\omega t - \frac{3\pi}{2} - \frac{\pi}{2} \cos \omega t - \sin \omega t\right]. \tag{4.95}$$

Supply DC current can be determined by performing the Fourier analysis on the drain current:

$$I_{DC} = \frac{1}{2\pi}\int_0^{2\pi} i(\omega t) d(\omega t) = \frac{V_{DD}}{\pi \omega L}. \tag{4.96}$$

The fundamental current in Eq. (4.90) is

$$I_m = \frac{\pi\sqrt{\pi^2 + 4}}{2} I_{DC} = 5.8499 I_{DC}. \tag{4.97}$$

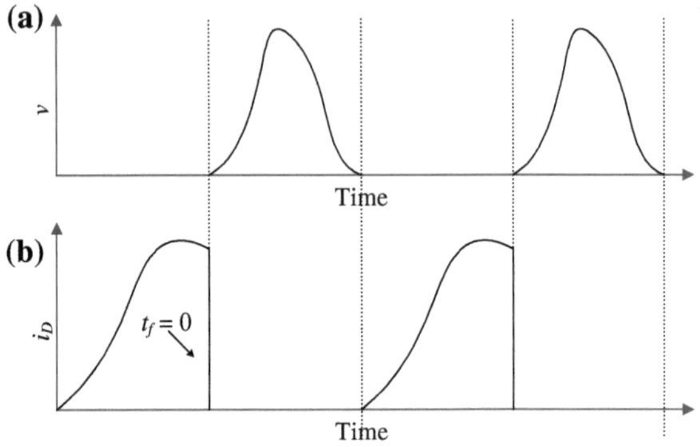

Fig. 4.10 Class-E ZCS power amplifier waveforms: **a** drain voltage and **b** drain current

Class-E ZCS waveforms are illustrated in Fig. 4.10.

As before, maximum values of drain voltage and current can be found by differentiation and substitution:

$$V_{\max} = 2.8621 V_{DD} \tag{4.98}$$

and

$$I_{\max} = 3.562 I_{DC}. \tag{4.99}$$

Thus the maximum values of the current and voltage waveforms of the Class-E ZCS power amplifier are swapped around when compared to those of the Class-E ZVS stage.Once again, by noting that $V_m = I_m R$, V_m can be calculated through Fourier analysis:

$$V_m = \frac{1}{\pi} \int_0^{2\pi} v \sin(\omega t + \phi) d(\omega t) = \frac{4 V_{DD}}{\pi \sqrt{\pi^2 + 4}} = 0.3419 V_{DD}. \tag{4.100}$$

The voltage of the capacitor is not sinusoidal, but the first harmonic can be expressed as

$$v_C(\omega t) = -V_C \cos(\omega t + \phi), \tag{4.101}$$

where V_C is the magnitude of the first harmonic, which can be calculated as

$$V_C = \frac{1}{\pi} \int_0^{2\pi} v \cos(\omega t + \phi) d(\omega t) = \frac{(\pi^2 + 12) V_{DD}}{4 \sqrt{\pi^2 + 4}} = 1.4681 V_{DD}. \tag{4.102}$$

Removing V_{DD} from Eq. (4.100) and (4.102) yields

$$V_m = \frac{8}{\pi(\pi^2+4)} \omega L I_m \tag{4.103}$$

and

$$V_C = \frac{\pi^2+12}{2(\pi^2+4)} \omega L I_m. \tag{4.104}$$

The phase shift between fundamental components of the capacitor and resistor voltages is

$$\tan\psi = -\frac{V_C}{V_m} = -\frac{1}{\omega RC}. \tag{4.105}$$

Thus, L and C can be calculated from

$$L = \frac{\pi(\pi^2+4)}{8}\frac{R}{\omega} = 5.446\frac{R}{\omega} = 0.8669\frac{R}{f} \tag{4.106}$$

and

$$C = \frac{16}{\pi(\pi^2+12)\omega R} = \frac{0.2329}{\omega R} = \frac{0.03707}{fR}. \tag{4.107}$$

As before, C_0 and L_0 are calculated in terms of the loaded Q-factor:

$$C_0 = \frac{1}{\omega R Q_L} = \frac{0.1592}{f R Q_L}, \tag{4.108}$$

$$L_0 = \left[Q_L - \frac{\pi(\pi^2+12)}{16}\right]\frac{R}{\omega} = (Q_L - 4.2941)\frac{R}{\omega} = 0.1592(Q_L - 4.2941)\frac{R}{f}. \tag{4.109}$$

The optimum load resistance R can be found in terms of supply voltage V_{DD} and output power as

$$R = \frac{8}{\pi^2(\pi^2+4)}\frac{V_{DD}^2}{P_O} - 0.05844\frac{V_{DD}^2}{P_O}. \tag{4.110}$$

4.2.2.2 Power and Efficiency

As stipulated before, the output power of the Class-E ZCS power amplifier is

$$P_O = 0.05844 \frac{V_{DD}^2}{R}, \tag{4.111}$$

which is about a tenth of the output power of the Class-E ZVS power amplifier. The output power capability is unchanged at 0.098.

Ideally, the efficiency can reach 100 %, but in practice, this efficiency is lower than that of Class-E ZVS power amplifier. Furthermore, the loaded Q-factor of the output network must be greater than 4.2941 for any practical results, as opposed to 1.1525 in the case of the Class-E ZVS amplifier, which makes the Class-E ZCS amplifier less realizable in practice. Furthermore, the assumption that the output capacitance of the transistor is zero is unrealistic. GaN HEMT devices, however, have small output capacitances, thus high-frequency low-output power applications can benefit from a Class-E ZCS power amplifier built around HEMT transistors [2]. One such example is listed at the end of this chapter. Alternatively, the resonant circuit can be replaced by the parallel resonant circuit and RFC and blocking capacitors can be introduced. Output power of 1.7337 V_{DD}^2/R can be reached in this case.

4.2.3 *Class-E ZVS Power Amplifier with Finite DC-Feed Inductance*

The Class-E ZVS power amplifier can also be implemented using a finite DC-feed inductance. Numerous circuit configurations are possible and the reader of this text is referred to [2] for a procedure on how to derive equations for a generalized network. In this chapter, however, we will only summarize the most important equations for the simplest configuration with one inductor and one capacitor, taken from the same text. The circuit diagram of this configuration is given in Fig. 4.11. In this figure C_B is the DC blocking capacitor.

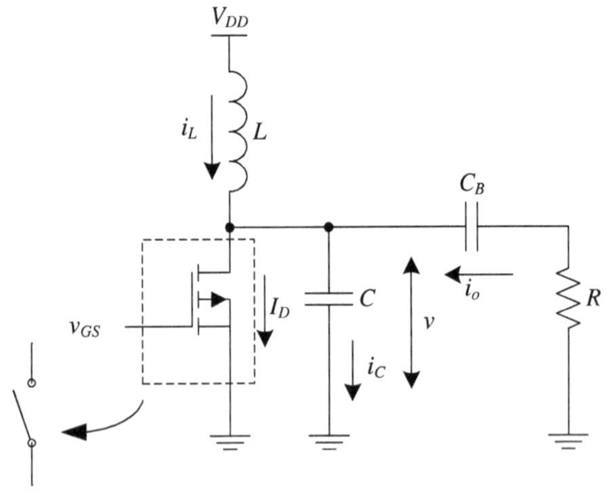

Fig. 4.11 Class-E ZVS power amplifier with one finite DC-feed inductance and one capacitor

The analysis can be performed by stating the usual assumptions about the driving waveform, transistor and passive components. Equations are derived from zero-voltage-switching conditions defined earlier (Eqs. (4.34 and 4.35)). When the transistor is turned on, there is no voltage across the switch, and the current, consisting of the total load current and inductor current, flows through the switch. When the transistor is turned off, this current flows through the shunt capacitor.

Using the same method for deriving equations used earlier in this chapter, the phase shift angle can be worked out to $\phi = 0.9552$ rad. The component values are

$$L = 0.41\frac{R}{\omega}, \tag{4.112}$$

$$C = \frac{1.025}{\omega R}, \tag{4.113}$$

and

$$R = 1.394\frac{V_{DD}^2}{P_O}. \tag{4.114}$$

The DC supply current is

$$I_{DC} = \frac{\pi V_{DD}}{4\omega L}. \tag{4.115}$$

Finally, the maximum values of drain voltage and current are

$$V_{\max} = 3.849V_{DD} \tag{4.116}$$

and

$$I_{\max} = 4I_{DC}. \tag{4.117}$$

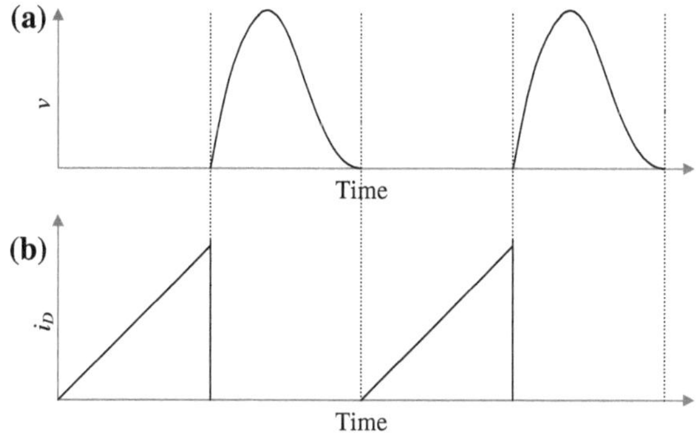

Fig. 4.12 The waveforms of the finite DC-feed Class E ZVS power amplifier: **a** drain voltage and **b** drain current

Since there is no simultaneous voltage and current, the ideal efficiency is once again 100 %.

Note that this configuration of Class-E power amplifier introduces a significant harmonic content to the load. For simplicity, only drain voltage and drain current waveforms are shown in Fig. 4.12.

4.3 Class-F Power Amplifier

The Class-F amplifier is similar to the Class E in the sense that it can also result in 100 % efficiency. However, at the same time, the output capability is similar to that of the Class-D amplifiers. This efficiency and power boost is a result of the presence of harmonic resonators in the output networks that shape drain (collector) waveforms in such a way that load appears to be short at even harmonics and open at odd harmonics, as shown in the general model in Fig. 4.13. As a result, the ideal drain voltage waveform approximates a square wave, while the collector current waveform approximates a half-sine wave. There is no overlap between the two waveforms.

In an inverse Class-F power amplifier, or Class F^{-1}, amplifier resonators are configured so that the drain voltage is shaped as a half-sinusoid and drain current is shaped as a square wave.

Shaping of waveforms can be done by means of transmission lines, as in the case of the Class-F amplifier with the quarter-wave transmission line. However, this is not a practical implementation at low-gigahertz frequencies. Instead, passive resonators are used. Class-F amplifiers would require an infinite number of resonators to shape output waveforms correctly. Figure 4.14 shows that the efficiency of the Class F amplifier increases when the number of harmonic traps increases. If only

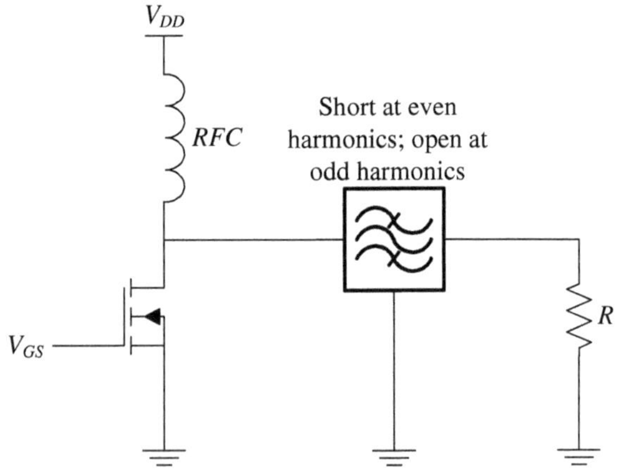

Fig. 4.13 General Class-F amplifier model

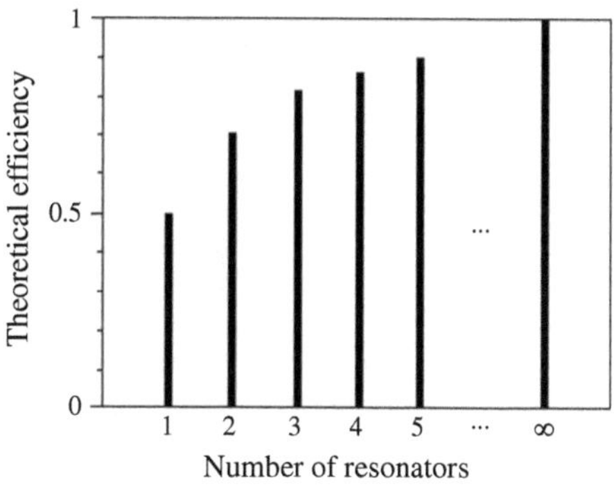

Fig. 4.14 Efficiency of a Class-F amplifier versus a number of resonators

the first harmonic trap is implemented, the amplifier behaves like a Class-A amplifier, with maximum efficiency of 50 %.

Most real-life integrated Class-F amplifier implementations deploy only a few resonator tanks, usually two or three. Maximally flat Class-F power amplifiers with resonators up to the third harmonic and maximally flat Class-F power amplifiers with resonators up to the fifth harmonic will be discussed in more detail, followed by two inverse Class-F configurations. Another group of Class-F power amplifiers, maximum drain efficiency Class-F power amplifiers, will not be discussed but the reader can consult reference [1]. Configuration with a quarter-wavelength line will also be considered.

4.3.1 Maximally Flat Class-F_3 Power Amplifier

The circuit diagram of the maximally flat Class-F_3 power amplifier consists of a driving transistor, an RFC, a blocking capacitor C_B, series resonant circuit L_3C_3 and shunt resonant circuit L_oC_o driving a load R, as shown in Fig. 4.15. The amplifier is driven with the sinusoidal wave and it is biased so that the transistor conducts at conduction angle 2θ.

4.3.1.1 Current and Voltage Waveform Analysis

In an ideal Class-F power amplifier, the drain-to-source voltage is shaped so that it contains only odd harmonics. Thus,

$$v_{DS} = V_{DD} - V_m \cos \omega_o t + \sum_{n=3,5,7...}^{\infty} V_{mn} \cos n\omega_o t, \tag{4.118}$$

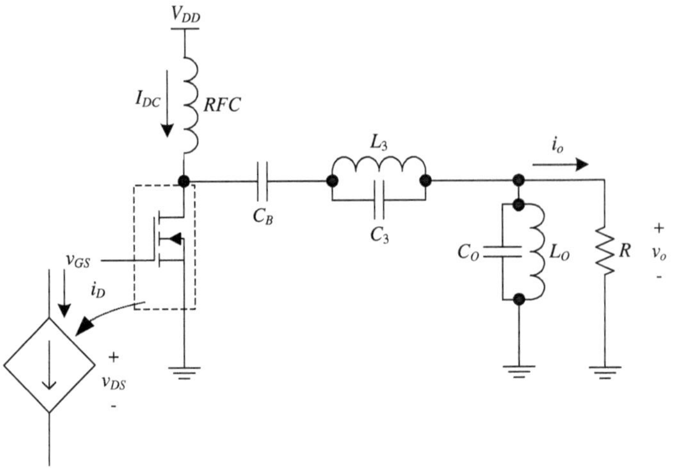

Fig. 4.15 Class-F_3 power amplifier circuit

where V_m is the amplitude of the fundamental harmonic and V_{mn} is the amplitude of the n-th harmonic. The current has only the fundamental and even harmonics:

$$i_D = I_{DC} + I_m \cos \omega_o t + \sum_{n=2,4,6...}^{\infty} I_{mn} \cos n\omega_o t, \tag{4.119}$$

where I_m is the amplitude and I_{mn} is the amplitude of the n-th harmonic. Variable $\gamma_1 = I_m/I_{DC}$ introduced in Chap. 3 can be formulated as

$$\gamma_1 = \frac{\theta - \sin\theta\cos\theta}{\sin\theta - \theta\cos\theta} \tag{4.120}$$

in terms of half of the conduction angle, θ, which allows for the DC current I_{DC} to be computed. Similarly, $\gamma_3 = I_{m3}/I_{DC}$ can be defined in terms of the magnitude of the third harmonic of the current:

$$\gamma_3 = \left.\frac{2(\sin n\theta\cos\theta - n\cos n\theta\sin\theta)}{n(n^2-1)(\sin\theta - \theta\cos\theta)}\right|_{n=3} = \frac{\sin 3\theta\cos\theta - 3\cos 3\theta\sin\theta}{12(\sin\theta - \theta\cos\theta)}. \tag{4.121}$$

Coefficients γ_1 and γ_3 are plotted in Fig. 4.16.

In the Class-F power amplifier with the resonant circuits only up to the third harmonic, drain-to-source voltage can be expressed as

$$v_{DS} = V_{DD} - V_m \cos\omega_o t + V_{m3} \cos 3\omega_o t. \tag{4.122}$$

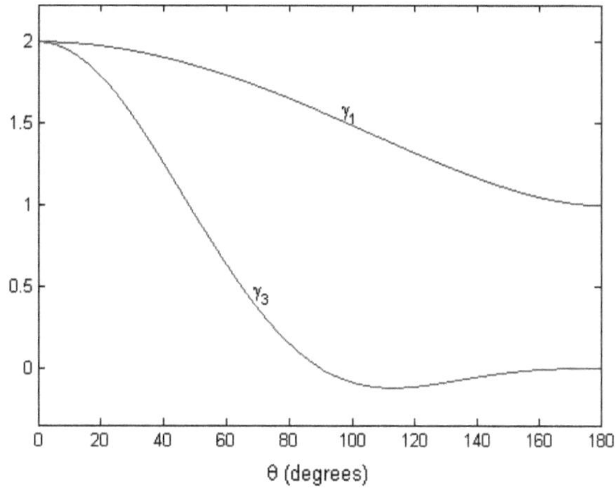

Fig. 4.16 γ_1 and γ_3 in terms of θ

Voltage over the load resistance R is sinusoidal owing to the presence of the shunt resonant tank and is equal to the fundamental harmonic of the drain-source voltage, thus

$$v_o = -V_m \cos \omega_o t. \quad (4.123)$$

To ensure that fundamental and third harmonic voltages are out of phase, γ_3 must be negative, which imposes a restriction that θ has to be greater than 90°.

The task of finding a Class-F_3 amplifier with maximally flat waveforms involves finding a correct ratio between V_m and V_{m3} resulting in maximum flatness at points where the waveform exhibits maxima and minima. Maxima and minima are found when the first derivative of the drain-source voltage is set to zero, and maximum flatness coefficients are obtained when all higher-order derivatives are zero. Thus, the first derivative is needed to get maxima and minima:

$$\frac{dv_{DS}}{d(\omega_0 t)} = V_m \sin \omega_o t - 3V_{m3} \sin 3\omega_o t = 0. \quad (4.124)$$

The solutions to Eq. (4.124) are

$$\omega_o t = 0,$$
$$\omega_o t - \pi,$$
$$\omega_o t = \pm \arcsin \sqrt{\frac{9V_{m3} - V_m}{12V_{m3}}}$$

and

$$\omega_o t = \pi \pm \arcsin\sqrt{\frac{9V_{m3} - V_m}{12V_{m3}}}, \tag{4.125}$$

where the last two solutions are real only for $V_{m3} \geq V_m/9$. To get maximum flatness, the second derivative is needed:

$$\frac{d^2 v_{DS}}{d(\omega_0 t)} = V_m \cos\omega_o t - 9V_{m3}\cos 3\omega_o t. \tag{4.126}$$

Substituting the first extreme location ($\omega_0 t = 0$) into this equation yields

$$V_m - 9V_{m3} = 0. \tag{4.127}$$

Thus a maximally flat waveform is when

$$\frac{V_{m3}}{V_m} = \frac{1}{9}. \tag{4.128}$$

To align the voltage waveform such that minimum is zero at $\omega_0 t = 0$,

$$v_{DS\min} = V_{DD} - V_m + \frac{V_m}{9} = 0, \tag{4.129}$$

which results in

$$V_m = \frac{9}{8} V_{DD}. \tag{4.130}$$

Thus, the drain-to-source voltage in terms of V_{DD} is

$$v_{DS} = V_{DD}\left(1 - \frac{9}{8}\cos\omega_o t + \frac{1}{8}\cos 3\omega_o t\right). \tag{4.131}$$

The maximum is reached when $\omega_0 t = \pi$:

$$v_{DSM} = V_{DD}\left(1 + \frac{9}{8} - \frac{1}{8}\right) = 2V_{DD}. \tag{4.132}$$

The current through the load is

$$i_o = \frac{v_o}{R} = -I_m \cos\omega_o t = -\frac{9}{8} V_{DD}\cos\omega_o t. \tag{4.133}$$

The DC input current is

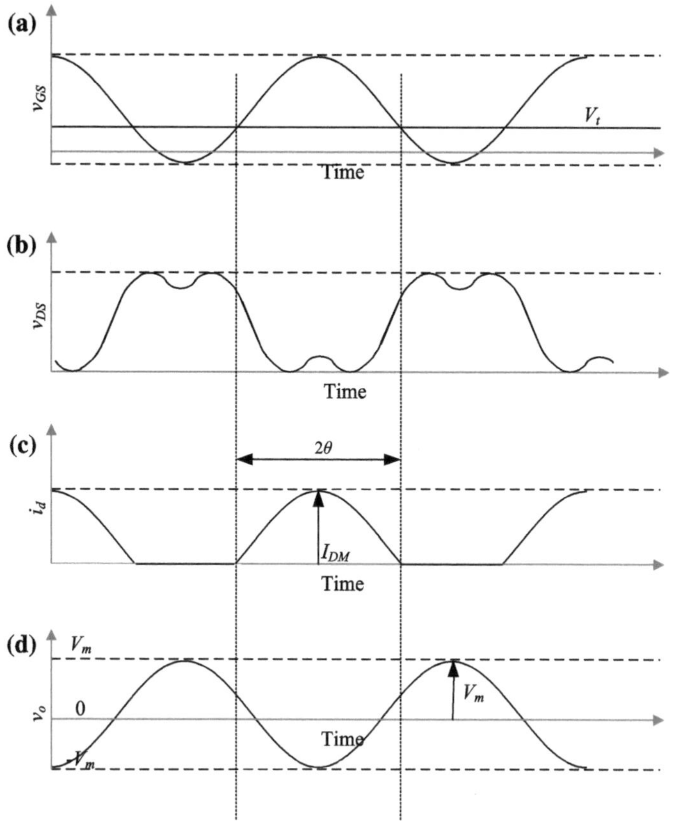

Fig. 4.17 Waveforms of the maximally flat Class-F_3 power amplifier: **a** gate-to-source voltage, **b** drain-to-source voltage, **c** drain current and **d** output voltage

$$I_{DC} = \frac{I_m}{\gamma_1} = \frac{9}{8}\frac{V_{DD}}{\gamma_1 R}. \tag{4.134}$$

Waveforms for the maximally flat Class-F_3 power amplifier are shown in Fig. 4.17, where the ripple in voltage waveform is exaggerated, while for the maximum flat waveform it is minimal.

4.3.1.2 Power and Efficiency

The output power is

$$P_O = \frac{V_m^2}{2R} = \frac{81}{128}\frac{V_{DD}^2}{R} = 0.6328\frac{V_{DD}^2}{R}. \tag{4.135}$$

The DC input power is

$$P_{DD} = V_{DD} I_{DC} = \frac{9 V_{DD}^2}{8 \gamma_1 R}. \tag{4.136}$$

Thus the drain efficiency is

$$\eta = \frac{P_O}{P_{DC}} = \frac{\frac{81}{128} \frac{V_{DD}^2}{R}}{\frac{9 V_{DD}^2}{8 \gamma_1 R}} = \frac{9}{16} \gamma_1 = 0.5625 \gamma_1 = \frac{9}{16} \frac{\theta - \sin\theta\cos\theta}{\sin\theta - \theta\cos\theta}. \tag{4.137}$$

The output power capability is

$$c_p = \frac{P_{out(\max)}}{I_{D\max} V_{DS(\max)}} = \frac{9}{32\pi} \frac{\theta - \sin\theta\cos\theta}{1 - \cos\theta}. \tag{4.138}$$

The efficiency and output power capability in terms of θ are shown in Fig. 4.18. Based on this figure, the conduction angle that results in the best compromise between the efficiency and output power capability is roughly

$$100^\circ < \theta < 120^\circ. \tag{4.139}$$

Useful θ-defined coefficients are listed in Table 4.1.

4.3.2 *Maximally Flat Class-F$_{35}$ Power Amplifier*

The Class-F$_{35}$ power amplifier has resonators up to the fifth harmonic, as depicted in Fig. 4.19.

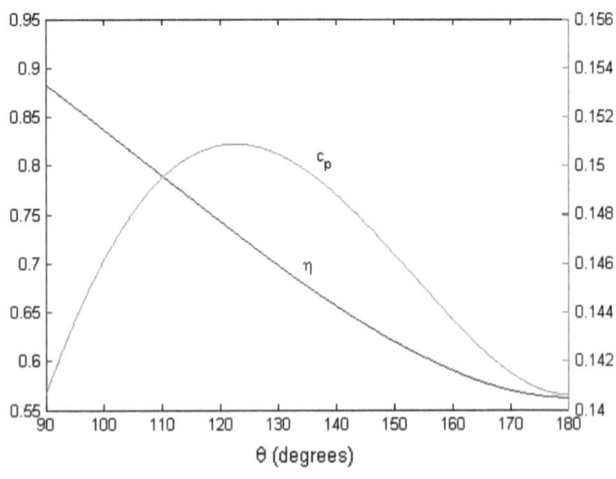

Fig. 4.18 Drain efficiency and output power capability of maximally flat Class-F$_3$ power amplifier

Table 4.1 Theta-defined relations for Class-F power amplifiers

θ (°)	$\frac{\theta - \sin\theta\cos\theta}{\sin\theta - \theta\cos\theta}$	$\frac{\theta - \sin\theta\cos\theta}{1 - \cos\theta}$
90	1.5708	1.5708
100	1.4880	1.6328
110	1.4040	1.6701
120	1.3210	1.6849
130	1.2413	1.6809
140	1.1675	1.6624
150	1.1025	1.6350
160	1.0498	1.6054
170	1.0137	1.5810
180	1.0000	1.5708

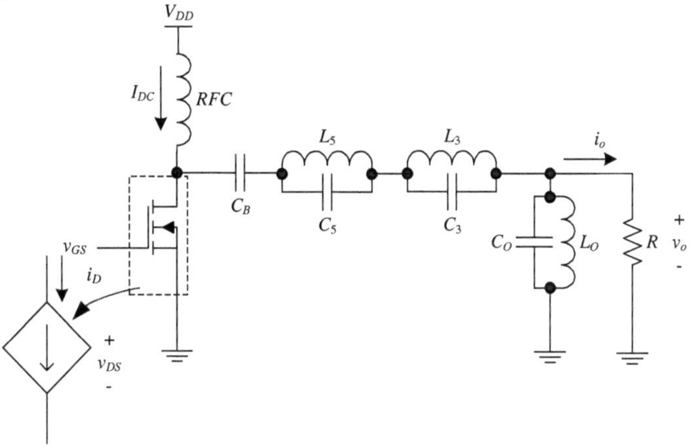

Fig. 4.19 Class-F_{35} power amplifier circuit

4.3.2.1 Current and Voltage Waveform Analysis

The drain-to-source waveform is defined by the following equation:

$$v_{DS} = V_{DD} - V_m \cos \omega_o t + V_{m3} \cos 3\omega_o t - V_{m5} \cos \omega_o t. \tag{4.140}$$

Half of the conduction angle has to be in the range $90^\circ \leq \theta \leq 127.76^\circ$. In order to calculate the coefficients, at least two derivative equations are needed. The first and third order derivatives do not give any equations, but the second and fourth order derivatives result in relations

$$V_{m3} = \frac{V_m}{6}, \tag{4.141}$$

and

$$V_{m5} = \frac{V_m}{50}. \tag{4.142}$$

The fundamental voltage magnitude V_m in terms of V_{DD} is

$$V_m = \frac{75}{64} V_{DD} = 1.1719 V_{DD} \tag{4.143}$$

and the maximum voltage V_{DSM} remains $2V_{DD}$. The waveforms are shown in Fig. 4.20, where the ripple in voltage waveform is exaggerated, while for the maximum flat waveform it is minimal.

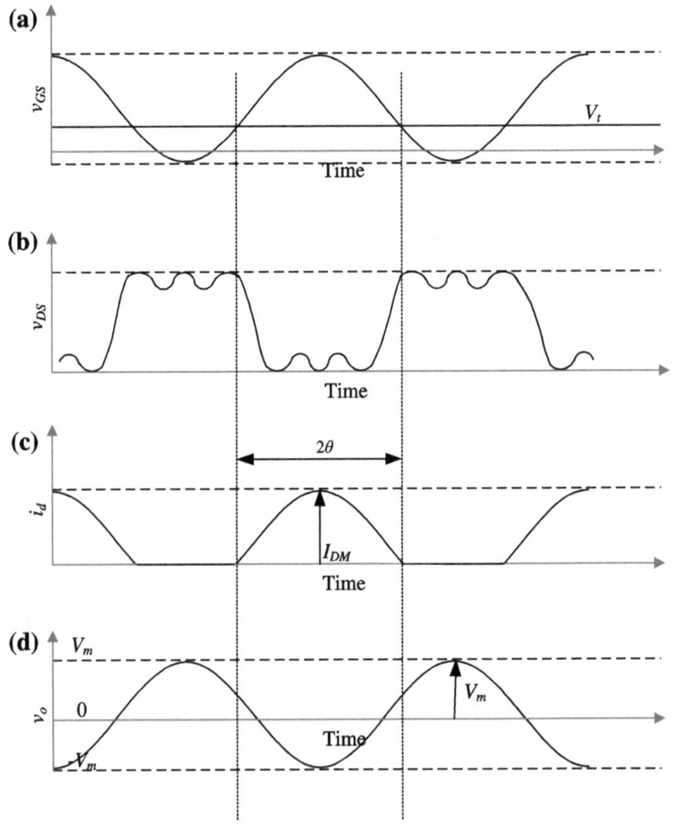

Fig. 4.20 Waveforms of the Class-F_{35} power amplifier: **a** gate-to-source voltage, **b** drain-to-source voltage, **c** drain current and **d** output voltage

4.3.2.2 Power and Efficiency

The output power is

$$P_O = \frac{5625}{8192}\frac{V_{DD}^{\,2}}{R} = 0.6866\frac{V_{DD}^{\,2}}{R}. \tag{4.144}$$

The DC input power is

$$P_{DD} = \frac{75V_{DD}^2}{64\gamma_1 R} = 1.1719\frac{V_{DD}^2}{\gamma_1 R}. \tag{4.145}$$

Thus the drain efficiency is

$$\eta = 0.5859\gamma_1 = \frac{75}{128}\frac{\theta - \sin\theta\cos\theta}{\sin\theta - \theta\cos\theta} \tag{4.146}$$

and the output power capability is

$$c_p = \frac{75}{256\pi}\frac{\theta - \sin\theta\cos\theta}{1 - \cos\theta}. \tag{4.147}$$

The efficiency and output power capability in terms of θ are shown in Fig. 4.21. Thus Class-F_{35} can achieve better efficiency and better output power capability than the Class-F_3. The output power capability is comparable to that of the Class-D amplifier.

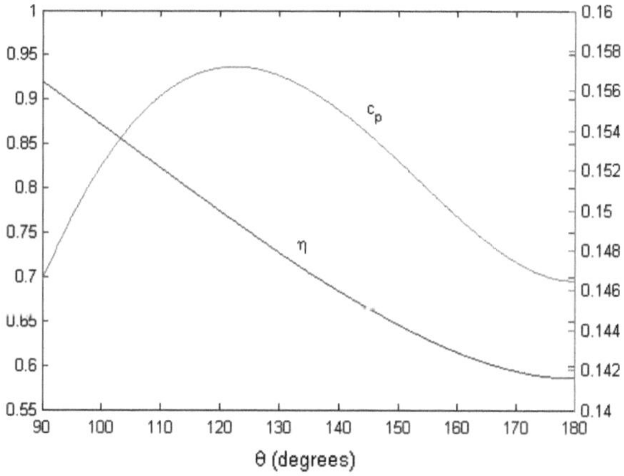

Fig. 4.21 Drain efficiency and output power capability of maximally flat Class-F_{35} power amplifier

4.3.3 Maximally Flat Class-F_2 and Maximally Flat Class-F_{24} Power Amplifier (Class-F^{-1} Power Amplifiers)

In an ideal Class-F^{-1} power amplifier, the drain-to-source voltage is shaped so that it contains only the fundamental and even harmonics. Thus,

$$v_{DS} = V_{DD} - V_m \cos \omega_o t + \sum_{n=2,4,6\ldots}^{\infty} V_{mn} \cos n\omega_o t. \tag{4.148}$$

The current has only the odd harmonics:

$$i_D = I_{DC} + I_m \cos \omega_o t + \sum_{n=3,5,7\ldots}^{\infty} I_{mn} \cos n\omega_o t = I_{DC}(1 + \gamma_1 \cos \omega_o t) + \sum_{n=3,5,7\ldots}^{\infty} \gamma_n \cos n\omega_o t, \tag{4.149}$$

where γ_n is defined as

$$\gamma_n = \frac{I_{mn}}{I_{DC}} = \frac{2 \sin n\theta}{n\theta}. \tag{4.150}$$

In a Class-F_2 power amplifier, depicted in Fig. 4.22, only the second harmonic tank is present in addition to the fundamental, and the voltage waveform is

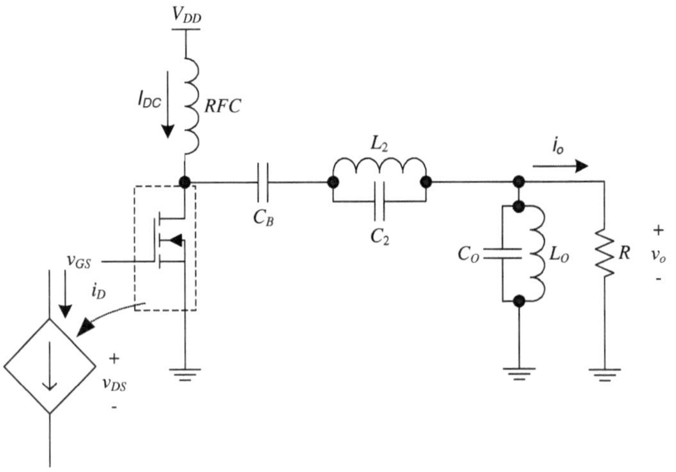

Fig. 4.22 Class-F_2 power amplifier circuit

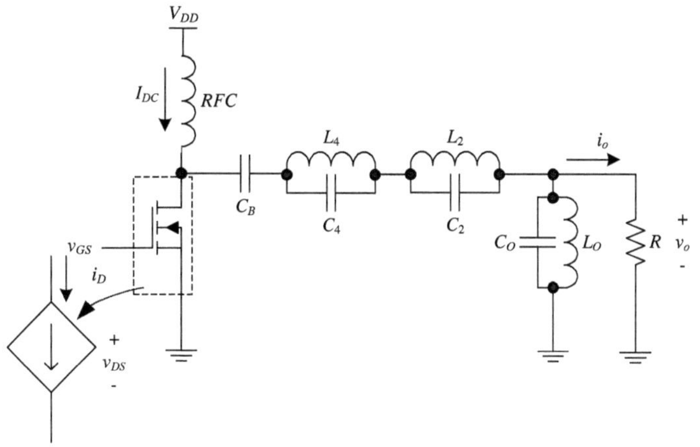

Fig. 4.23 Class-F_{24} power amplifier circuit

Table 4.2 Amplitude coefficients, power and efficiency of Class-F_2 and Class-F_{24} power amplifiers

Parameter/class	F_2	F_{24}
θ	$90° \leq \theta \leq 180°$	$90° \leq \theta \leq 135°$
V_m/V_{DD}	4/3	64/45
$V_{m}2/V_{DD}$	1/3	4/9
$V_{m}4/V_{DD}$	0	1/45
P_{DD}	$4/3\ \theta V_{DD}^2/R\sin\theta$	$64/45\ \theta V_{DD}^2/R\sin\theta$
P_O	$8/9\ V_{DD}^2/R$	$2048/2025\ V_{DD}^2/R$
η	$4/3\ \sin\theta/\theta$	$64/45\ \sin\theta/\theta$
c_p	$\sin\theta/2\pi$	$\sin\theta/2\pi$

$$v_{DS} = V_{DD} - V_m \cos \omega_o t + V_{m2} \cos 2\omega_o t \tag{4.151}$$

and in a Class-F_{24} power amplifier, depicted in Fig. 4.23, the second and fourth harmonic tanks are present in addition to the fundamental, and the voltage waveform is

$$v_{DS} = V_{DD} - V_m \cos \omega_o t + V_{m2} \cos 2\omega_o t + V_{m4} \cos 4\omega_o t. \tag{4.152}$$

A summary of voltage amplitude coefficients, power and efficiency is presented in Table 4.2. The waveforms of Class-F_2 and Class-F_{24} are similar and are shown in Fig. 4.24, where the ripple in voltage waveform is exaggerated, while for maximum flat waveform it is minimal.

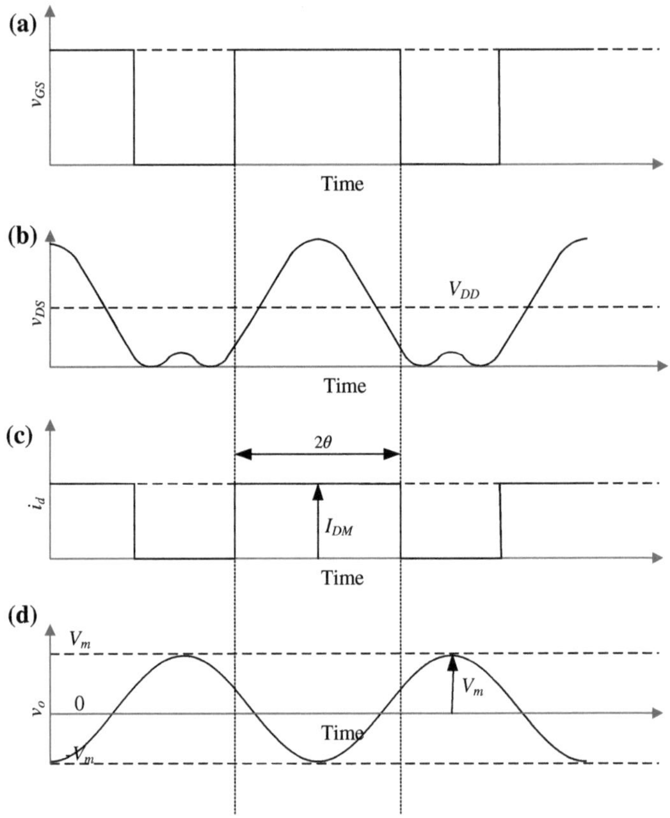

Fig. 4.24 Waveforms of the Class-F_2 power amplifier: **a** gate-to-source voltage, **b** drain-to-source voltage, **c** drain current and **d** output voltage

4.3.4 Class-F Power Amplifier with Quarter-Wavelength Transmission Line

The Class-F power amplifier with quarter-wavelength transmission line and parallel-resonant circuit, depicted in Fig. 4.25, is the ideal Class-F power amplifier. Because of the mechanism described in Chap. 5, the quarter-wavelength transmission line acts as an infinite number of resonators, thus the Class-F power amplifier with quarter-wavelength transformer is sometimes also denoted a Class-F_{∞} amplifier. At 2.4 GHz, the length of the quarter-wavelength transmission line is about 3 cm.

If the amplifier is operated as a switch with saturation resistance r_{sat}, then the power dissipated in the transistor is

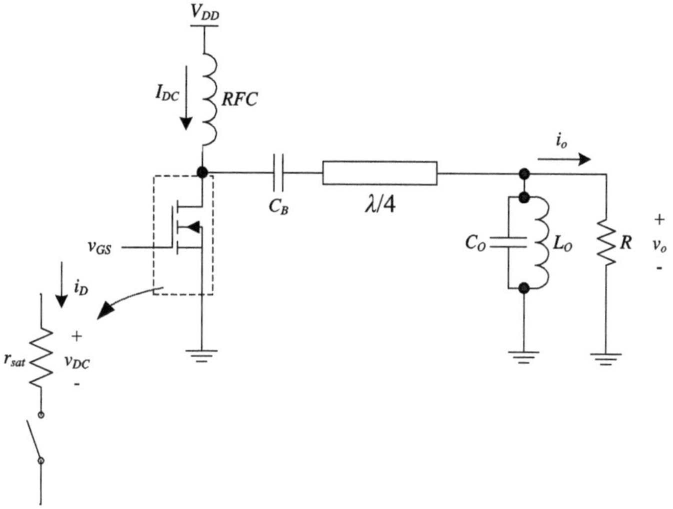

Fig. 4.25 Class-F$_\infty$ power amplifier circuit

$$P_{rsat} = \frac{r_{sat} I_m^2}{4}, \tag{4.153}$$

and the output power is

$$P_O = \frac{R_i I_m^2}{2}, \tag{4.154}$$

where R_i is the input impedance seen by the drain and source terminals at fundamental frequency f_0, and is defined as

$$R_i = \frac{Z_0^2}{R}, \tag{4.155}$$

where Z_0 is the characteristic impedance of the transmission line and R is load resistance. The characteristic impedance of the transmission line will be defined in Chap. 5, in which the derivation of Eq. (4.155) will be also be presented.The efficiency is

$$\eta = \frac{P_O}{P_O + P_{rsat}} = \frac{R_i}{R_i + \frac{r_{sat}}{2}}. \tag{4.156}$$

The efficiency can reach 100 % when $r_{sat} = 0$. The output power capability can reach the figure of $c_p = 0.159$, that of the Class-D amplifier. With $\theta = 90°$, the following describing quantities can be derived. The peak drain-to-source voltage is

$$V_{DSM} = 2V_{DD}, \tag{4.157}$$

and the amplitude of its fundamental component is

$$V_m = \frac{4}{\pi} V_{DD}. \tag{4.158}$$

The DC input current is given by

$$I_{DC} = \frac{8V_{DD}}{\pi^2 R_i}. \tag{4.159}$$

The waveforms of the Class-F$_\infty$ power amplifier are shown in Fig. 4.26 for $\theta = 90°$.

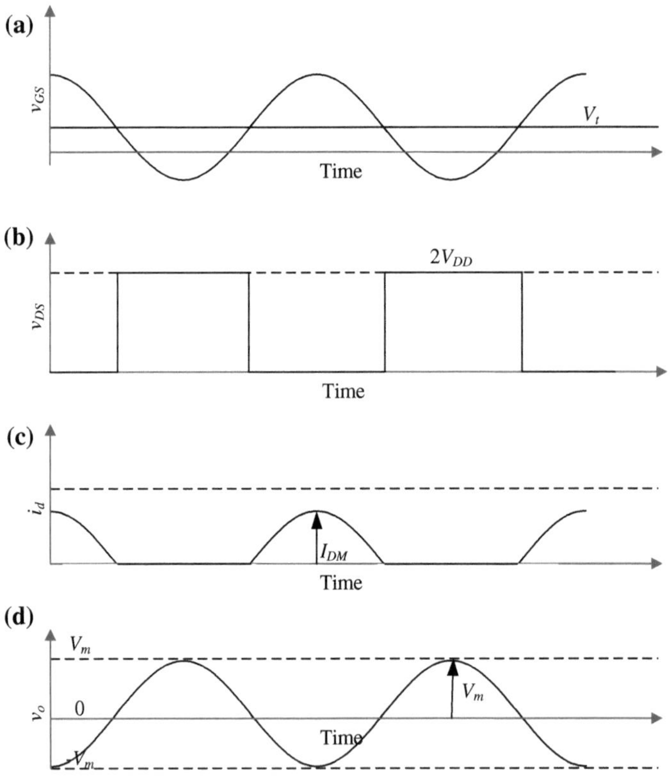

Fig. 4.26 Waveforms of the Class-F$_\infty$ power amplifier: **a** gate-to-source voltage, **b** drain-to-source voltage, **c** drain current and **d** output voltage

4.4 Other Power Amplifier Classes

Several other power amplifier classes are also possible. These classes are usually a combination of classes already mentioned in this chapter.

Class-DE combines Class-E ZVS switching conditions with the low-voltage stress of Class-D output stage, thus effectively extending the Class-D power amplifier operation to higher frequencies. This stage is also known as the Class-D_{ZVS} power amplifier. The power amplifier is driven so that there are time intervals when both Class-D driving transistors are off. The circuit is similar to the Class-D power amplifier circuit, but with capacitors in shunt with the transistors.

Similarly, the Class-FE power amplifier is achieved when Class-E ZVS conditions are applied to the Class-F power amplifier. This is accomplished by inserting a shunt capacitor and an additional series inductor into the Class-F circuit.

The Class-E/F power amplifier is a Class-E amplifier with inverse Class-F (Class-1/F) harmonic tuning. This is accomplished by inserting additional Class-F^{-1} resonators into the Class-E power amplifier circuit.

Class-E_M is essentially Class-E optimized for higher frequencies, where efforts are made to minimize switching time that becomes more prominent as frequency increases. This entails finding a solution that removes the instantaneous jump in the drain current during turn-off. The simplest Class-E_M power amplifier consists of a main power amplifier that consumes about three-quarters of load power and converts this power and power generated by the auxiliary amplifier to the power at output frequency f_0 with a smaller auxiliary power amplifier that generates a quarter of the load power at frequency $2f_0$, and is phased and locked to the main amplifier.

Recently, a continuous Class-F power amplifier was introduced [21]. It is similar to the Class-F power amplifier the sense that both the normal Class-F and continuous Class-F versions have the same half-sine current waveform. The voltage waveform of the continuous Class-F amplifier is, however, more complicated. The Class-F amplifier has been one of the most popular broadband high-efficiency power amplifiers, with theoretical efficiency of 90 %. The mechanism and performance of the Class-F amplifier is similar to that of Class J.

4.5 Recent State-of-the-Art Examples

As in Chap. 3, in this section we list several state-of-the art examples of switch-mode power amplifiers published in recent years.

Aflaki et al. described a current-mode Class-D power amplifier for *S*-band [3]. The amplifier operates at 2.35 GHz, has 68 % efficiency and is able to deliver 39 dBm power using a GaN transistor.

Leng et al. demonstrated a GaN Class-E^{-1} power amplifier with a finite DC-feed inductance [4]. This amplifier uses transmission lines and it can deliver 40.7 dBm output power at 2.87 GHz while accomplishing drain efficiency of 70.6 %.

Thian et al. designed a Class-FE amplifier with 91 % drain efficiency operating at a frequency of 2.22 GHz [5]. The amplifier is a transmission-line based discrete implementation around a GaN HEMT and delivers 39.5 dBm of power.

Calvillo-Cortes et al. presented a Chireix outphasing RF power amplifier based on a Class-E configuration at 2.2 GHz [6]. Using a wideband bondwire-based transformer, this power amplifier reaches 81 % and 53.5 % efficiency from a 20 and 28 V supply voltage respectively.

Barton and Perreault demonstrated a microstrip-based Chireix outphasing power amplifier based on the Class-F^{-1} output stage [7]. It can deliver 14.1 W of power at 55.6 % efficiency.

Ji et al. presented a 10-W GaN Class-F power amplifier [8]. The power amplifier uses digital predistortion for linearity improvement and delivers 33.3 dBm of power at 2.12 GHz with an efficiency of 37.8 %.

4.6 Summary

In this chapter we presented the theory of operation of switch-mode power amplifiers, specifically variations of Class D, Class E and Class F. Some basic defining concepts of other classes of switch-mode power amplifiers (typically combinations of the above-mentioned three classes) were also mentioned, but the theory was not presented in great detail.

Switch-mode power amplifiers have high efficiency (it can reach 100 % in theory), and some have the highest output-power capability reported to date. High efficiency is typically due to the shape of output waveforms, where the basic idea is that output voltage and current waveforms never overlap, thus none of the power is dissipated in the active device. At the end of this chapter, we listed some power amplifiers with fairly high efficiencies achieved in practice.

In Chap. 3 and in this chapter, we realized that each power amplifier, in addition to an active component (transistor), requires a different number of passive components, depending on the output stage. Design of passives is presented in the next chapter.

References

1. Kazimierczuk MK. RF power amplifiers. 2nd ed. Chiechester: Wiley; 2015.
2. Grebennikov A, Sokal NO, Franco MJ. Switchmode RF and microwave power amplifiers. 2nd ed. Burlington: Elsevier; 2012.
3. Aflaki P, Negra R, Ghannouchi FM. Enhanced architecture for microwave currentmode Class-D amplifiers applied to the design of an S-band GaN-based power amplifier. IET Microwave Antennas Propagation. 2009;3(6):997–1006.
4. Leng Q, Zeng Y, Zhang GL, Peng W, Zhang LJ, Peng YT, Guan J, Lu H, Zheng ZQ. An inverse class E power amplifier with finite DC-feed inductance. Microw J. 2013;56(8):86–100.

5. Thian M, Barakat A, Fusco V. High-efficiency harmonic-peaking Class-EF power amplifiers with enhanced maximum operating frequency. IEEE Trans Microw Theory Tech. 2015;63 (2):659–71.
6. Calvillo-Cortes DA, van der Heijden MP, Acar M, de Langen M, Wesson R, van Rijs F, de Vreede LCN. A package-integrated chireix outphasing RF switch-mode high-power amplifier. IEEE Trans Microw Theory Tech. 2013;61(10):3721–32.
7. Barton TW, Perreault DJ. Four-way microstrip-based power combining for microwave outphasing power amplifiers. IEEE Trans Circuits Syst I Regul Pap. 2014;61(10):2987–98.
8. Ji L, Xu Z, Zhou J, Zhai J. Highly efficient 10 W GaN class F power amplifier using DPD. Microw J. 2013;56(10):120–30.
9. Rogers J, Plett C. Radio frequency integrated circuit design. 2nd ed. Boston: Artech House; 2010.
10. Raab FH, Asbeck P, Cripps S, Kenington PB, Popovic ZB, Pothecary N, Sevic JF, Sokal NO. Power amplifiers and transmitters for RF and microwave. IEEE Trans Microw Theory Tech. 2002;50(3):814–26.
11. Raab FH. Maximum efficiency and output of Class-F power amplifiers. IEEE Trans Microw Theory Tech. 2001;49(6):1162–6.
12. Sokal NA, Sokal AD. Class-E-A new class of high-efficiency tuned single-ended switching power amplifiers. IEEE J Solid-State Circuits. 1975;10(3):168–76.
13. Gao S. High-efficiency Class F RF/microwave power amplifiers. IEEE Microwave Mag. 1055;2006(7):40–8.
14. Hayati M, Kazimierczuk MK, Sekiya H. Modeling and analysis of Class-E amplifier with a shunt inductor at sub-nominal operation for any duty ratio. IEEE Trans Circuits Syst I Regul Pap.
15. Thian M, Fusco VF. Transmission-line Class-E power amplifier with extended maximum operating frequency. IEEE Trans Circuits Syst II Express Briefs. 2011;58(4):195–9.
16. Wei X, Sekiya H, Kuroiwa S, Suetsugu T, Kazimierczuk K. Design of Class-E amplifier with MOSFET linear gate-to-drain and nonlinear drain-to-source capacitances. IEEE Trans Circuits Syst I Regul Pap. 2011;58(10):2556–65.
17. Zhang R, Acar M, van der Heijden MP, Apostolidou M, Leenaerts DMW. Generalized semi-analytical design methodology of Class-E outphasing power amplifier. IEEE Trans Circuits Syst I Regul Pap. 2014;61(10):2951–60.
18. Wei X, Nagashima T, Kazimierczuk MK, Sekiya H, Suetsugu T. Analysis and design of Class-EM power amplifier. IEEE Trans Circuits Syst I Regul Pap. 2014;61(4):976–86.
19. Raab FH. Class-E, Class-C, and Class-F power amplifiers based upon a finite number of harmonics. IEEE Trans Microw Theory Tech. 2001;49(12):1462–8.
20. Juhas A, Novak LA. Closed form of optimal current waveform for Class-F PA up to fourth harmonic. Sadhana. 2015;40(2):425–36.
21. Peng C, He S. A 0.4–2.3 GHz broadband power amplifier extended continuous Class-F design technology. Int J Electron. 2015;102(8):1320–33.

Chapter 5
Passives for Power Amplifiers

In this chapter passive components, paramount for power amplifier design, are discussed. As the frequency increases towards the RF and microwave bands defined in Chap. 2, the quality (Q) factor of passive devices becomes more dominant in the power dissipation and efficiency of a power amplifier, and it is thus investigated in detail. At higher frequencies, modeling of passive devices also becomes more complex. Special focus will be placed on different types of inductor implementations, as they tend to exhibit low Q-factors even at lower frequencies and their design is therefore an important step in power amplifier design. Integrated inductors allow for smaller floorplans and more compact packages and are also discussed in some detail. MEMS inductors are discussed as a promising technology for the design of inductors. The chapter will also cover resistors, discrete and integrated capacitors, discrete inductors such as toroids and solenoids, RFCs and transformers. It concludes with the discussion of a quarter-wave transformer and its importance in power amplifier design.

5.1 Resistors

The purpose of the resistor in the electronic circuit is to provide an adequate voltage drop [1]. This is accomplished by converting some of the electrical energy into heat. Both integrated and discrete resistors are used.

Several types of discrete resistors are generally used in RF and microwave electronics, specifically with power amplifiers:

- Carbon composite resistors,
- Wire-wound resistors,
- Metal-film resistors,
- Thin film resistors.

In integrated circuitry, resistors are fairly easy to fabricate by using layers with different (high or low) resistivity, depending on the required resistance value. In design kits, the resistivity of the layer is given in Ω/square. The designer can reach the required resistance by designing for a required number of squares. If carefully

M. Božanić and S. Sinha, *Power Amplifiers for the S-, C-, X- and Ku-bands*,
Signals and Communication Technology, DOI 10.1007/978-3-319-28376-0_5

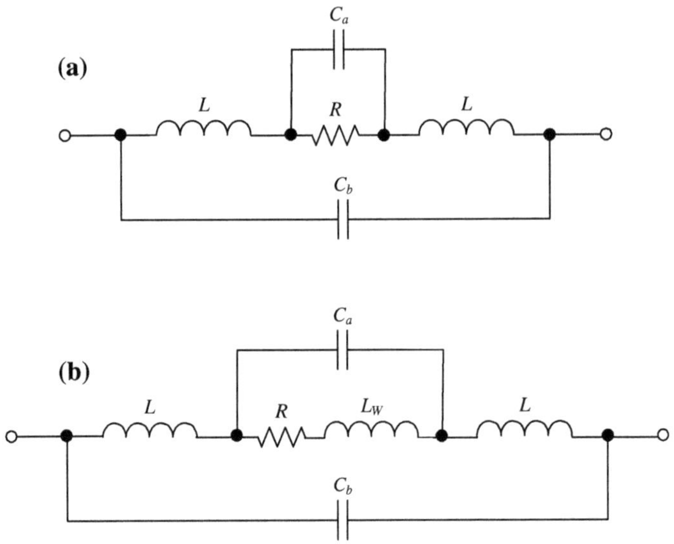

Fig. 5.1 Circuit model of **a** general resistor and **b** wire-wound resistor

controlled resistance is needed, integrated resistors can be divided into segments that can be matched to remove the influence of the resistivity gradient that may appear during layer fabrication.

By definition, the impedance of the resistor is constant at all frequencies. However, as frequency increases, depending on the type of the resistors, parasitics start playing a more prominent role. Figure 5.1a shows a general model of a resistor. In this model, R is the actual resistance of the resistor, L is the resistor lead inductance and C_a and C_b are various parasitic capacitances. The model of the wire-wound resistor is slightly more complex, with additional inductance L_W due to the influence of the wire windings that cause the device to act as an inductor. The additional inductance is illustrated in Fig. 5.1b.

Some of the inductor parameters can be calculated as indicated below. The resistance of the wire-wound resistor is given by

$$R = \frac{l}{2\pi a \delta \sigma}, \tag{5.1}$$

where a is the radius of the cylindrical wire, $\sigma = 1/\rho$ is the material conductivity and δ is the skin depth related to frequency f via relation

$$\delta = \sqrt{\frac{\rho}{\pi \mu f}}. \tag{5.2}$$

The inductance in microhenries of the leads can be determined via the equation for the inductance of the straight piece of wire [2]:

$$L = 0.002\,l\left[2.3\log\left(\frac{4l}{d} - 0.75\right)\right], \tag{5.3}$$

where l is the length of the spiral and d is the wire diameter. Parasitic capacitances depend on the exact resistor implementation, thus there is no single determining equation that can be mentioned here. Wire-wound inductance, on the other hand, can be calculated; however, the equations will be discussed in the part of the chapter that deals with inductors.

A typical frequency response of a thin-film resistor may look something like what is illustrated in Fig. 5.2a, where the capacitive effect becomes more prominent as the frequency increases, but with the inductor effect becoming even more prominent above the resonant frequency $f_r = \frac{1}{2\pi\sqrt{LC}}$. Wire-wound resistors are more strongly influenced by inductive parasitics, resulting in an impedance characteristics as shown in Fig. 5.2b. Depending on the type of resistor, even multiple resonance points may exist.

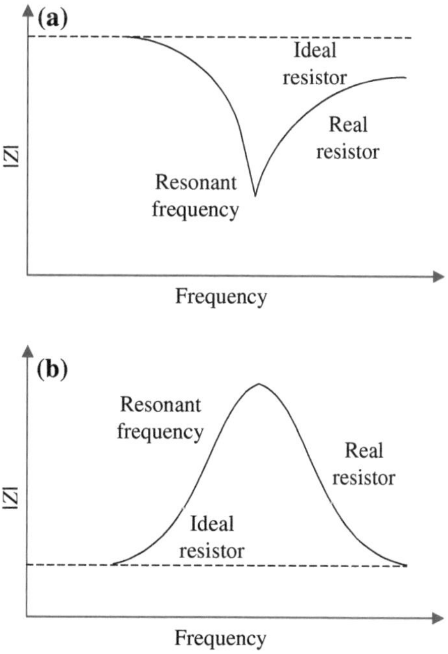

Fig. 5.2 Frequency response of the impedance of ideal and real resistors: **a** thin-film resistor and **b** wire-wound resistor

5.2 Capacitors

The role of a capacitor in an electronic circuit is to store the energy electrostatically in an electric field. A typical capacitor consists of two parallel plates separated by an insulator layer (air, ceramic, plastic glass, metalized film, etc.). If the surface area A of the plates is large compared to its separation distance d, the capacitance can be calculated by [1]

$$C = \varepsilon_0 \varepsilon_r \frac{A}{d}, \tag{5.4}$$

where ε_0 is the absolute permittivity defined in Chap. 2 and ε_r is the relative permittivity of the dielectric used. Mica capacitors typically have a dielectric constant of about 6, making them fairly large [2]. On the other hand, these capacitors have a good temperature coefficient. Ceramic capacitors have dielectric constants up to 10,000 but worse temperature performance.

5.2.1 *Discrete Capacitors*

A high-frequency model of a typical discrete capacitor is illustrated in Fig. 5.3. Capacitance C is the capacitor (design) capacitance calculated by Eq. (5.4). R_S and L represent the losses in both lead conductors and can be calculated with resistor equations (Eqs. 5.1 and 5.3). R_e is the dielectric loss resistance, defined as

$$R_e = \frac{d}{\sigma_{diel} A}, \tag{5.5}$$

where σ_{diel} is the conductivity of the dielectric. The frequency response of the impedance of ideal and real capacitors is shown in Fig. 5.4.

The Q-factor of a device is defined as 2π times the ratio of energy stored in the device and energy lost in one oscillation cycle. If Z is the impedance of a capacitor, then its Q-factor is given by

$$Q = \frac{\mathrm{Im}(Z)}{\mathrm{Re}(Z)}. \tag{5.6}$$

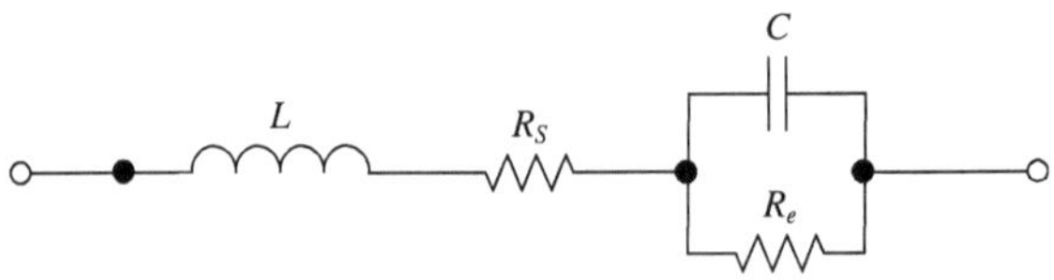

Fig. 5.3 High-frequency model of a capacitor

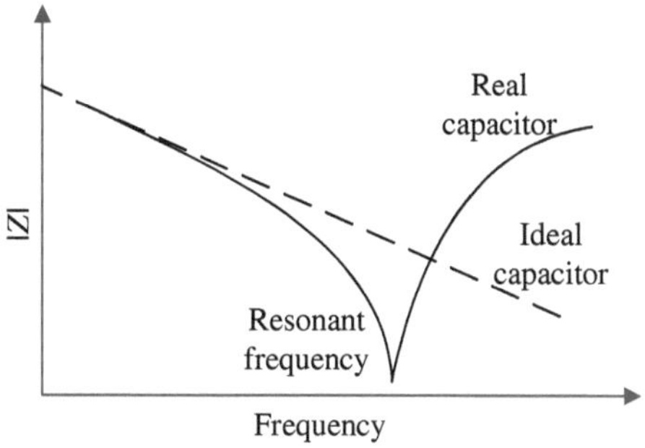

Fig. 5.4 Frequency response of the impedance of ideal and real capacitors

Thus the Q-factor of a capacitor can be defined as

$$Q = \frac{X_C}{R_{total}}, \quad (5.7)$$

where X_C is the capacitor reactance and R_{total} is the total capacitor equivalent resistance. The Q-factor of discrete capacitors is generally not a problem at RF and microwave. At microwave frequencies, interdigital capacitors with high Q-factors but limited capacitance values [3] can also be used.

5.2.2 *Integrated Capacitors*

In integrated circuits, capacitors are implemented by utilizing two polysilicon (poly) layers or two metal layers and the insulating material between the two layers. Typically, poly-insulator-poly (PIP) and metal-insulator-metal (MIM) capacitors are used. A MIM capacitor is illustrated in Fig. 5.5.

A model of the MIM (or PIP) capacitor needs to cater for the process substrate effects in addition to the parasitics mentioned with the discrete capacitors. A typical MIM model is shown in Fig. 5.6 [4]. In this model, parasistics due to two leads need to be separated into parasitics for the bottom and the top plates. Furthermore, C_{ox} represents the capacitance due to the oxide below the bottom plate, and C_{sub}

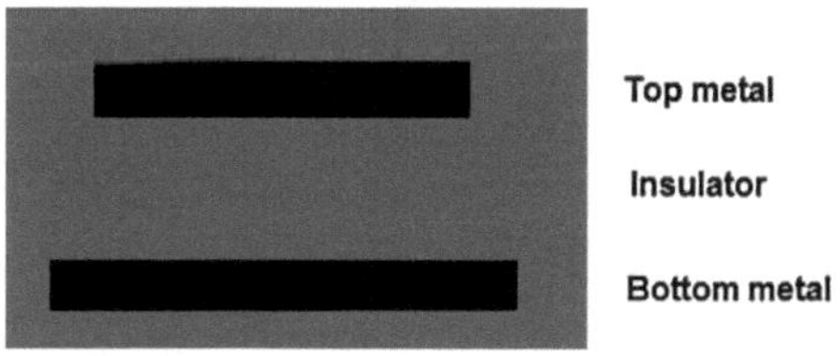

Fig. 5.5 A MIM capacitor

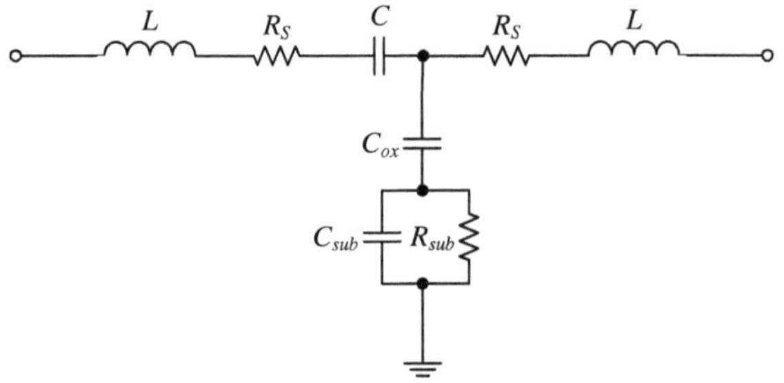

Fig. 5.6 Model of the MIM capacitor

and R_{sub} are the capacitance and resistance due to the substrate, respectively. The oxide and substrate parasitics are approximately proportional to the area of the capacitor, but are also highly dependent on the conductivity of the substrate and the operating frequency. Integrated inductors also suffer from oxide and substrate parasitics, so detailed oxide and substrate calculations will be presented in the part of the chapter dealing with inductors.

The capacitance value of an integrated capacitor is deterministic. This means that by choosing the width and the length of the capacitor, a carefully controlled capacitance value can be obtained in an IC implementation. Also, accurate models of integrated capacitors are typically provided by the foundry, where the designer can change the width and length of the capacitor until the required capacitance is reached.

The frequency response of the capacitance and the Q-factor of a typical MIM capacitor are shown in Fig. 5.7. As expected, the Q-factor decreases with frequency, from about 20 at 20 GHz to less than 5 at 100 GHz in a 130 nm process, so for power amplifiers up to *Ku*-band, it is in an acceptable range, and thus should not be of a great concern to the designer [4].

MIM and PIP capacitors can be considered passive capacitors. Active capacitors, such as the single-ended active capacitor (SAC) configuration shown in Fig. 5.8, could present an advantage at mm-wave frequencies.

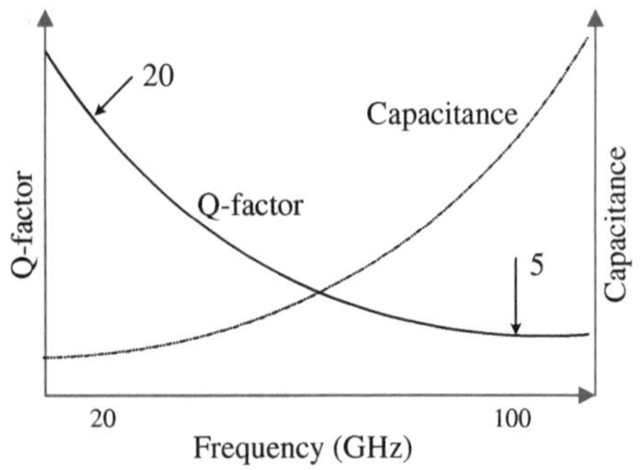

Fig. 5.7 Capacitance and Q-factor versus frequency of a typical MIM capacitor

Fig. 5.8 An active capacitor

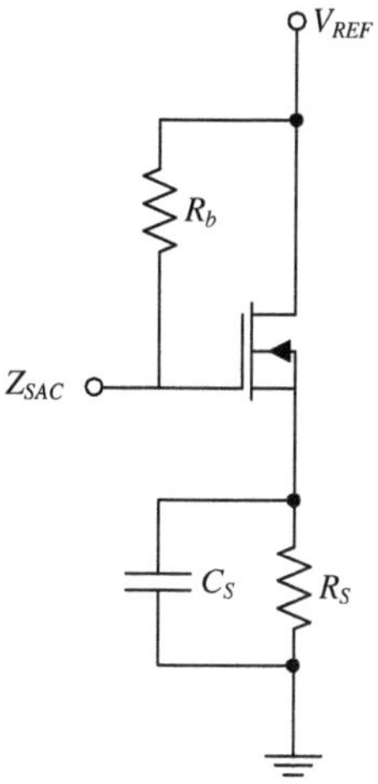

5.3 Inductors

The inductor or a coil is a passive electronic device that, stores energy in its magnetic field when current passes through the device. A real inductor is usually modeled as an ideal inductor L_S in series with a resistor R_S, both in parallel with capacitor C_S, as shown in Fig. 5.9 [1, 5]. Inclusion of the series resistor and parallel capacitor is necessary to model the losses of the inductor even at frequencies below RF.

While the impedance of the ideal inductor increases with constant slope for all frequencies, the impedance of a non-ideal inductor exhibits a slope value dependent on frequency, as shown in Fig. 5.10. The frequency where the magnitude of impedance ($|Z|$) peaks is the resonant frequency of an inductor. The resonant frequency, $f_r = \frac{1}{2\pi\sqrt{L_S C_S}}$, should ideally peak at infinity, but the finite value of the peak is due to the resistance R_S. Similarly, capacitance C_S is the reason the inductor exhibits capacitive instead of inductive behavior at frequencies above the resonance.

The Q-factor of an inductor can be defined similarly to the Q-factor of the capacitor. Thus

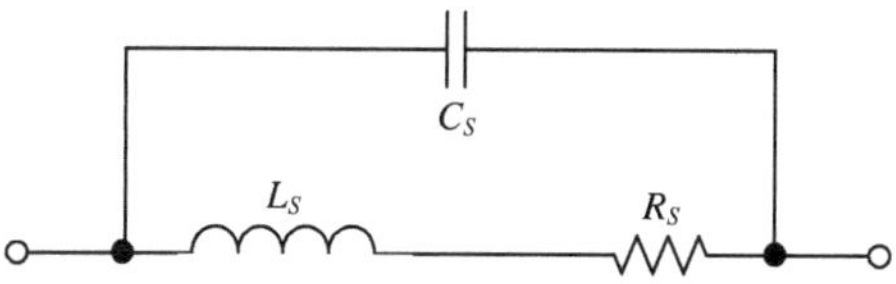

Fig. 5.9 General high-frequency model of an inductor

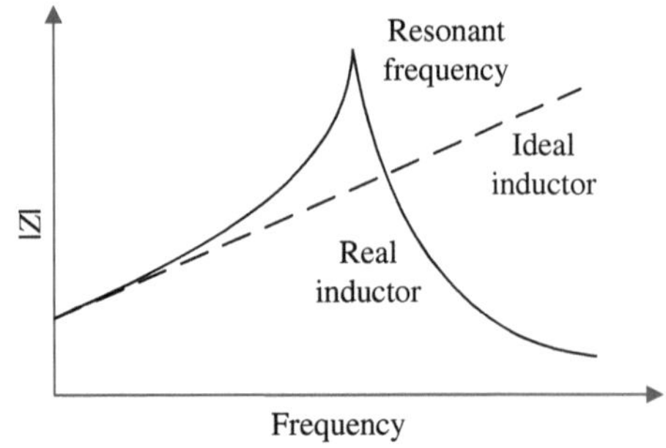

Fig. 5.10 Frequency response of the impedance of ideal and real inductors

$$Q = \frac{X_L}{R_S}, \tag{5.8}$$

where X_L is the total reactance of the inductor. The Q-factor of an inductor is much more heavily dependent on the frequency than that of the capacitor. Thus careful design of the inductor is needed.

Various factors, such as inductor size and its Q-factor, lead to numerous inductor implementation options. These include:

- Discrete inductors,
- Active integrated inductors,
- Bond wires,
- Passive integrated inductors (spiral inductors),
- MEMS inductors, and
- Other on-chip or on-package/in-package implementations.

Each of the above options is discussed in more detail in the sections that follow.

5.3.1 Discrete Inductors

Discrete or off-chip inductors are usually implemented as a solenoidal coil or a toroid, as illustrated in Fig. 5.11.

The simplest solenoidal configuration involves an air core around which a single layer of wire is wound. For radius r (in cm), length l (in cm) and number of turns N, the resulting inductance in μH is [2]

$$L = \frac{0.394r^2N^2}{9r + 10l}. \tag{5.9}$$

The parasitic resistance can be calculated from the wire-resistance equation (Eq. 5.1).

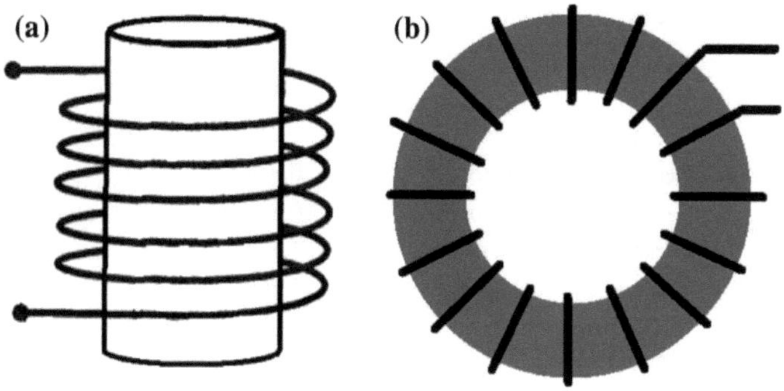

Fig. 5.11 Diagrams of **a** solenoidal, **b** toroidal inductors

Air core inductors are impractical in RF and microwave applications because of their size. Thus most RF inductors are designed with cores manufactured with magnetic materials, such as iron on ferrite. They allow for smaller size and increased Q-factor owing to the lower number of turns required to achieve the same inductance as air-gap inductors.

Toroid inductors are magnetic-core solenoidal inductors, curved such that they assume a ring or doughnut shape. The main advantage of toroids is their self-shielding capability. Toroidal inductor inductance in nH is given by

$$L = \frac{0.004\,\pi N^2 \mu_i A_c}{l_e}, \tag{5.10}$$

where A_c is the cross-sectional area of the core in cm^2, l_e is the effective length of the core in cm and μ_i is the initial magnetic permeability (permeability at low magnetic fields), a relative quantity typically greater than 500 for ferrite. Sometimes all quantities other than the number of turns are combined into a single constant for the inductor core given by the core manufacturer, called the inductance index, A_L (in nH/turn^2). In terms of inductance index, the toroidal inductance in nH is then

$$L = N^2 A_L. \tag{5.11}$$

The usage of discrete inductors at high frequencies also implies careful PCB modeling and design. The frequency of the Q-factor peak (typically in the range of hundreds) is predefined and is typically located in either the high-megahertz or the low-gigahertz range. Certain inductor manufacturers, such as Coilcraft Inc., provide an easy interface to look for inductors optimized for a specific frequency [6]. Tables 5.1 and 5.2 illustrate typical discrete inductor sizes, corresponding Q-factors and self-resonance frequencies (SRF) that can be obtained from this supplier for applications at 1 GHz and 3 GHz respectively. Nominal inductances of 1 nH, 2 nH and 5 nH were searched for.

Table 5.1 Commercially available high-Q inductors (1 GHz application)

Part number	Inductor family	Q factor	Inductance (nH)	Nominal L (nH)	SRF (GHz)
0604HQ-1N1	Ceramic chip inductors	96	1.21	1.2	6
0402CS-1N0	Ceramic chip inductors	52	0.83	1	12.7
0402HP-1N0	Ceramic chip inductors	48	0.81	1	16
0603CT-1N0	Ceramic chip inductors	44	1.13	1	16
0603CT-1N2	Ceramic chip inductors	39	1.36	1.2	16
0906-2	Micro Spring™ air core inductors	199	1.47	1.7	10
0403HQ-1N9	Ceramic chip inductors	70	2.02	1.9	11.84
0603CT-2N3	Ceramic chip inductors	64	2.42	2.3	11
0402HP-2N2	Ceramic chip inductors	63	2.01	2.2	15.1
0402CS-2N2	Ceramic chip inductors	61	1.87	2.2	10.8
0402CS-2N0	Ceramic chip inductors	60	1.92	2	11.1
0402HP-2N0	Ceramic chip inductors	60	1.79	2	15.2
0603CT-2N0	Ceramic Chip Inductors	60	1.98	2	12
0402HP-2N4	Ceramic chip inductors	59	2.25	2.4	14
0402CS-1N8	Ceramic chip inductors	57	1.82	1.8	12
0403HQ-2N1	Ceramic chip inductors	56	2.24	2.1	12.4
0402CS-1N9	Ceramic chip inductors	55	1.64	1.9	11.3
0402CS-2N4	Ceramic chip inductors	55	2.4	2.4	10.5
0603CS-1N6	Ceramic chip inductors	53	1.42	1.6	12.5
0603HC-1N6	Ceramic chip inductors	53	1.35	1.6	12.5
0805HT-1N8	Ceramic chip inductors	52	2.07	1.8	9.4
1508-5N5	Mini Spring™ air core inductors	214	5.61	5.5	5
0906-5	Micro Spring™ air core inductors	150	5.07	5.4	7
1606-6	Micro Spring™ air core inductors	130	5.18	5.6	6.5
0603HP-5N6	Ceramic chip inductors	111	5.53	5.6	6.6
0604HQ-4N5	Ceramic chip inductors	104	4.5	4.5	5.8
0604HQ-5N0	Ceramic chip inductors	103	5.03	5	5.3
0805HQ-5N6	Ceramic chip inductors	102	5.62	5.6	6
0603HP-5N1	Ceramic chip inductors	98	4.96	5.1	8.9
0603HP-6N0	Ceramic chip inductors	96	6.19	6	6
0806SQ-6N0	Ceramic chip inductors	96	6.26	6	5.2
0805CS-050	Ceramic chip inductors	94	5.5	5.6	5.9
0403HQ-5N5	Ceramic chip inductors	85	5.59	5.5	8.6
0603CT-5N6	Ceramic chip inductors	83	5.46	5.6	5.05
0805HT-4N7	Ceramic chip inductors	79	4.68	4.7	5.5

Table 5.2 Commercially available high-Q inductors (3 GHz application)

Part number	Inductor family	Q factor	Inductance (nH)	Nominal L (nH)	SRF (GHz)
0604HQ-1N1	Ceramic chip inductors	201	1.22	1.2	6
0402CS-1N0	Ceramic chip inductors	116	0.82	1	12.7
0402HP-1N0	Ceramic chip inductors	114	0.81	1	16
0603CT-1N2	Ceramic chip inductors	89	1.34	1.2	16
0603CT-1N0	Ceramic chip inductors	88	1.12	1	16
0201DS-1N2	Ceramic chip inductors	69	1.19	1.2	17.9
0402CS-1N2	Ceramic chip inductors	67	1	1.2	12.9
0906-2	Micro Spring™ air core inductors	10,513	1.47	1.7	10
0603CT-2N3	Ceramic chip inductors	179	2.49	2.3	11
0603CS-1N6	Ceramic chip inductors	163	1.42	1.6	12.5
0402HP-2N2	Ceramic chip inductors	153	2.02	2.2	15.1
0603CT-2N0	Ceramic chip inductors	148	1.99	2	12
0402HP-2N0	Ceramic chip inductors	147	1.8	2	15.2
0403HQ-1N9	Ceramic chip inductors	137	2.05	1.9	11.84
0805HT-1N8	Ceramic chip inductors	136	2.11	1.8	9.400
0402CS-2N0	Ceramic chip inductors	133	1.95	2	11.1
0402CS-2N2	Ceramic chip inductors	133	1.89	2.2	10.8
0402HP-2N4	Ceramic chip inductors	128	2.3	2.4	14
1606-6	Micro Spring™ air core inductors	1397	5.33	5.6	6.5
1508-5N5	Mini Spring™ air core inductors	810	6.39	5.5	5
0906-5	Micro Spring™ air core inductors	533	5.63	5.4	7
0603HP-5N1	Ceramic chip inductors	471	5.14	5.1	8.9
0603HP-5N6	Ceramic chip inductors	325	5.91	5.6	6.6
0805HQ-5N6	Ceramic chip inductors	305	6.22	5.6	60
0805HT-4N7	Ceramic chip inductors	242	5.01	4.7	5.5
0603CT-5N6	Ceramic chip inductors	237	5.73	5.6	5.05
0403HQ-5N5	Ceramic chip inductors	231	6.14	5.5	8.6
0806SQ-6N0	Ceramic chip inductors	228	8.25	6	5.2
0603HP-4N3	Ceramic chip inductors	221	4.21	4.3	7.5
0603CT-5N1	Ceramic chip inductors	217	5.24	5.1	5.5
0603CS-4N3	Ceramic chip inductors	191	4.16	4.3	5.9
0604HQ-4N5	Ceramic chip inductors	181	4.85	4.5	5.8
0805CS-050	Ceramic chip inductors	165	6.07	5.6	5
0603CT-4N3	Ceramic chip inductors	16	4.62	4.3	6.3
0604HQ-5N0	Ceramic chip inductors	163	5.44	5	5.3

5.3.2 Integrated Active Inductors

As seen before, for frequencies up to the *Ku*-band, there is no real advantage to active integrated capacitors because passive capacitors exhibit acceptable Q-factors. High-Q integrated active inductors, on the other hand, are a good alternative to passive integrated inductors because passive inductors typically have low Q-factors. Typical Q-factors that can be obtained for active configurations are between 10 and 100, which are up to ten times those of spiral inductors [7]. Active inductors can also take up a smaller area on the chip than spiral inductors. The main disadvantages of active inductors include increased power consumption, the presence of electrical noise from active devices and limited dynamic range.

A design requiring only six transistors has been demonstrated [8], which makes active inductors much more compact compared to traditional designs requiring ten transistors or more. The active die area of this topology is only 30 μm by 65 μm, and it achieves a Q-factor of about 20 at 1 GHz while exhibiting a differential inductance of 20 nH.

Some active inductors employ a gyrator approach [9]. An ideal gyrator is a linear two-port network that does not absorb or dissipate energy. A linearized model of an ideal gyrator is shown in Fig. 5.12. At low frequencies the effective inductance and series resistance can be accurately modeled in terms of capacitances (C_1, C_0 and C), output resistance (R_0) and transconductances (g_1 and g_2) represented in Fig. 5.12. At higher frequencies, the parasitics become more prominent, thus more complex models are required and the analysis of active inductors becomes specific to a given configuration. Some examples of active inductor designs are included in [7, 8].

5.3.3 Bond Wires

Bond wires are very thin metals used to connect the bonding pads of an IC to the pins (leads) of the IC package [10]. Typically, they are made of gold, copper, or aluminum. They normally present a parasitic quantity for signals transmitted between systems inside and outside the packaged device because of their inductive

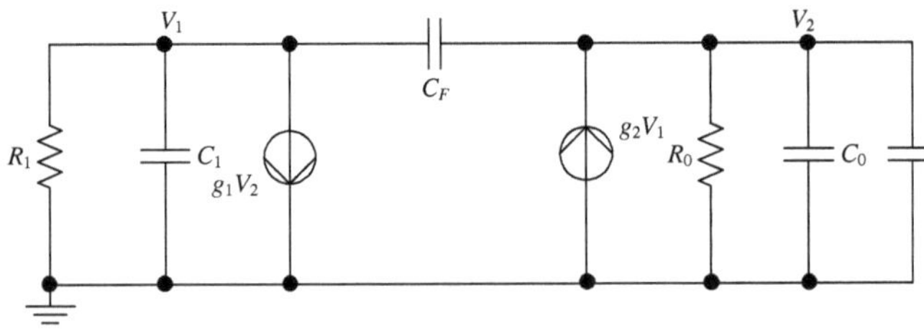

Fig. 5.12 Two-port equivalent model of a gyrator

behavior [11]. The fact that the bond wire is inductive can be used as an advantage in RF and microwave design.

Electrical characteristics of bond wires depend on the type of material of which they are made, the wire cross-section (or radius), the horizontal length and the pitch between the adjacent wires, if more than one bond wire is used [12]. Inductance in henries of the bond wire is approximated as [13]:

$$L = \frac{\mu_0 l}{2\pi}\left(\ln\frac{2l}{a} - 0.75\right), \tag{5.12}$$

where l is the length of the bond wire, a is the radius of the wire and μ_0 is the absolute permeability ($4\pi\ 10^{-7}$ H/m). The mutual inductance of adjacent bond wires is approximated as

$$M = \frac{\mu_0 l}{2\pi}\left(\ln\frac{2l}{s} - \frac{s}{l} - 1\right) \tag{5.13}$$

and the resistance of the wire is calculated using the wire formula in Eq. (5.1).

With the inductance and resistance of the bond wire calculated, the Q-factor of the bond wire can be calculated by assuming the series inductance-resistance model and Eq. (5.8). The parasitics of the bond wire can be decreased by placing the wire well above any conducting planes. Thus high Q-factors and high SRF can be obtained. Although bond wires with Q-factors of 50 have been reported, their inductances will typically be less than 1 nH [12]. This limits their feasibility for the gigahertz range where well-controlled inductances of 1 nH and more are often needed.

5.3.4 *Spiral Inductors*

Spiral integrated inductors present a viable option for practical RF and microwave implementation. This is due to the fact that a number of deterministic models exist that can be used to predict the inductance value and Q-factors of any inductive structure on chip with varying degrees of accuracy, given the process parameters and geometry of that inductive structure.

Several spiral inductor geometries are commonly used. These include square, circular, hexagonal and octagonal inductors [14]. The square spiral has traditionally been more popular, since some IC processes constrained all angles to 90° [15], but it generally has a lower Q-factor than the circular spiral, which most closely resembles the common off-chip solenoidal inductors. The layout process of the circular inductor is somewhat complicated, so a polygon (hexagonal and octagonal) spiral is sometimes implemented as a compromise. Drawings of these geometries are shown in Fig. 5.13.

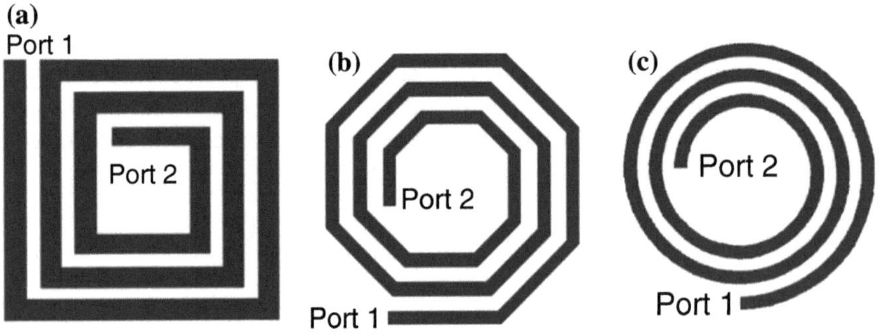

Fig. 5.13 Square (**a**), polygonal (octagonal) (**b**) and circular (**c**) spiral inductors

The geometries shown in Fig. 5.13 require only a single metal layer for fabrication and are consequently asymmetric. Additional layers are only needed to lay out an underpass, a wire that serves to bring the signal lines from the center to the edge of an inductor. Symmetrical inductors require more than one underpass, also known as a metal-level interchange, as shown in Fig. 5.14a [15]. Alternatively, the second metal layer can be used as part of the core of the inductors. An example of a multi-layer geometry is a two-layer square inductor shown in Fig. 5.14b [16]. The multi-layer spirals can deliver higher quality factors than a single layer inductor due to mutual inductance coupling of different layers.

Another common inductor geometry is a taper geometry, where inner spirals of inductors decrease in width in respect of the outer spirals [17] (Fig. 5.15). Tapering is done to suppress eddy current losses in the inner turns in order to increase the Q-factor, but it is most effective when substrate losses are negligible, which is possible with the MEMS approach discussed later.

For a given geometry, a spiral inductor is fully specified by the number of turns (n), the turn width (w) inner diameter (d_{in}) and outer diameter (d_{out}), as shown in Fig. 5.16 for the square and circular inductors. The average diameter, d_{avg} is

Fig. 5.14 Symmetrical (**a**) and two-layer (**b**) spiral inductor

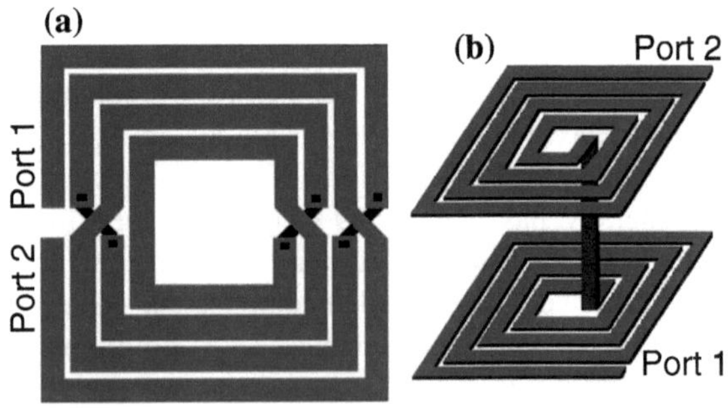

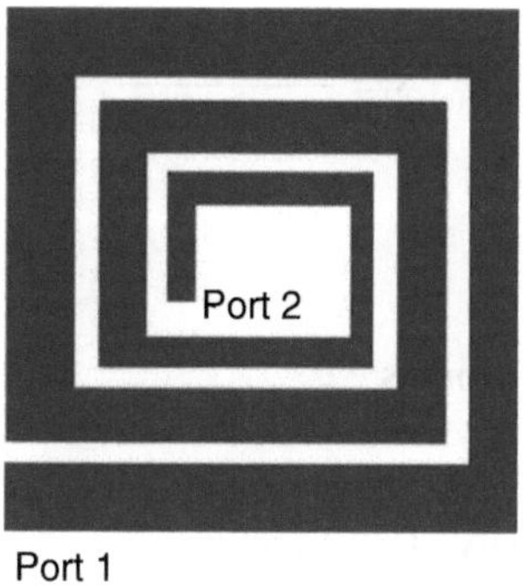

Fig. 5.15 A taper spiral inductor

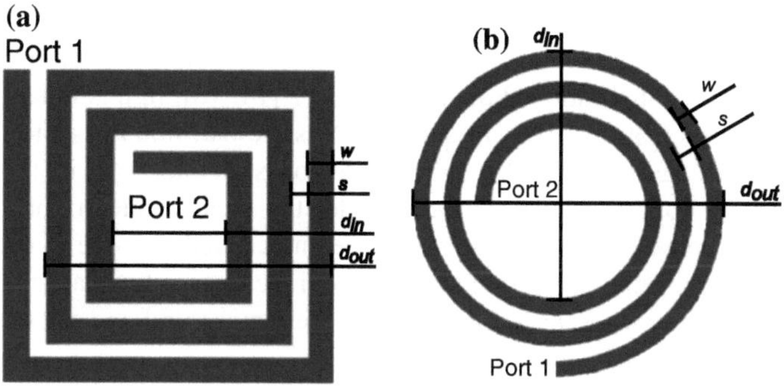

Fig. 5.16 Geometry parameters of the **a** square and **b** circular spiral inductors

$$d_{avg} = \frac{d_{in} + d_{out}}{2} \tag{5.14}$$

and the fill ratio is defined as

$$\rho_{fill} = \frac{d_{out} - d_{in}}{d_{out} + d_{in}}. \tag{5.15}$$

The total length of a spiral is dependent on inductor geometry and for a square inductor, it can be calculated as

$$l = 4(d_{in} + w) + 2n(2n - 1)(s + w), \tag{5.16}$$

where s is the pitch between the runs of the spiral, and it can be calculated for given d_{in}, d_{out}, w and n as

$$s = \frac{\frac{d_{out} - d_{in}}{2} - nw}{n - 1}. \tag{5.17}$$

for $n \geq 2$.

5.3.4.1 Spiral Inductor Models

Several spiral inductor models have found widespread use, depending on the required modeling complexity. These include:

- Single-π model,
- Segmented model,
- Double-π model and
- Third-order models.

Single-π Model The most commonly used model is a lumped single-π nine-component configuration shown in Fig. 5.17 [14, 18]. In this model, L_S is the (design) inductance at the given frequency, R_S is the parasitic resistance and C_S is the parasitic capacitance of the spiral inductor structure. C_{ox} is the parasitic capacitance due to oxide layers directly under the metal inductor spiral. Finally, C_{sub} and R_{sub} represent the parasitic resistance and capacitance due to the substrate (typically silicon). This topology does not model the distributive capacitive effects, but it models correctly for parasitic effects of the metal spiral and the oxide below the spiral, as well as for substrate effects.

Segmented Model A somewhat more complicated model is the segmented model presented by Koutsoyannopoulos and Papananos [19]. Each segment of the inductor is modeled separately with a circuit shown in Fig. 5.18. In this model, parasitics C_{ox}, C_{sub} and R_{sub} represent parasitics of only one inductor segment, L_S and R_S represent the inductance and parasitic capacitance of one segment coupled to all segments. Finally, C_{f1} and C_{f2} are added to represent coupling to adjacent segment nodes.

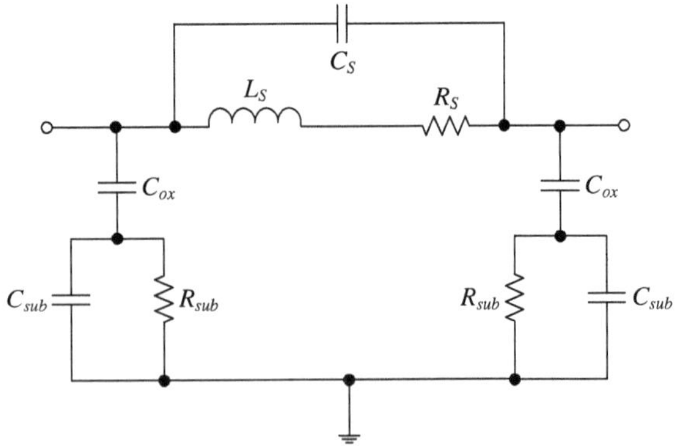

Fig. 5.17 A commonly used nine-component spiral inductor model

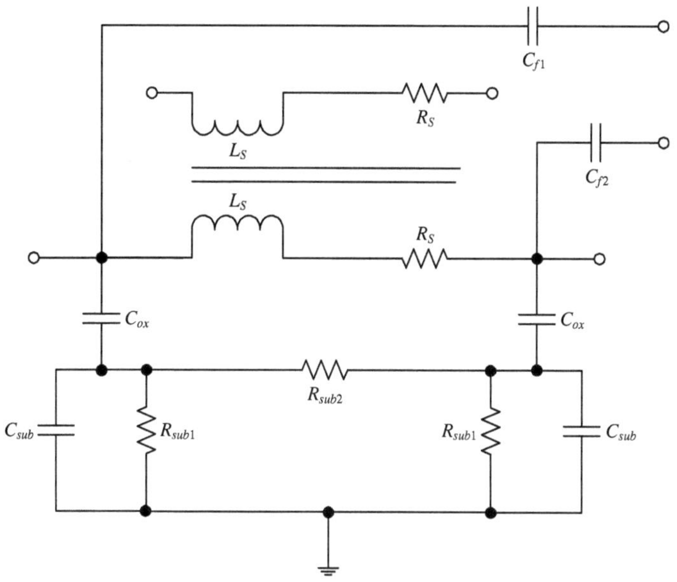

Fig. 5.18 An equivalent two-port model for one segment of a spiral inductor

Distributed Double-π Model The standard single-π model can also be extended into a second-order, distributed double-π model shown in Fig. 5.19 [18, 20]. A second-order ladder (with third grounded branch) is used to model the distributive characteristics of metal windings. The interwinding capacitance (C_w) is included to model the capacitive effects between metal windings of the inductor. The transformer loops (M_{S1} and M_{S2}) represent the effects of frequency-dependent series loss.

Third-Order Transmission-Line Model The second-order model shown in Fig. 5.19 is valid for the inductor up to the first resonance frequency. If a third-order model is used, it is possible to predict inductor behavior accurately, even beyond the resonant frequency. One such model is presented by [21]. An equivalent circuit diagram for this configuration is shown in Fig. 5.20. Extrinsic admittances are used and all circuit components are self-explanatory from this figure.

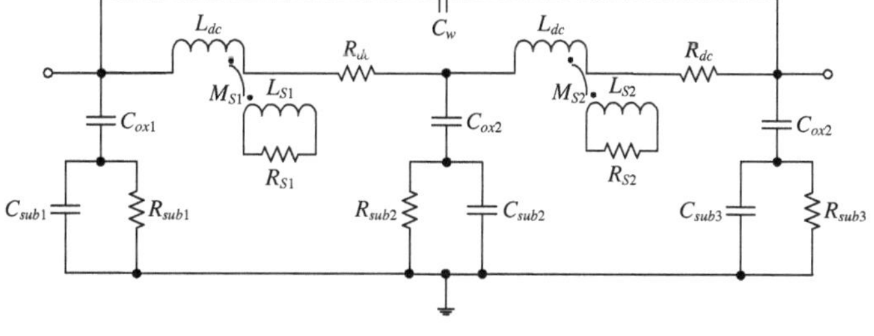

Fig. 5.19 A double-π distributed inductor model

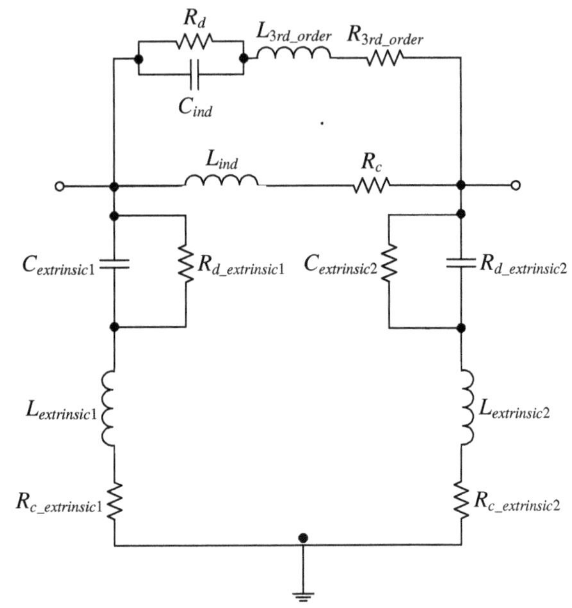

Fig. 5.20 A complete third-order inductor model

5.3.4.2 Inductance and Parasitic Component Values for the Single-π Inductor Model

The single-π inductor model of Fig. 5.17 is sufficient to model spiral inductors accurately for frequencies below resonance [22].

Series Inductance (L_S) Several equations are commonly used in literature to represent the series inductance of spiral inductors with various levels of accuracy. The modified Wheeler equation, current-sheet equation, Bryan's equation and monomial expression are four equations that are commonly used.

Inductance according to the modified Wheeler equation is [14]:

$$L_{mw} = K_1 \mu_0 \frac{n^2 d_{avg}}{1 + K_2 \rho_{fill}}. \tag{5.18}$$

Here, n is the number of turns, d_{avg} is the average diameter of the spiral and x_{fill} is the fill ratio, all defined previously. K_1 and K_2 are geometry-dependent coefficients with values defined in Table 5.3 and $\mu = \mu_0 \mu_r$ is the absolute magnetic permeability of the metal layer defined previously.

The expression obtained by approximating the sides of the spiral by symmetrical current sheets of equivalent current is [14]:

$$L_{gmd} = \mu \frac{n^2 d_{avg} c_1}{2} \left[\ln \frac{c_2}{\rho} + c_3 \rho_{fill} + c_4 \rho_{fill}^2 \right]. \tag{5.19}$$

Table 5.3 Coefficients for the modified Wheeler expression

Layout	K_1	K_2
Square	2.34	2.75
Octagonal	2.33	3.82
Hexagonal	2.25	3.55

Table 5.4 Coefficients for the current sheet expression

Layout	c_1	c_2	c_3	c_4
Square	1.27	2.07	0.18	0.13
Hexagonal	1.09	2.23	0.00	0.17
Octagonal	1.07	2.29	0.00	0.19
Circular	1.00	2.46	0.00	0.20

Here, c_1, c_2, c_3 and c_4 are geometry-dependent coefficients with values defined in Table 5.4. This expression exhibits a maximum error of 8 % for $s \leq 3w$.

The inductance expressed by Bryan's equation is [23]

$$L = 0.00241\left(\frac{d_{out} + d_{in}}{4}\right) n^{\frac{5}{3}} \ln\left(\frac{4}{\rho_{fill}}\right). \tag{5.20}$$

Finally, the data-fitted monomial expression results in an error smaller than seen in the expressions given above (typically less than 3 %). It is based on a data-fitting technique. Inductance in nH is calculated as [14, 24]:

$$L_{mon} = \beta d_{out}^{\alpha_1} w^{\alpha_2} d_{avg}^{\alpha_3} n^{\alpha_4} s^{\alpha_5}, \tag{5.21}$$

where coefficients β, α_1, α_2, α_3, α_4 and α_5 are once again geometry-dependent, as presented in Table 5.5.

The monomial expression has been developed by curve fitting over a family of 19,000 inductors [14]. It has better accuracy and higher simplicity than the equations described above, and is the equation of choice.

Parasitic Resistance (R_S) Parasitic resistance is dependent on the frequency of operation. At DC, this value is mostly determined by the sheet resistance of the material of which the wire is made. At higher frequencies, this value is surpassed by the resistance that arises from the formation of eddy currents. It is governed by the resistivity of the metal layer in which the inductor is laid out (ρ),

Table 5.5 Coefficients for the spiral inductor inductance calculation

Layout	β	α_1 (d_{out})	α_2 (w)	α_3 (d_{avg})	α_4 (n)	α_5 (s)
Square	1.62×10^{-3}	−1.21	−0.147	2.40	1.78	−0.030
Hexagonal	1.28×10^{-3}	−1.24	−0.174	2.47	1.77	−0.049
Octagonal	1.33×10^{-3}	−1.21	−0.163	2.43	1.75	−0.049

the total length of the spiral (l), the width of each turn (w) and the effective thickness of the spiral (t_{eff}) [24]:

$$R_S = \frac{\rho l}{w t_{eff}}. \tag{5.22}$$

The effective thickness, t_{eff}, is dependent on the actual thickness of the metal layer, t:

$$t_{eff} = \delta(1 - e^{-t/\delta}), \tag{5.23}$$

where δ is the frequency-dependent skin depth defined earlier.

Parasitic Capacitance (C_S) Parasitic capacitance is the sum of all the capacitances created between the spiral and the underpass. If there is only one underpass with the same width as the spiral, then the total capacitance is equal to [24]

$$C_S = nw^2 \frac{\varepsilon_{ox}}{t_{oxM1-M2}}, \tag{5.24}$$

where $t_{oxM1\text{-}M2}$ is the oxide thickness between the spiral and the underpass and ε_{ox} is the dielectric constant of the oxide layer between the two metals.

Oxide and substrate parasitics (C_{ox}, C_{sub} and R_{sub}) In order to calculate the oxide capacitance C_{ox} and substrate capacitance C_{sub}, the effective thickness (t_{eff}) and effective dielectric constant (ε_{eff}) of either oxide or substrate must be determined. The effective thickness is calculated as [25]

$$t_{eff} = \begin{cases} w\left[\frac{w}{t} + 2.42 - 0.44\frac{t}{w} + \left(1 - \frac{t}{w}\right)^6\right]^{-1}, & \text{for } \frac{t}{w} \leq 1 \\ t_{eff} = \frac{w}{2\pi}\ln\left(\frac{8t}{w} + \frac{4w}{t}\right), & \text{for } \frac{t}{w} \geq 1 \end{cases} \tag{5.25}$$

for both oxide and substrate. The effective dielectric constant is determined as

$$\varepsilon_{eff} = \frac{1+\varepsilon}{2} + \frac{\varepsilon - 1}{2}\left(1 + \frac{10t}{w}\right)^{-\frac{1}{2}}. \tag{5.26}$$

Then,

$$C_{ox} = \frac{wl\varepsilon_0\varepsilon_{effox}}{t_{effox}}, \tag{5.27}$$

$$C_{sub} = \frac{wl\varepsilon_0\varepsilon_{effsub}}{t_{effoxsub}} \tag{5.28}$$

and

$$R_{sub} = \frac{t_{eff\,sub}}{\sigma_{eff} w l}, \tag{5.29}$$

where the σ_{eff} is the effective conductivity of the substrate, obtained from

$$\sigma_{eff} = \sigma \left[\frac{1}{2} + \frac{1}{2} \left(1 + \frac{10t}{w} \right)^{-\frac{1}{2}} \right], \tag{5.30}$$

where $\sigma = 1/\rho$ represents the substrate conductivity.

5.3.4.3 Quality Factor and Resonance Frequency for Single-π Inductor Model

The Q-factor of the inductor modeled by the single-π model is [26]:

$$Q = \frac{\omega L_S}{R_S} \cdot \frac{R_P}{R_P + \left[\left(\frac{\omega L_S}{R_S} \right)^2 + 1 \right] R_S} \cdot \left[1 - (C_P + C_S) \cdot \left(\omega^2 L_S + \frac{R_S^2}{L_S} \right) \right], \tag{5.31}$$

where R_P and C_P are defined as

$$R_P = \frac{1}{\omega^2 C_{ox}^2 R_{sub}} + \frac{R_{sub} (C_{ox} + C_{sub})^2}{C_{ox}^2} \tag{5.32}$$

and

$$C_P = C_{ox} \cdot \frac{1 + \omega^2 (C_{ox} + C_{sub}) C_{sub} R_{sub}^2}{1 + \omega^2 (C_{ox} + C_{sub})^2 R_{sub}^2}, \tag{5.33}$$

respectively and $\omega = 2\pi f$.

The Q-factor is influenced by three types of loss [27], thus the Q-factor equation can be split into three factors. The first factor

$$F_1 = \omega L_S / R_S \tag{5.34}$$

is the intrinsic (nominal) Q-factor of the overall inductance. The second factor,

$$F_2 = \frac{R_P}{R_P + \left[(\omega L_S / R_S)^2 + 1 \right] R_S}, \tag{5.35}$$

models the substrate loss in the semiconducting silicon substrate. The last factor,

$$F_3 = 1 - (C_P + C_S) \cdot \left(\omega^2 L_S + R_S^2 / L_S\right), \tag{5.36}$$

models the self-resonance loss due to total capacitance C_P + C_S. This resonant frequency can be isolated by equating the last factor to zero, and solving for ω. This results in the formula for the SRF of the spiral inductor modeled by the single-π model:

$$f_r = \frac{\omega_o}{2\pi} = \frac{1}{2\pi}\sqrt{\frac{1}{L_S \cdot (C_P + C_S)} - \left(\frac{R_S}{L_S}\right)^2} \tag{5.37}$$

Each factor restricts the performance of the inductor of the Q-factor at a different frequency range. At lower frequencies, the loss of the metal line (F_1) has the most prominent effect [28]. In high-frequency ranges, the loss of substrate (F_2) prevails as the restricting factor. F_2 is greatly dependent on the resistivity of the substrate. As resistivity decreases at a fixed frequency, the skin depth of the substrate increases, leading to an increase of eddy currents in the substrate, resulting in a decrease of the Q-factor of the inductor. Heavily doped substrates are usually used in a submicron process, with substrate resistivity usually lying in the range of 10–30 Ω cm. As a result, in the traditional processes, the performance of spiral inductors is limited by the substrate. Inductors fabricated in MEMS processes strive to minimize the effects of this limitation. Figure 5.21 shows the analysis of factors F_1, F_2 and F_3 for 1 nH and 5 nH sample spiral inductors optimized at different frequencies for their highest quality operation. It can be observed that, although the nominal Q-factor (F_1) increases with frequency, F_2 and F_3 decrease in the same range, resulting in a decrease in the overall Q-factor (Q) at frequencies above 1 GHz.

Close to resonant frequency, the frequency has some effect on the apparent inductance value, which can be calculated from

$$L_{eff} = \frac{\mathrm{Im}(Z)}{2\pi f_r}, \tag{5.38}$$

where Z is the total impedance of the single-π-modeled inductor with its one port grounded. This impedance shift effect can also be observed for discrete inductors in Tables 5.1 and 5.2.

5.3.4.4 Guidelines for Integrating Spiral Inductors

Hastings [29] isolates some general guidelines that can assist in reducing parasitics, eddy current losses and undesired coupling, as well increasing the desired magnetic coupling. Adhering to the guidelines allows for fairly high-quality inductors to be

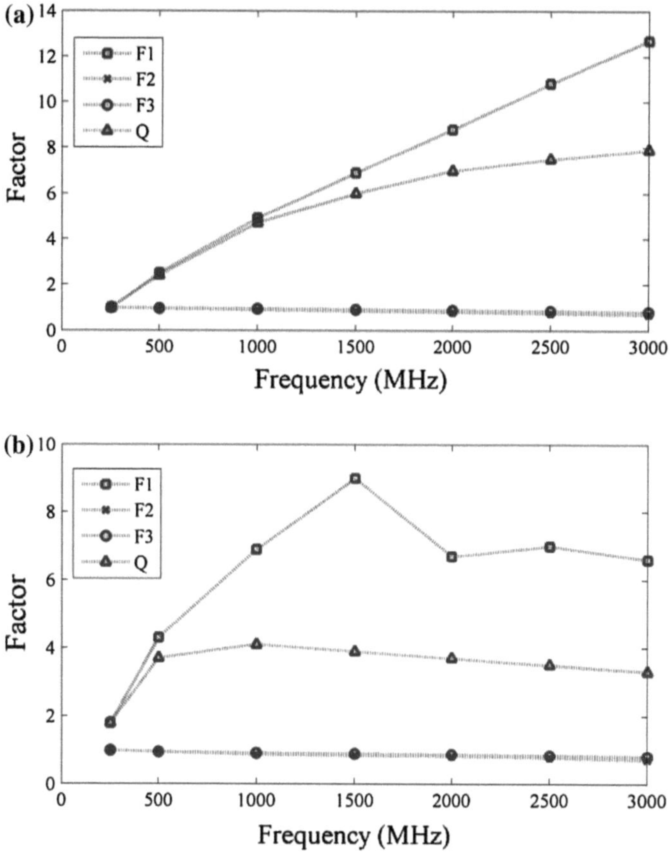

Fig. 5.21 Analysis of the determining factors of the Q-factor equation for **a** 1 nH inductor and **b** 5 nH inductor

designed, irrespective of the geometry of the inductor. These guidelines are also model-independent.

- Where possible, the highest resistivity substrate available should be used. This will reduce the eddy losses that reduce the Q-factor.
- Inductors should be placed on the highest possible metal layers. In this way, substrate parasitics will play a less prominent role because the inductor will be further away from the silicon.
- If necessary, parallel metal layers for the body of the inductor may be used to reduce the sheet resistance.
- Unconnected metal should be placed at least five turn widths away from inductors. This is another technique that helps to reduce eddy current losses.

- Exceptionally wide or narrow turn widths should be avoided. Narrow turns have high resistances, and wide turns are vulnerable to current crowding.
- The narrowest spacing between the turns, as allowed by the process design rules, should be maintained. Narrow spacing enhances magnetic coupling between the turns, resulting in higher inductance and Q-factor values.
- Filling the entire inductor with turns should be avoided. Inner turns are susceptible to the magnetic field, once more resulting in eddy current losses.
- Placing of unrelated metal plates above or under inductors should be avoided. Ungrounded metal plates will also aid the build-up of eddy currents.
- Placing of junctions beneath the inductor should be avoided. The presence of a junction close to the inductor can produce undesired coupling of AC signals.
- Short and narrow inductor leads should be used. The leads will unavoidably produce parasitics of their own.

5.3.5 MEMS Inductors

MEMS require an IC fabrication technique that allows conventional two-dimensional (2-D) circuits to expand into the third dimension (3-D) [30]. This principle becomes particularly useful in inductor fabrication, because the influence of substrate parasitics can be reduced significantly. Earlier in the section on inductors, it was noted that the substrate losses are the main contributors to the low Q-factor of integrated passive inductors.

Different techniques for producing MEMS devices, including bulk and surface micromachining, effectively replace silicon below the inductor by air or another material that has lower relative permittivity, which allows the Q-factor and resonant frequency to approach the values of off-chip inductors. Typical obtainable Q-factors range from 10 to 30 for 1 nH inductors at SHF. An example of a high-Q silicon-based inductor using polymer cavity can be found in [31]. This inductor achieves a Q-factor of 38 for a 1.61 nH inductor around 10 GHz.

As an alternative to spiral MEMS inductors, solenoidal inductors suspended on-chip can be used with various degrees of chip stability [32]. Several advantages over conventional spiral inductors can be identified, which include a lower stray capacitance because only part of the inductor is lying on the silicon substrate, a simple design equation and greater possibilities for flexible layout. Advanced micromachining techniques for solenoidal inductors have been proposed, including 3-D laser lithography, multiple-trenched sidewalls, the U-shaped solenoidal shape and a concave-suspending MEMS process.

Out-of-plane inductors [33] are similar to MEMS inductors, but their coils are fabricated using stress-engineered thin films. The stress gradient is induced by changing the ambient pressure during film deposition. When released, a stress-graded film curls up in a circular trajectory. The typical Q-factor of this configuration is over 70 at 1 GHz.

Although MEMS devices present an attractive alternative to conventional passive inductors, particularly because of the high Q-factors, their fabrication requires process changes or modifications to the wafer after fabrication, which influences repeatability [34] and the cost of fabrication.

5.3.6 Other On-Chip Inductor Implementations

Some inductor implementations are mentioned in a separate section because of their fundamental differences from standard inductor types.

The 2-D and 3-D meanders are discussed in [35]. A 2-D meander is a flat passive inductor consisting of a long piece of metal that is not wound as in the case of the spiral inductor, which will be described in detail later, but meanders similarly to rivers in their lower watercourses. This inductor occupies a small area and no underpass is needed, but its measured Q-factor is quite low (about 2.1 for inductance of 1.3 nH). Such a trade-off between the area and Q-factor is acceptable for matching network applications. The second type of inductor is a snake inductor that meanders into the third dimension. Both of these inductors are shown in Fig. 5.22. The inductance of a meander inductor can also be calculated using a data-fitted monomial expression, similar to that of the spiral inductor introduced earlier.

Electrically tunable solenoidal on-chip inductors are discussed in [36]. Other tunable inductors are commonly implemented in active configuration, such as in the case of the inductor in [37].

Toroid inductors can also be implemented on-chip by means of micromachining [38].

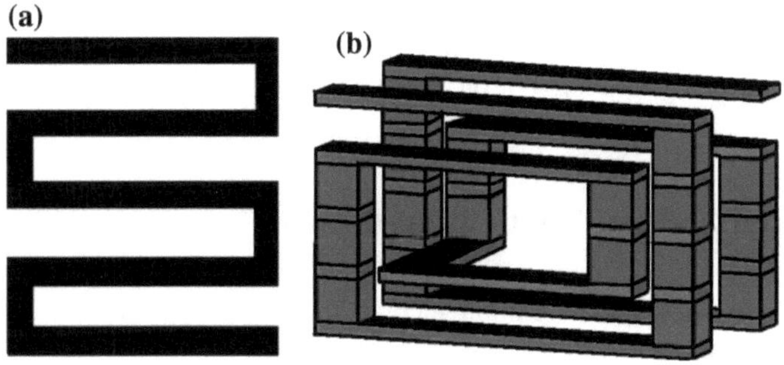

Fig. 5.22 Meander inductors: **a** 2-D meander inductor and **b** 3-D meander (*snake*) inductor

5.4 RF Chokes

RFC is the device that is used to suppress the AC signals while passing the DC signal. Ideally, it is an inductor with infinite inductance value and zero resistance. In practice, however, this is unreachable, and as seen in Chaps. 3 and 4, an RFC is designed for an acceptable amount of ripple. Depending on the amplifier design topology and design parameters, the required inductance of the RFC remains quite large. For example, for a Class-E power amplifier operating at 2.4 GHz, with a load resistance of 5 Ω, an 18 nH RFC inductor is needed. Thus, even integrated power amplifiers typically use off-chip RFCs. Since the inductance of the RFC need not be carefully controlled, the additional reactance of the tracks and bond wires is not a cause for concern.

RFCs are typically manufactured in the form of toroidal inductors with ferrite cores, because of their high inductance index.

5.5 Transformers

A transformer is a device that uses the principle of electromagnetic induction to transfer energy between two or more circuits. When used with power amplifiers, they are usually used for impedance transformation, to connect balanced and unbalanced source and load (balun), and for power combination or splitting [2]. A transformer is essentially two or more inductors placed closely together (called primary and secondary winding), thus the inductor design principles presented earlier in this chapter apply. Figure 5.23 shows a balun, n:1 (4:1) transformer, and a power combiner. Another way to design an n:1 transformer is to have a different number of turns between the primary and the secondary winding.

Typical integrated transformers are similar to spiral inductors treated in this chapter in some detail. Figure 5.24 shows two types of spiral integrated transformers. The type shown in Fig. 5.24a is the interleaved 1:1 transformer (balun). The type shown in Fig. 5.24b is a 1:1 transformer where the primary and the secondary windings are laid out on two different metal layers.

5.6 Quarter-Wavelength Transformer

Transmission lines tend to exhibit increased Q-factors with the increase in frequency, making them extremely useful components in RF electronics. A special type of transmission line, the quarter-wavelength transformer, was already mentioned in previous chapters, where it was used as an impedance converter for Doherty amplifiers and for impedance termination in case of Class-F power amplifiers. Furthermore, it can also act as an RFC, or it can be used for impedance

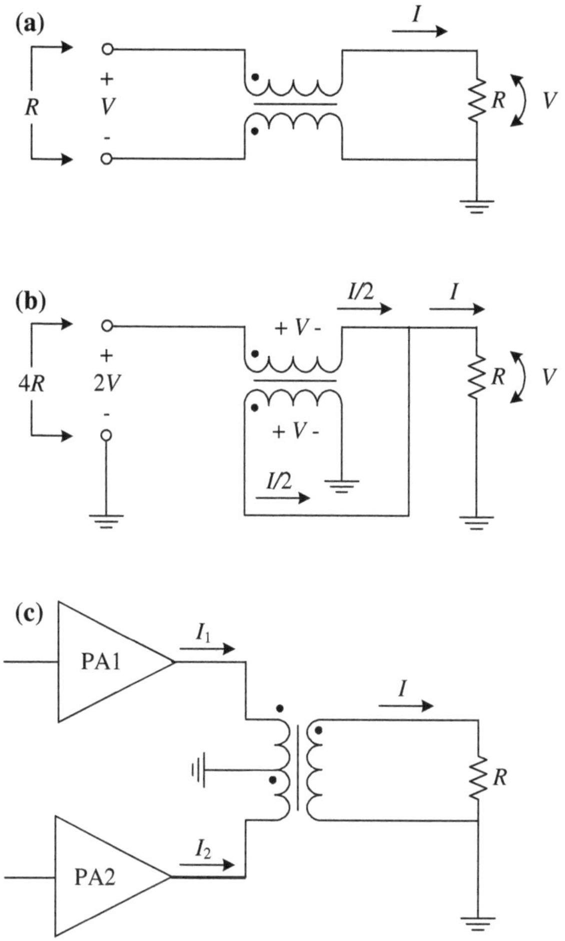

Fig. 5.23 Three types of transformer applications: **a** balun, **b** n:1 transformer (4:1 ratio shown) and **c** power combiner or splitter (power combiner shown)

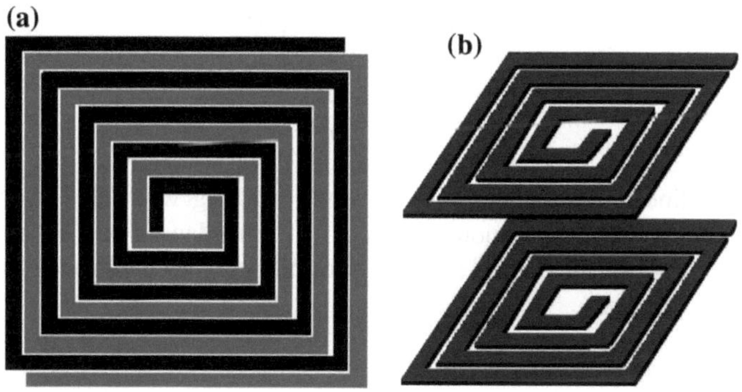

Fig. 5.24 Integrated transformers: **a** one-layer configuration and **b** two-layer configuration

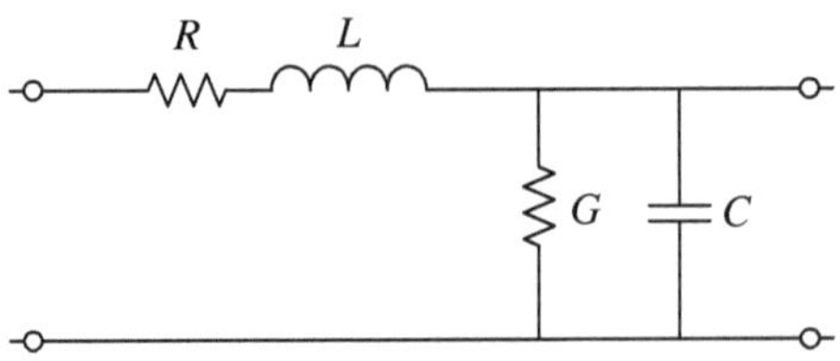

Fig. 5.25 General transmission line model

matching of real loads (impedance matching will form the topic of Chap. 6). Because of the importance of the quarter-wave transformer, its impedance and bandwidth will be discussed in some detail in this section. However, it is first necessary to present the model of a general transmission line and define the concept of its characteristic impedance.

5.6.1 General Transmission Line

A transmission line is a structure specialized to carry RF and microwave signals. This can be achieved by several configurations, including two-wire, coaxial, parallel-plate and microstrip lines. The transmission line is fully defined by four parameters: resistance R, inductance L, conductance G and capacitance C [1]. The model of a general transmission line is shown in Fig. 5.25. Each parameter depends on the geometry of the particular line and material properties. Assuming that R, L, G and C are all known or can be calculated (examples of calculations for different transmission lines can be found in [1]), the characteristic impedance of the lossy transmission line can be obtained with the following complex equation:

$$Z_0 = \sqrt{\frac{R + j\omega L}{G + j\omega C}} \tag{5.39}$$

For short line segments, the transmission line can, however, be considered lossless, meaning that $R = 0$ and $G = 0$, thus Eq. (5.39) simplifies to

$$Z_0 = \sqrt{\frac{L}{C}}. \tag{5.40}$$

A microstrip line is particularly popular when working with RF and microwave power amplifiers, because various characteristic impedances can be obtained by changing the substrate height h and the transmission line width w. Approximate equations can be obtained for narrow microstrip lines ($w/h < 1$) and wide lines ($w/h > 1$), if the thickness of the conductor forming the line is negligible compared to the substrate height. If we define

$$Z_f = \sqrt{\frac{\mu_0}{\varepsilon_0}} \approx 376.8\,\Omega, \tag{5.41}$$

then the characteristic impedance of the narrow line is

$$Z_0 = \frac{Z_f}{2\pi\sqrt{\varepsilon_{eff}}} \ln\left(\frac{8h}{w} + \frac{w}{4h}\right) \tag{5.42}$$

and the characteristic impedance of the wide line is

$$Z_0 = \frac{Z_f}{\sqrt{\varepsilon_{eff}}\left[1.393 + \frac{w}{h} + \frac{2}{3}\ln\left(\frac{w}{h} + 1.444\right)\right]}, \tag{5.43}$$

where the effective dielectric constant is given by

$$\varepsilon_{eff} = \frac{\varepsilon_r + 1}{2} + \frac{\varepsilon_r - 1}{2}\left[\left(1 + \frac{12h}{w}\right)^{-\frac{1}{2}} + 0.04\left(1 - \frac{w}{h}\right)^2\right]. \tag{5.44}$$

for narrow line and

$$\varepsilon_{eff} = \frac{\varepsilon_r + 1}{2} + \frac{\varepsilon_r - 1}{2}\left(1 + \frac{12h}{w}\right)^{-\frac{1}{2}}. \tag{5.45}$$

for wide line. Note the similarity between Eqs. (5.26) and (5.44).

The input impedance of a lossless transmission line with an arbitrary length l and terminated with load resistance Z_L and with characteristic impedance Z_0 is

$$Z_{in} = Z_0 \frac{Z_L + jZ_0 \tan \beta l}{Z_1 + jZ_0 \tan \beta l}, \tag{5.46}$$

where β is the propagation constant of the transmission line, dependent on the wavelength λ by relation

$$\beta = \frac{2\pi}{\lambda}. \tag{5.47}$$

5.6.2 Input Impedance of the Quarter-Wave Transformer

With the characteristic impedance of the general transmission line known, we can proceed to calculate the input impedance of the terminated quarter-wave transformer. A quarter-wave transformer with characteristic impedance Z_1, connected to another transmission line with characteristic impedance Z_0, is illustrated in Fig. 5.26 [39].

In case of a quarter-wave transformer, $l = \lambda/4$ at frequency f_0, thus

$$\beta l = \frac{\pi}{2}, \tag{5.48}$$

and the input impedance of the quarter-wave transformer simplifies to

$$Z_{in} = \lim_{\beta l \to \frac{\pi}{2}} Z_1 \frac{\frac{RZ_L}{\tan \beta l} + jZ_1}{\frac{Z_1}{\tan \beta l} + jZ_L} = \frac{Z_1^2}{Z_L}. \tag{5.49}$$

Therefore, at the fundamental frequency, the quarter-wave transformer acts as an impedance inverter.

5.6.3 Bandwidth of the Quarter-Wave Transformer

The reflection coefficient on the input side of the transformer at any frequency is

$$\Gamma = \frac{Z_{in} - Z_0}{Z_{in} + Z_0}. \tag{5.50}$$

By substituting Eqs. (5.49) into (5.50) and grouping real and imaginary parts,

$$\Gamma = \frac{Z_1(Z_L - Z_0) + j(Z_1^2 - Z_0 Z_L) \tan \beta l}{Z_1(Z_L + Z_0) + j(Z_1^2 + Z_0 Z_L) \tan \beta l}. \tag{5.51}$$

Near f_0, $Z_1 = Z_{in} = Z_0 Z_L$, thus

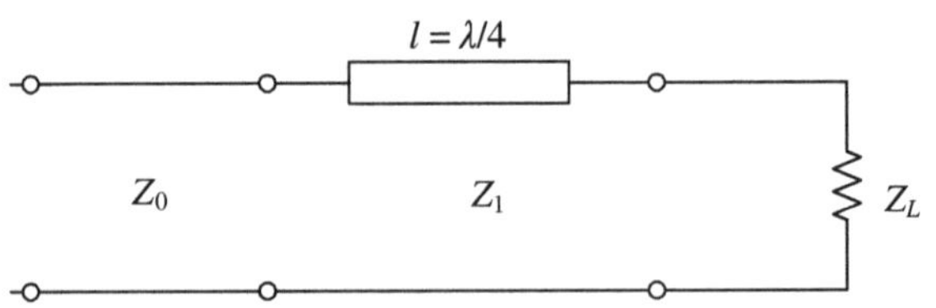

Fig. 5.26 The quarter-wave transformer

$$\Gamma = \frac{Z_L - Z_0}{Z_L + Z_0 + j2(\sqrt{Z_0 Z_L})\tan\beta l}. \tag{5.52}$$

The magnitude of Γ is then

$$|\Gamma| = \frac{1}{\sqrt{1 + \frac{4Z_0 Z_L}{(Z_L - Z_0)^2 \cos^2 \beta l}}}. \tag{5.53}$$

Bandwidth can now be defined for the maximum value of the reflection coefficient Γ_m. We start by finding the maximum value of βl, $(\beta l)_m$, as illustrated in Fig. 5.27. From Eq. (5.53):

$$\frac{1}{\Gamma_m^2} = 1 + \left[\frac{2\sqrt{Z_0 Z_L}}{(Z_L - Z_0)\cos(\beta l)}\right]^2. \tag{5.54}$$

Thus,

$$(\beta l)_m = \arccos\left(\frac{\Gamma_m}{\sqrt{1-\Gamma_m}}\frac{2\sqrt{Z_0 Z_L}}{|Z_L - Z_0|}\right). \tag{5.55}$$

If we incorporate the wavelength equation, Eq. (2.1), into Eq. (5.48), where f_m is the frequency corresponding to Γ_m, we obtain

$$(\beta l)_m = \frac{2\pi f_m}{v}\frac{v}{4f_0} = \frac{\pi f_m}{2f_0}. \tag{5.56}$$

and

$$f_m = \frac{2(\beta l)_m}{\pi} f_0. \tag{5.57}$$

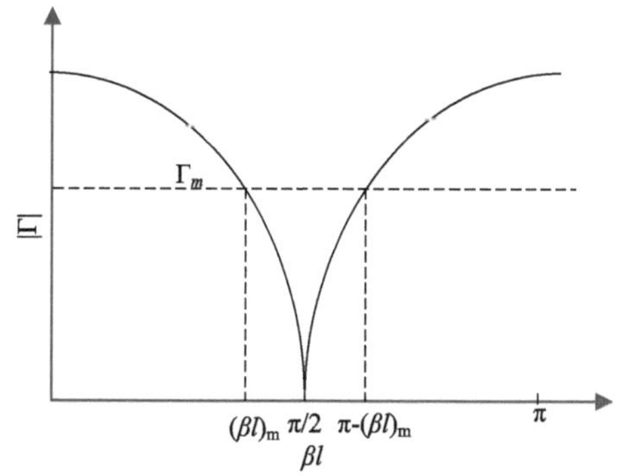

Fig. 5.27 Behavior of the magnitude of the reflection coefficient in terms of βl

Finally, the bandwidth is

$$BW = 2(f_0 - f_m) = 2f_0\left[1 - \frac{2}{\pi}(\beta l)_m\right]$$
$$= 2f_0\left\{1 - \frac{2}{\pi}\arccos\left(\frac{\Gamma_m}{\sqrt{1-\Gamma_m}}\frac{2\sqrt{Z_0 Z_L}}{|Z_L - Z_0|}\right)\right\}. \quad (5.58)$$

5.6.4 Impedance of the Quarter-Wave Transformer Terminated with a Resonant Tank

When used with a Class-F power amplifier, the quarter-wave transformer is terminated with a parallel resonant tank, as shown in Fig. 5.28. This causes the transformer to appear with resistance R at the fundamental frequency f_0, with zero impedance at even harmonics and with infinite impedance at odd harmonics.

At the fundamental frequency, the parallel resonant tank appears to have resistance R, thus with $Z_1 = R$, Eq. (5.49) simplifies to

$$Z_{in} = \frac{R^2}{R} = R. \quad (5.59)$$

At even harmonics ($n = 2k$), $\beta l = 2k(\pi/2) = k\pi$, and $\tan k\pi = 0$, thus

$$Z_{in} = Z_1\frac{Z_L + jZ_1 \cdot 0}{Z_1 + jZ_L \cdot 0} = Z_L, \quad (5.60)$$

and the quarter-wave acts as an impedance repeater. According to Chap. 2, the impedance of the resonant tank far from the resonant frequency f_0 is almost zero, thus Eq. (5.60) simplifies to

$$Z_{in} \approx 0. \quad (5.61)$$

At odd harmonics, Eq. (5.49) is applicable again, but the impedance of the resonant tank still approaches zero, therefore in the limit

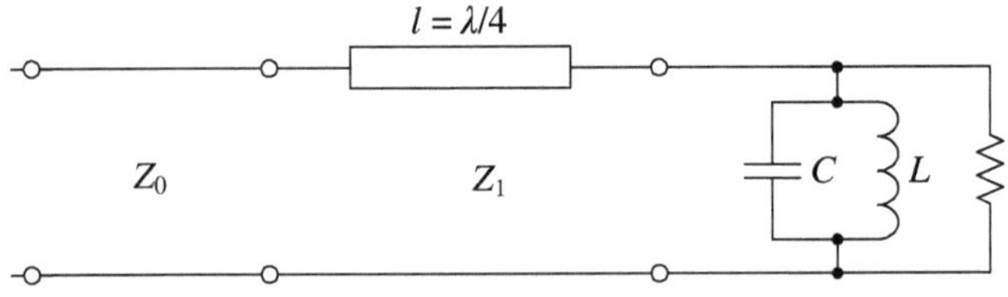

Fig. 5.28 Quarter-wave transformer terminated with a parallel resonant tank

$$Z_{in} = \lim_{Z_L \to 0} \frac{Z_1^2}{Z_L} = \infty, \tag{5.62}$$

as required for close to ideal Class-F power amplifier operation.

An interesting advantage of the Class-F power amplifier with the quarter-wave transmission line is that it can be designed such that the transmission line simultaneously performs impedance matching.

5.7 Summary

In this chapter, we carried out an extensive discussion on passives typically used with power amplifiers. Resistors, capacitors, inductors, transformers and transmission lines were all discussed, from the viewpoint of both integrated and discrete implementations. Defining equations were analyzed, and various models for different component types were given in some detail. The importance of parasitics with increased frequency was discussed; it was noted that the inductor is typically the passive that suffers the largest losses. We also defined a Q-factor, an important quantity for describing passives.

The discussion of passives carries on to Chap. 6 where they are also used to build matching networks, another important block in the power amplifier design.

References

1. Ludwig R, Bretchko P. RF circuit design: theory and applications. 1st ed. Upper Saddle River: Prentice Hall; 2000.
2. Bowick C, Blyler J, Ajluni C. RF circuit design. 2nd ed. Burlington: Elsevier; 2008.
3. Bahl IJ. Lumped elements for RF and microwave circuits. 1st ed. Norwood: Artech House; 2003.
4. Ghadiri A, Moez K. High-quality-factor active capacitors for millimeter-wave applications. IEEE Trans Microw Theory Tech. 2012;60(12):3710–8.
5. Fakhfakh M, Tlelo-Cuautle E, Siarry P. Computational intelligence in analog and mixed-signal (AMS) and radio-frequency (RF) circuit design. 1st ed. Cham: Springer; 2015.
6. Coilcraft, Inc. Highest Q finder. [Internet]. [cited 2015 June 8]. Available from: http://www.coilcraft.com/apps/hiq/hiQ.cfm.
7. Uyanik H, Tarim N. Compact low voltage high-Q CMOS active inductor suitable for RF applications. Analog Integr Circ Sig Process. 2007;51:191–4.
8. Ler CL, A'ain AKB, Kordesh AV. CMOS source degenerated differential active inductor. Electron Lett. 2008;44(3):196–7
9. Bakken T, Choma J. Gyrator-based synthesis of active on-chip inductances. Analog Integr Circuit Sign Process. 2003;34(3):171–91.
10. Chen A, Lo HY. Semiconductor packaging: materials interaction and reliability. 1st ed. Boca Ranton: CRC Press; 2012.

11. Murad SAZ, Pokharel RK, Kanaya H, Yoshida K, Nizhnik O. A 2.4-GHz 0.18-μm CMOS Class E single-ended switching power amplifier with a self-biased cascode. Int J Electr Commun 2010;64(9):813–818.
12. Khatri H, Gudem PS, Larson LE. Integrated RF interference suppression filter design using bond-wire inductors. IEEE Trans Microw Theory Tech. 2008;56(5):1024–34.
13. Kazimierczuk MK. RF power amplifiers. 2nd ed. Chiechester: Wiley; 2015.
14. Mohan SS, Hershenson M, Boyd SP, Lee TH. Simple accurate expressions for planar spiral inductances. IEEE J Solid-State Circuits. 1999;34(10):1419–24.
15. Niknejad AM, Meyer RG. Design, simulation and application of inductors and transformers for Si RF ICs. 1st ed. New York: Springer; 2000.
16. Xu X, Li P, Cai M, Han B. Design of novel high-Q-factor multipath stacked on-chip spiral inductors. IEEE Trans Electron Dev. 2012;59(8):2011–8.
17. Pei S, Wanrong Z, Lu H, Dongyue J, Hongyun X. Improving the quality factor of an RF spiral inductor with non-uniform metal width and non-uniform coil spacing. CIE J Semiconductors. 2011;32(6):1–5.
18. Wang TP, Li ZW, Tsai HY. Performance improvement of a 0.18-μm CMOS microwave amplifier using micromachined suspended inductors: theory and experiment. IEEE Trans Electron Dev. 2013;60(5):1738–44.
19. Koutsoyannopoulos YK, Papananos Y. Systematic analysis and modeling of integrated inductors and transformers in RF IC Design. IEEE Trans Circ Syst II Analog Digit Sig Process 2000;47(8):699–713.
20. Watson AC, Melendy D, Francis P, Hwang K, Weisshaar A. A comprehensive compact-modeling methodology for spiral inductors in silicon-based RFICs. IEEE Trans Microw Theory Tech. 2004;52(3):849–57.
21. Lee KY, Mohammadi S, Bhattacharya PK, Katehi LPB. Compact models based on transmission-line concept for integrated capacitors and inductors. IEEE Trans Microw Theory Tech. 2006;54(12):4141–8.
22. Wang H, Sun L, Yu Z, Gao J. Analysis of modeling approaches for on-chip spiral inductors. Int J RF Microwave Comput Aided Eng. 2012;22(3):377–86.
23. Musunuri S, Chapman PL, Zou J, Liu C. Design issues for monolithic DC–DC converters. IEEE Trans Power Electron. 2005;20(3):639–49.
24. Yue CP, Wong SS. Physical modeling of spiral inductors on silicon. IEEE Trans Electron Devices. 2000;47(3):560–8.
25. Huo X, Chan PCH, Chen KJ, Luong HC. A Physical model for on-chip spiral inductors with accurate substrate modeling. IEEE Trans Electron Dev. 2006;53(12):2942–9.
26. Lee CY, Chen TS, Deng JDS, Kao CH. A simple systematic spiral inductor design with perfected Q improvement for CMOS RFIC application. IEEE Trans Microw Theory Tech. 2005;53(2):523–8.
27. Sun H, Liu Z, Zhao J, Wang L, Zhu J. The enhancement of Q-factor of planar spiral inductor with low-temperature annealing. IEEE Trans Electron Dev. 2008;55(3):931–6.
28. Xue C, Yao F, Cheng B, Wang Q. Effect of the silicon substrate structure on chip spiral inductor. Front Electr Electron Eng China. 2008;3(1):110–5.
29. Hastings A. The art of analog layout. 2nd ed. Upper Saddle River: Prentice Hall; 2006.
30. De Los Santos HJ. MEMS for RF/microwave wireless applications. Microwave J. 2001;44 (3):20–41.
31. Khoo YM, Lim TG, Ho SW, Li R, Xiong YZ, Zhang X. Enhancement of silicon-based inductor Q-factor using polymer cavity. IEEE Trans Compon Packag Manuf Technol. 2012;2 (12):1973–9.
32. Gu L, Li X. High-Q Solenoid inductors with a CMOS-compatible concave-suspending MEMS process. J Microelectromech Syst. 2007;16(5):1162–72.
33. Chua LC, Fork DK, Van Schuylenbergh K, Lu JP. Out-of-plane high-Q inductors on low-resistance silicon. J Microelectromech Syst. 2003;12(6):989–95.
34. Foty D. Prospects for nanoscale electron devices: some little-recognized problems. In: Proceedings of the international semiconductor conference (CAS); 2008; Sinaia.

35. Masu K, Okada K, Ito H. RF passive components using metal line on Si CMOS. Trans Electr. 2006;89(6):681–91.
36. Vroubel M, Zhuang Y, Rejaei B, Burghartz JN. Integrated tunable magnetic RF inductor. IEEE Electr Dev Lett. 2004;25(12):787–9.
37. Seo S, Ryu N, Choi, H., Jeong Y. Novel high-Q inductor using active inductor structure and feedback parallel resonance circuit. In: Proceedings of IEEE radio frequency integrated circuits symposium; 2007; Honolulu.
38. Zine-El-Abidine I, Okoniewski M. High quality factor micromachined toroid and solenoid inductors. In: Proceedings of the 37th European microwave conference; 2007; Munich.
39. Pozar M. Microwave engineering. 4th ed. Hoboken: Wiley; 2012.

Chapter 6
Impedance Matching

In Chap. 1, impedance matching was introduced in the context of power amplifier design as a means of adding a power amplifier between the modulator and the antenna with minimum insertion loss. In general, impedance matching is necessary in any RF and microwave design in order to avoid unnecessary power losses that arise because signal wavelength decreases to a size where it becomes comparable to the characteristic circuit component length. A block diagram of a power amplifier showing input and output matching networks was given in Fig. 1.6. It was described that an input impedance network had to be designed with biasing of the driving transistor in mind, and that output impedance matching had to be performed with the output impedance under large drive conditions in mind. Finally, a few generic two-component and three-component matching networks were illustrated.

This chapter will expand the matching design theory that was touched on in Chap. 1 by undertaking a detailed analysis of two- and three-component networks, with emphasis on the mathematics of their designs. Since the latter part of this book focuses on power amplifier design automation, most of this chapter will focus on laying the groundwork for numerical (analytic) matching network design, which is practical for the case when the two values that are to be matched are both real. For complex loads, matching using existing MATLAB matching functions will be introduced. The Smith chart, introduced in Chap. 2, is a very useful graphical tool for RF and microwave design and will also be reconsidered here from the matching perspective. Although in the era of computer design the tasks of RF matching can be accomplished faster by using specialized software functions, such as the ones presented in MATLAB, the Smith chart remains an excellent tool for visualizing the matching steps, particularly when multiple elements are used. Finally, a part of this chapter will also deal with transmission-line matching.

6.1 Importance of Impedance Matching

A general impedance-matching block diagram is illustrated in Fig. 6.1a [1]. The role of the matching network is to match the source impedance Z_S to the load impedance Z_L. In the case of power amplifiers, matching can be at the input of the

M. Božanić and S. Sinha, *Power Amplifiers for the S-, C-, X- and Ku-bands*,
Signals and Communication Technology, DOI 10.1007/978-3-319-28376-0_6

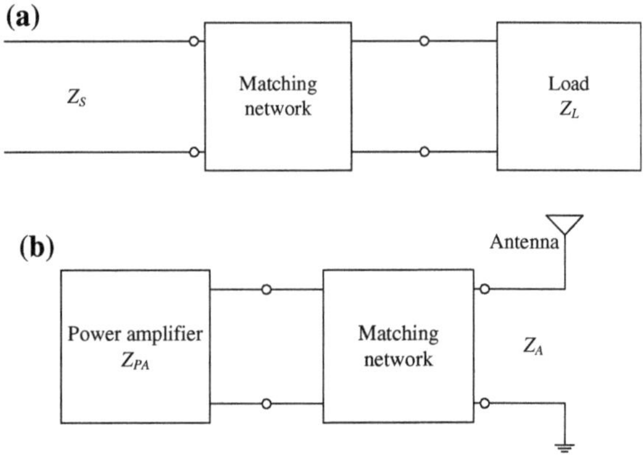

Fig. 6.1 A lossless impedance-matching network (**a**) and power-amplifier-to-antenna matching network (**b**)

power amplifier or at its output, or it can even be required for connecting various amplification stages (inter-stage matching). For the common case of matching at the output of the power amplifier, Fig. 6.1a simplifies to Fig. 6.1b, [2] where Z_{PA} is the output impedance of the power amplifier and Z_A is the input impedance of the antenna. To retain generality, however, anything before the matching network will be denoted as source with impedance Z_S and anything following the matching network will be denoted as load with impedance Z_L, unless specified otherwise.

The process of impedance matching is sometimes referred to as impedance tuning. Impedance tuning is important for several reasons:

- The maximum power can only be delivered when the load is matched to the source;
- The signal-to-noise ratio of the matched system is improved; and
- Matching reduces amplitude and phase errors.

Many types of matching networks are possible, irrespective of whether the matching is performed with lumped components or transmission lines. The choice of adequate matching networks might not always be clear, but it is typically dependent on one of the following factors: complexity, bandwidth, implementation and adjustability. In this chapter, lumped-element matching networks will be discussed, because of their practicality in power amplifier designs up to the *Ku*-band. Transmission-line matching will be mentioned as well, which can be used for discrete power amplifier implementations at frequencies of several GHz [1–6].

6.2 Load-Pull Characterization

When doing power amplifier output impedance matching, the load impedance of the matching network is the impedance of the antenna, with a typical value of 50 Ω. On the other hand, the impedance of the source, in this case a power amplifier, is not always known. A technique to obtain the power amplifier output impedance at

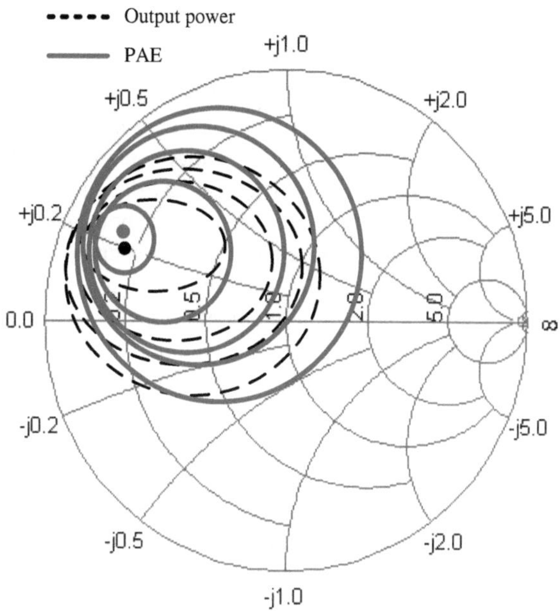

Fig. 6.2 Constant power and constant PAE contours on the Smith chart

different drive levels is called load-pull characterization [7]. Typically, the end result of load-pull characterization is a series of contours on the Smith chart representing constant output power, with all inner contours having increased power. This means that the highest power is reachable only at one point on the chart. The contours are drawn for several frequencies of interest and they represent the loci of transistor output impedance at specified power levels. Similarly, contours for specific PAE can also be drawn and they are generally not located at the same position on the graph as the power contours. Both power and PAE contours are illustrated on a single Smith chart in Fig. 6.2. Typically, the contours vary between elliptical and circular shapes. With both power contours and PAE contours plotted, the designer of the power amplifier system can choose an adequate power and efficiency level and read the required amplifier output impedance.

Unfortunately, the load-pull characterization process is expensive and time-consuming to realize in practice. It is performed using input and output tuners together with a vector network analyzer or it can be performed with specialized equipment. Alternatively, load-pull results acquired by the power transistor manufacturers can be used. They typically use an automated procedure and specialized software and report the results in the transistor datasheets for specified output power, supply voltage and frequency range.

When designing integrated solutions, load pull cannot be done experimentally and is typically done by means of extensive simulations of a system comprising the amplifier and the load, with real elements at the frequency of interest. Most commercial simulators support load-pull analysis and the setup is typically straightforward.

6.3 Lumped-Element Matching

As seen in Chap. 1 and illustrated in Figs. 1.7 and 1.8, lumped-element matching is accomplished by two-element or three-element networks. In practice, however, matching using any number of components is possible, but this increases matching network complexity. Any type of network can give a perfect match (zero reflection) at a single frequency, therefore to cover both narrowband and wideband networks, it is sufficient to explore only two-element (wideband) and three-element (narrowband) networks. Multiple-element networks are fairly complex to analyze analytically, thus multiple-element tuning is usually performed graphically with the aid of Smith charts. The main advantage of introducing multiple-element networks is that with these networks, which can be understood as a combination of several simpler networks, higher bandwidths can be achieved.

For DC circuits, the maximum power theorem states that the maximum power is transmitted from the source to the load if the load resistance equals the source resistance. For AC (RF or microwave) circuits, the same theorem states that the maximum power is transferred when the load impedance is equal to the complex conjugate of the source impedance. So, if the source impedance is $Z_S = R + jX$, the load needs to be made to look like $Z_S^* = R{-}jX$, resulting in the process called conjugate matching.

When source and load are both real quantities, an analytical solution to the matching problem is fairly simple. If either the source or the load is complex, one way of tackling the matching problem is ignoring the imaginary parts and performing the matching as if both source and load were real, after which the imaginary component (capacitive or inductive stray reactance) of the source or the load is either absorbed in the matching network, or resonated with an equal and opposite reactance. Using both techniques in the same system is also possible.

If absorption needs to take place, the matching network has to be chosen correctly, that is, if the stray reactance is inductive, the matching network needs to be chosen so that the inductor is in series with the stray inductance, and the stray inductance is subtracted from the design inductance values, provided that it is sufficiently large. Similarly, if stray capacitance is present, the matching network is chosen so that the capacitance is in parallel with the stray capacitance. If the design reactance ends up being smaller than the stray reactance, resonance is employed.

A proper analytical solution to matching of one real to one complex value without absorption or resonance does, exist, but it requires solving a system of two equations, so it will be demonstrated only for a two-component network. Three-component network matching will only be done for real load and source, assuming that any reactance in either the source or the load can be either absorbed or resonated. Alternatively, one may opt to design an L-network to transform the complex quantity into a real quantity.

6.3.1 Wideband Two-Element Networks (L-Networks)

6.3.1.1 Matching of Real Sources and Loads

The simplest matching network is the L-network, with four variations shown in Fig. 6.3, where each network consists of one capacitor and one inductor, one in shunt (parallel) and the other one in series with either the source or the load. If we define Q_s as the Q-factor of the series element and Q_p as the Q-factor of the shunt respectively, with

$$Q_s = \frac{X_s}{R_s} \tag{6.1}$$

and

$$Q_p = \frac{R_p}{X_p}, \tag{6.2}$$

where R_p is the parallel resistance and R_s is the series resistance, then a generic matching circuit of Fig. 6.4 can be used. Then, for $R_p > R_s$,

$$Q_S = Q_p = \sqrt{\frac{R_p}{R_s} - 1} \tag{6.3}$$

is a design equation that can be used to calculate the required Q-factor of the whole network, and reactances of each of the two matching elements. Recall that if the element is an inductor,

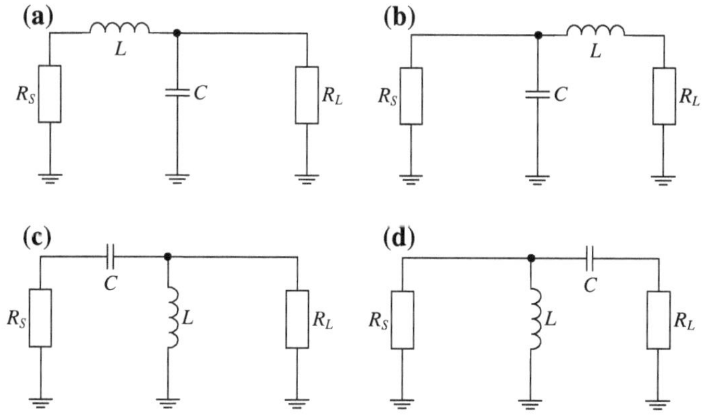

Fig. 6.3 Four variations of the L-type matching network: **a** and **b** are low-pass, and **c** and **d** are high-pass networks

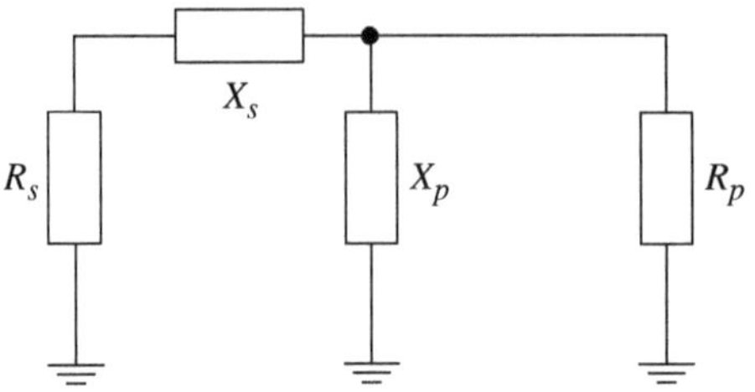

Fig. 6.4 Generic matching circuit with source and load resistances R_S and R_L replaced with series and parallel (shunt) reactances

$$L = \frac{X}{2\pi f_0} \tag{6.4}$$

and if the element is a capacitor,

$$C = \frac{1}{2\pi f_0 X}, \tag{6.5}$$

where f_0 is the operating frequency. Note that the condition $R_p > R_s$ means that for a solution to be possible, a parallel component needs to be placed next to the larger resistance value, be it the source or the load. Another interesting thing to note here is that the designer has no control over the quality factor of the network. If there is need for a precise Q-factor value, one of the three-component (narrowband) networks is used, described in the following section.

6.3.1.2 Matching of a Complex Load to a Complex Source

Now consider the case when one or both of the impedances being matched are complex. Let the source impedance be $Z_S = R_S + jX_L$ and the load impedance be $Z_L = R_L + X_L$, as illustrated for two networks in Fig. 6.5. The Q-factor analysis we used for real sources and loads now becomes tedious, thus we elect to find the equivalent impedance looking into the matching circuit. Then we look for matching components so that the conjugate match is found, i.e. so that $Z_{in} = Z_S^*$. For the circuit of Fig. 6.5a, $R_L > R_S$ and the input impedance looking from the source is

$$Z_{in} = Z_s + \frac{Z_p Z_L}{Z_p + Z_L}, \tag{6.6}$$

where $Z_s = jX_s$ and $Z_p = jX_p$. By rearranging Eq. (6.6) and by grouping real and imaginary parts, a system of two equations can be obtained. The two solutions of this system are:

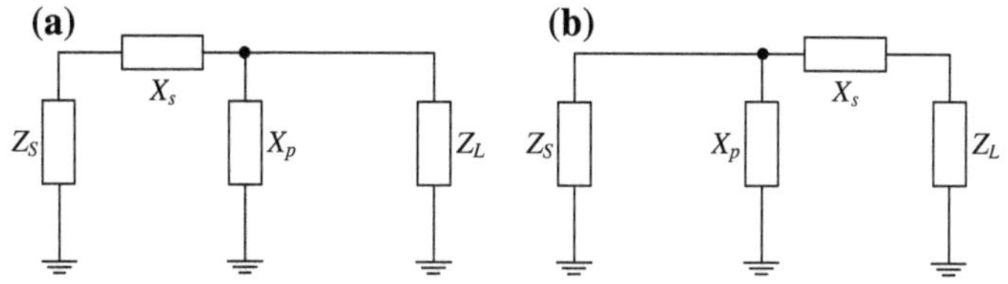

Fig. 6.5 L-networks for matching of a complex load to a complex source: **a** shunt component next to the load and **b** shunt component next to the source

$$(X_p)_{1,2} = \frac{X_L \pm R_L Q}{\frac{R_L}{R_S} - 1} \tag{6.7}$$

and

$$(X_s)_{1,2} = -(X_S \pm R_S Q), \tag{6.8}$$

where Q can be defined as

$$Q = \sqrt{\frac{R_L}{R_S} - 1 + \frac{X_L^2}{R_S R_L}}. \tag{6.9}$$

From the above analysis, it is obvious that the analytic solution is impractical for any manual calculations and only makes sense for computer analysis. Note that if $X_S = 0$ and $X_L = 0$, Eqs. (6.7)–(6.9) can be rewritten as Eqs. (6.1)–(6.3) with $R_s = R_S$ and $R_p = R_L$

For the circuit of Fig. 6.5b, $R_S > R_L$ and the input impedance looking from the source is

$$Z_{in} = \frac{Z_p(Z_s + Z_L)}{Z_p + Z_s + Z_L} \tag{6.10}$$

As before, by grouping the real and imaginary parts, a system of two equations is obtained, yielding solutions

$$(X_p)_{1,2} = \frac{X_S \pm R_S Q}{\frac{R_S}{R_L} - 1}. \tag{6.11}$$

and

$$(X_s)_{1,2} = -(X_L \pm R_L Q), \tag{6.12}$$

where

$$Q = \sqrt{\frac{R_S}{R_L} - 1 + \frac{X_S^2}{R_S R_L}}. \tag{6.13}$$

Note again that if $X_S = 0$ and $X_L = 0$, Eqs. (6.11)–(6.13) can be rewritten as Eqs. (6.1)–(6.3) with $R_s = R_L$ and $R_p = R_S$.

As before, the definitions for reactances of inductors and capacitors (6.4 and 6.5) can be used to obtain the actual values of the inductors and capacitors.

6.3.2 Narrowband Three-Element Networks (T- and Π-Networks)

Three-element matching networks assume the shape of the letter T or Greek letter Π, and are therefore called the T-networks and Π-networks respectively. Inductor-capacitor-inductor and capacitor-inductor-capacitor network variations of T- and Π-networks are the simplest to analyze and are shown in Fig. 6.6.

6.3.2.1 T-Networks

T-networks can be understood as two L networks separated by a virtual resistance R_V as shown in Fig. 6.7. In this figure, X_{S1} and X_{P1} are reactances of the first L-network, and X_{S2} and X_{P2} are reactances of the second L-network. The two L-networks can be analyzed separately and then combined into a single T-network by removing the virtual resistance.

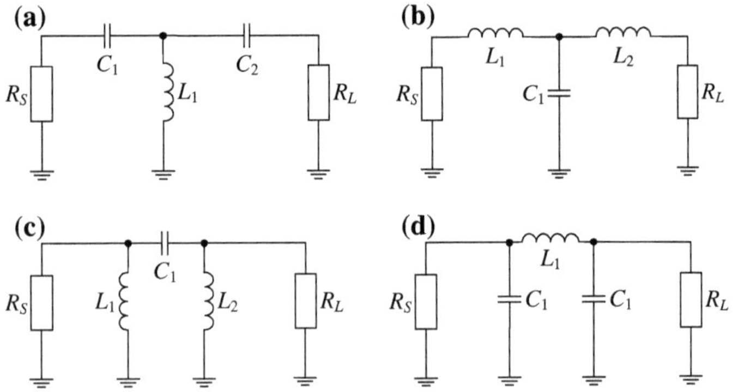

Fig. 6.6 Four matching networks: **a** capacitor-inductor-capacitor T-network, **b** inductor-capacitor-inductor T-network, **c** inductor-capacitor-inductor Π-network and **d** capacitor-inductor-capacitor Π-network

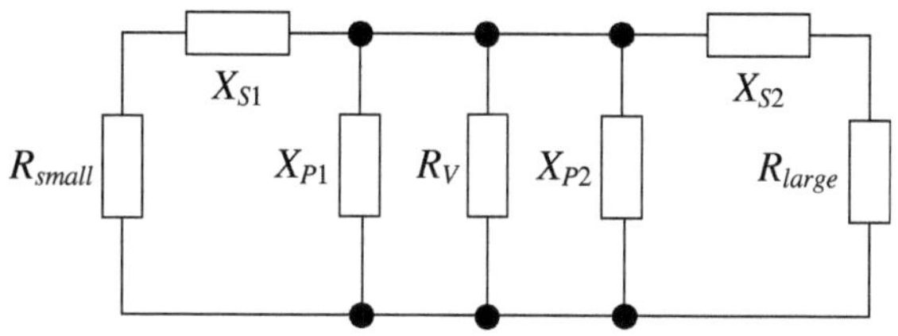

Fig. 6.7 Two L-networks separated by a virtual resistance, forming a T-network

The virtual resistance is found using

$$R_V = R_{small}(Q^2 + 1), \tag{6.14}$$

where $R_{small} = \min(R_S, R_L)$, and $Q = f_0/BW$ is the desired Q-factor of the matching network calculated from the operating frequency f_0 and bandwidth BW. Series reactance X_{S1} is then

$$X_{S1} = QR_S \tag{6.15}$$

and parallel reactance X_{P1} is

$$X_{P1} = \frac{R_V}{Q} \tag{6.16}$$

For the second L-network, a separate Q-factor needs to be calculated:

$$Q_2 = \sqrt{\frac{R_V}{R_L} - 1} \tag{6.17}$$

and X_{S2} and X_{P2} are then

$$X_{S2} = Q_2 R_L \tag{6.18}$$

and

$$X_{P2} = \frac{R_V}{Q_2} \tag{6.19}$$

respectively.

After all the reactances have been calculated and the virtual resistance has been removed, X_{P1} and X_{P2} can then be combined in parallel to get the final reactance X_P:

$$X_P = \frac{X_{P1} X_{P2}}{X_{P1} + X_{P2}}. \tag{6.20}$$

However, (6.20) is only applicable if both components (X_{P1} and X_{P1}) are of the same type i.e. both are inductors or both are capacitors, which will be the case for inductor-capacitor-inductor and capacitor-inductor-capacitor networks. Finally, the definitions of reactances of inductors and capacitors (6.4 and 6.5) can be used to obtain the actual values of the inductors and capacitors.

6.3.2.2 Π-Networks

A similar procedure can be applied to the Π-networks, with the difference that in this case serial reactances X_{S1} and X_{S2}, instead of parallel reactances X_{P1} and X_{P2}, are separated by a virtual resistor, as shown in Fig. 6.8. Virtual resistance is given by

$$R_V = \frac{R_{large}}{Q^2 + 1}, \tag{6.21}$$

where R_{large} = max(X_S, X_L). X_{S2} and X_{P2} are then

$$X_{S2} = QR_V \tag{6.22}$$

and

$$X_{P2} = \frac{R_L}{Q}. \tag{6.23}$$

The Q-factor of the first L network is

$$Q_1 = \sqrt{\frac{R_S}{R_V} - 1}. \tag{6.24}$$

and X_{S1} and X_{P1} are then

$$X_{S1} = Q_1 R_V \tag{6.25}$$

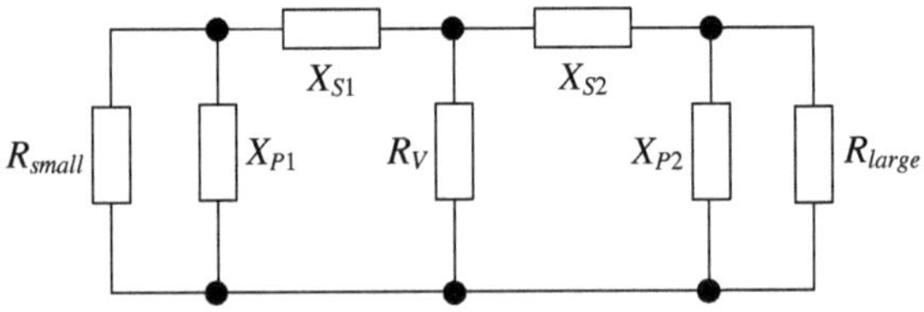

Fig. 6.8 Two L networks separated by virtual resistance forming a Π network

and

$$X_{P1} = \frac{R_P}{Q_1}. \tag{6.26}$$

After all the reactances have been calculated and the virtual resistance has been removed, X_{S1} and X_{S2} can then be simply added, but only if both components are of the same type, which will be the case for inductor-capacitor-inductor and capacitor-inductor-capacitor networks. Once again, the values of the inductors and capacitors are obtained from (6.4) and (6.5) respectively.

6.4 Lumped Matching Using Smith Charts

To illustrate the process of impedance matching using Smith charts, because of the nature of the problem, we will have to focus on several particular matching tasks:

- Matching of two real values using an L-section;
- Matching of two complex values using an L-section;
- Matching of two complex values using a T-network; and
- Matching of two complex values using a Π-network.

For each matching type, we will present one solution to the matching problem. The four examples should be sufficient for the reader to understand the Smith chart matching concept. Before proceeding to the examples, we note that positive movement on the impedance chart will indicate a series inductor and negative movement will indicate a series capacitor, while positive movement on the admittance chart will indicate a shunt capacitor and negative movement will indicate a shunt inductor.

Task 1 Using an L-network, match a source of $R_S = 50\ \Omega$ to a load of $R_L = 250\ \Omega$.

Solution: To accomplish this task, we realize that we need one shunt and one series component to complete the L-network. Here we notice that it is easier to deal with impedances when describing series components and that it is easier to deal with admittances when dealing with shunt components. One approach is to use impedance-to-admittance conversion, as demonstrated in Chap. 2, but this seems tedious, since we need to convert from impedance to admittance or vice versa twice. The second idea is to introduce a Smith chart with both impedance and admittance coordinates, such as the one shown in Fig. 6.9. Since the admittance chart is the impedance chart rotated by 180°, the negative values in the admittance graph are in the top part of the graph.

With the correct Smith chart chosen, the first step is to normalize the impedances and plot them on the graph. Since one of the impedances is 50 Ω, we choose 50 as the normalization factor. The impedances plotted are therefore 50/50 = 1 Ω and 250/50 = 5 Ω, also shown in this figure. The second step is to look for a path on the

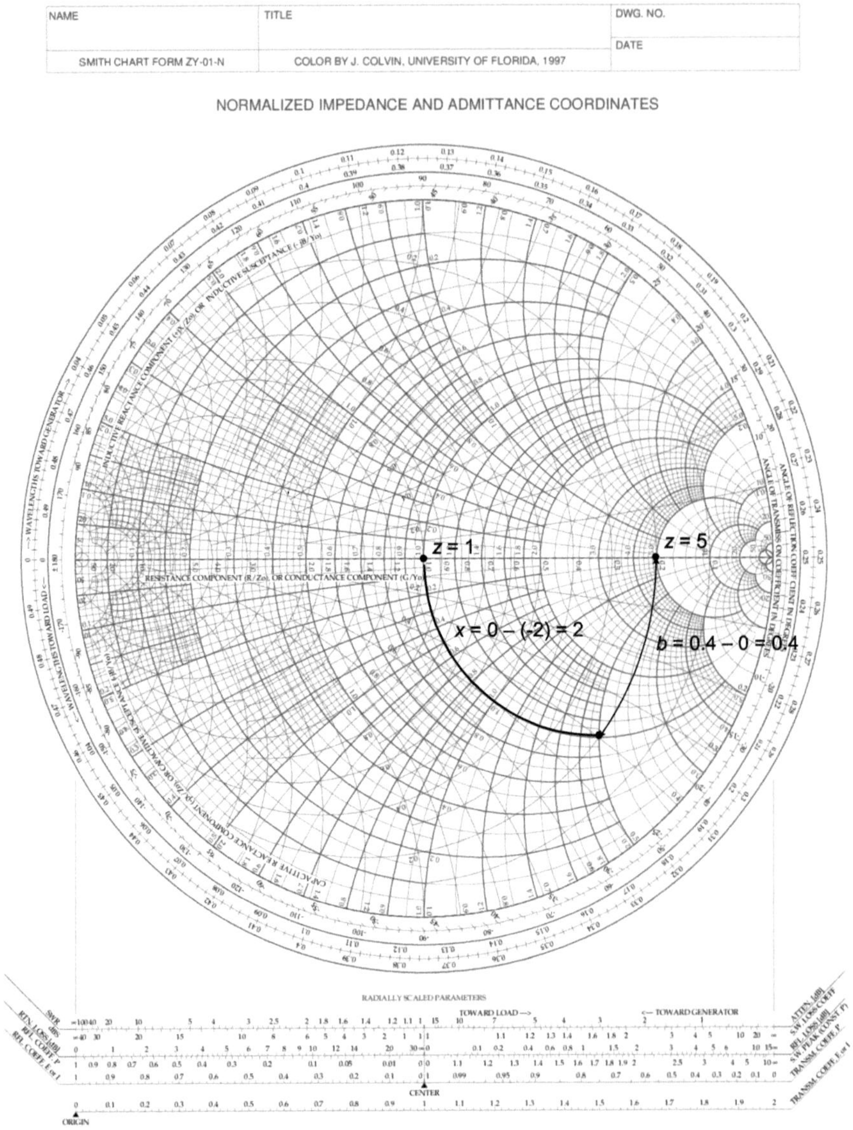

Fig. 6.9 Matching procedure of task 1

combined admittance-impedance chart that will connect the two plotted impedances. We identify that one way to connect these two points is to move on the constant admittance graph closer to the load and on the constant impedance circle closer to the source. The intersection point is also illustrated in the figure.

Next we check the "distance traveled" on each circle: we moved 0.4 S on the admittance circle, and 2 Ω on the impedance circle. This means that we need a

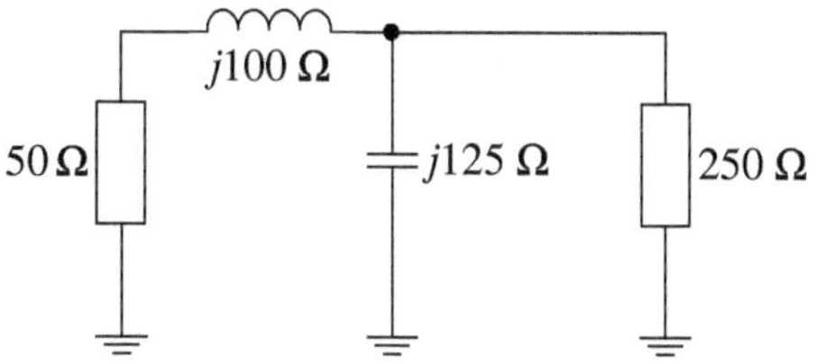

Fig. 6.10 Network designed in task 1

shunt capacitor with impedance 50 × (1/0.4) = 125 Ω next to the load and an inductor of 50 × 2 = 100 Ω in series with the source. This basically completes the matching task, and the only task left is to calculate the exact capacitance value, which we will omit. The resulting circuit is shown in Fig. 6.10. Note that this is one of two possible solutions, the other one sitting on the top part of the Smith chart, where the inductor is in shunt and the capacitor is in series.

The fact that we chose to match two real impedances allows us to check our calculations by using the analytic equations. We first note that $R_L > R_S$, meaning that the shunt component is next to the load. Thus using Eq. (6.3), we can calculate the inherited Q-factors:

$$Q_S = Q_p = \sqrt{\frac{250}{50} - 1} = \sqrt{4} = 2.$$

From Eqs. (6.4) and (6.5), the required impedances are

$$X_s = 2 \times 50 = 100\,\Omega$$

and

$$X_p = \frac{250}{2} = 125\,\Omega,$$

which are exactly the same values as obtained graphically. This allows us to gain confidence that the Smith chart method is correct so that we can proceed with matching two complex impedances.

Task 2 Using an L-network, match a source impedance of (30 − j20) Ω to a load of (75 − j50) Ω.

Solution: We again start by plotting the values on the Smith chart. Note that now that both the impedances are complex, we need to implement conjugate matching. Thus we need to make the load look like the conjugate of the source, which is 30 + *j*20 Ω. Again, we can normalize by 50, and plot (0.6 + *j*0.4) Ω and (1.5 − *j*1) Ω. The solution then follows the same steps of Task 1: we "move" on the admittance chart away from the load and on the impedance chart away from the conjugate of the source, until we find the intersection point. This movement again

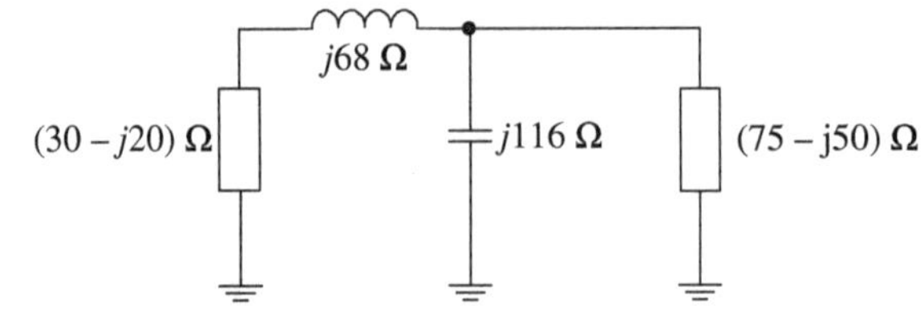

Fig. 6.11 Network designed in task 2

results in a capacitor parallel with the load and an inductor in series with the source. The distance travelled on the admittance graph is about 0.75 − 0.32 = 0.43 S and the distance travelled on the impedance graph is about 0.4 − (−0.95) = 1.35 Ω. Finally, we normalize back to yield a capacitor with reactance of ~116 Ω and an inductor with reactance of ~68 Ω. The resulting network is illustrated in Fig. 6.11 and the Smith chart for analysis of the problem is illustrated in Fig. 6.12.

Task 3 Repeat task 2, but use a Π-network.

Before attempting the solution to this task, we have to recall three things. First, when matching using three elements, we have control of the Q-factor of the matching network. Second, in the case of T-networks, Q is controlled on the side of the smaller resistance, and in the case of Π-networks, Q is controlled on the side of the larger resistance. Finally, without specifying the Q-factor, an infinite number of three-element matching networks can accomplish the matching task.

Therefore, to limit the number of solutions to four, a value of Q can be specified. Then two intersection points on the Smith chart are needed, which can be determined as follows:

1. For the T-network where $R_S > R_L$, the first intersection is determined by the resistance circle on which R_L is located and the Q-factor arc (two options), and the second intersection is then between the resistance circle on which R_S is located and the admittance circle on which the first intersection is located (two more options).
2. For the T-network where $R_S < R_L$, the first intersection is determined by the resistance circle on which R_S is located and the Q-factor arc (two options), and the second intersection is then between the resistance circle on which R_L is located and the admittance circle on which the first intersection is located (two more options).
3. For the Π-network where $R_S > R_L$, the first intersection is determined by the admittance circle on which R_S is located and the Q-factor arc (two options), and the second intersection is then between the admittance circle on which R_L is located and the resistance circle on which the first intersection is located (two more options).
4. For the Π-network where $R_S < R_L$, the first intersection is determined by the admittance circle on which R_L is located and the Q-factor arc (two options), and

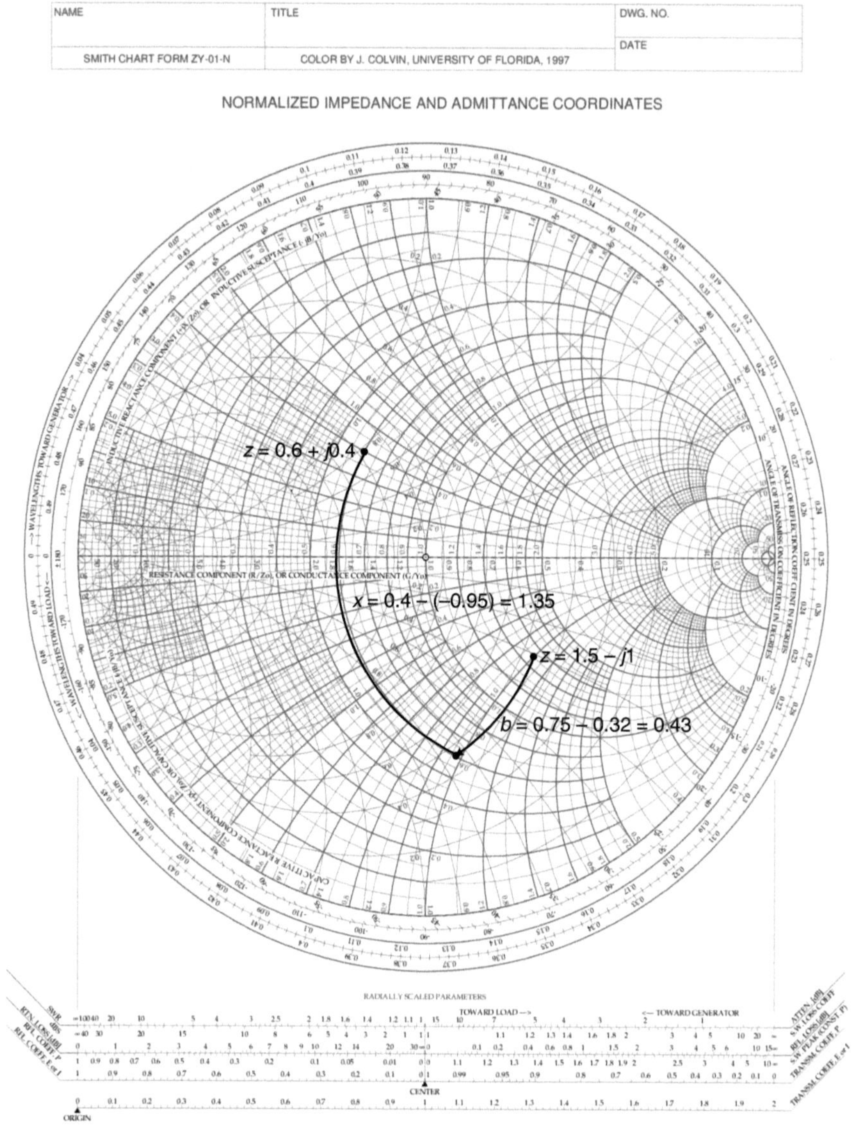

Fig. 6.12 Matching procedure of task 2

the second intersection is then between the admittance circle on which R_S is located and the resistance circle on which the first intersection is located (two more options).

The Q-factor arc can be found by plotting all impedances on the impedance Smith chart that satisfy the equation

$$Q = \frac{X}{R}. \tag{6.27}$$

The graph always passes through (0, 0) and (∞, ∞) points, with the infinite Q arc defining the outer circle of the Smith chart. In practice, it is sufficient to plot only a few extra points in order to plot the arc of the particular Q-factor value when working on paper.

We now have enough information to proceed to the matching task.

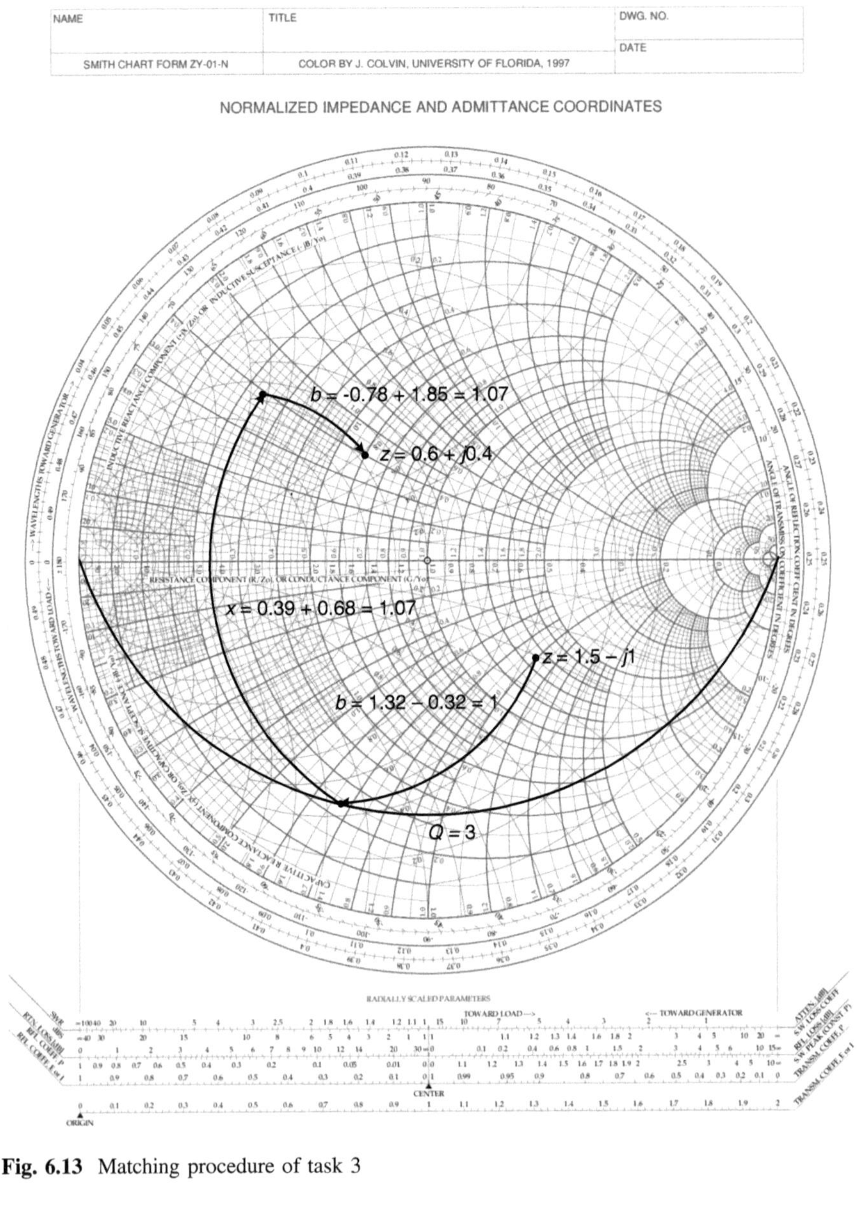

Fig. 6.13 Matching procedure of task 3

Solution: We begin by normalizing the impedance with a factor of 50 and plot the load impedance and the conjugate of the source impedance on the Smith chart, as illustrated in Fig. 6.13. Secondly, we need to decide on the Q-factor of the matching network and plot the Q-arc. Let us assume that $Q = 3$ will be sufficient for the bandwidth requirement (which is not specified). Also, let us always move in clockwise direction on the chart.

Now, since $R_S < R_L$, we move from Z_L on the admittance circle until we hit the intersection with the Q-arc, and note that the distance traveled is $1.32 - 0.32 = 1$ S. Then we move on the resistance circle until we find the intersection with the admittance circle coming from Z_S^* and note that the distance traveled is $0.39 + 0.68 = 1.07\ \Omega$. Finally, we note that the distance traveled from this intersection to Z_S^* is $-0.78 + 1.85 = 1.07$ S. Thus the denormalized impedances are 50, 53.5 and 46.7 Ω, arranged as shown in Fig. 6.14. The network is capacitor-inductor-capacitor network, which is determined by the direction of movement on the Smith chart, as explained before.

Task 4 Repeat task 2, but use a T-network.

Solution: As in the previous example, we first normalize the impedances, and plot them together with the Q-arc of $Q = 3$. Now, since $R_S < R_L$, we find the intersection of the resistance circle coming from Z_S^* and the Q-factor arc, and then trace the admittance circle from this intersection until we find the second intersection with the resistance circle coming from Z_L. Then when moving from Z_L towards Z_S^*, we note that the distances are $-2.7 + 1 = -1.7\ \Omega$, $0.5 - 0.28 = 0.32$ S and $0.4 + 0.8 = 2.2\ \Omega$. The resulting denormalized impedances (looking from the load) are −85, 156 and 110 Ω. The minus sign next to the 85 Ω impedance indicates negative movement, thus the series component is a capacitor, and the network is an inductor-capacitor-capacitor network, as shown in Fig. 6.15. The Smith chart movement is illustrated in Fig. 6.16.

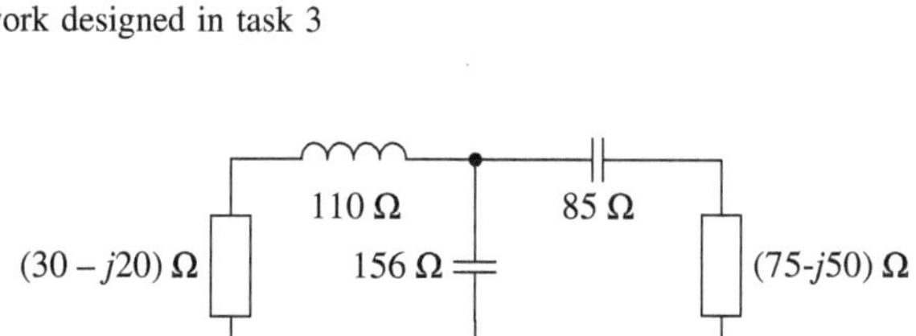

Fig. 6.14 Network designed in task 3

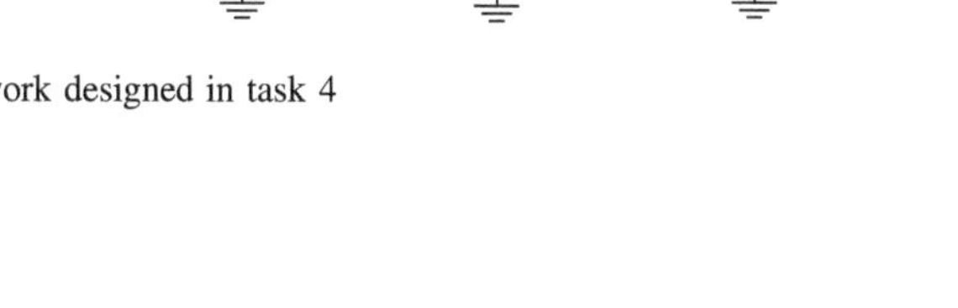

Fig. 6.15 Network designed in task 4

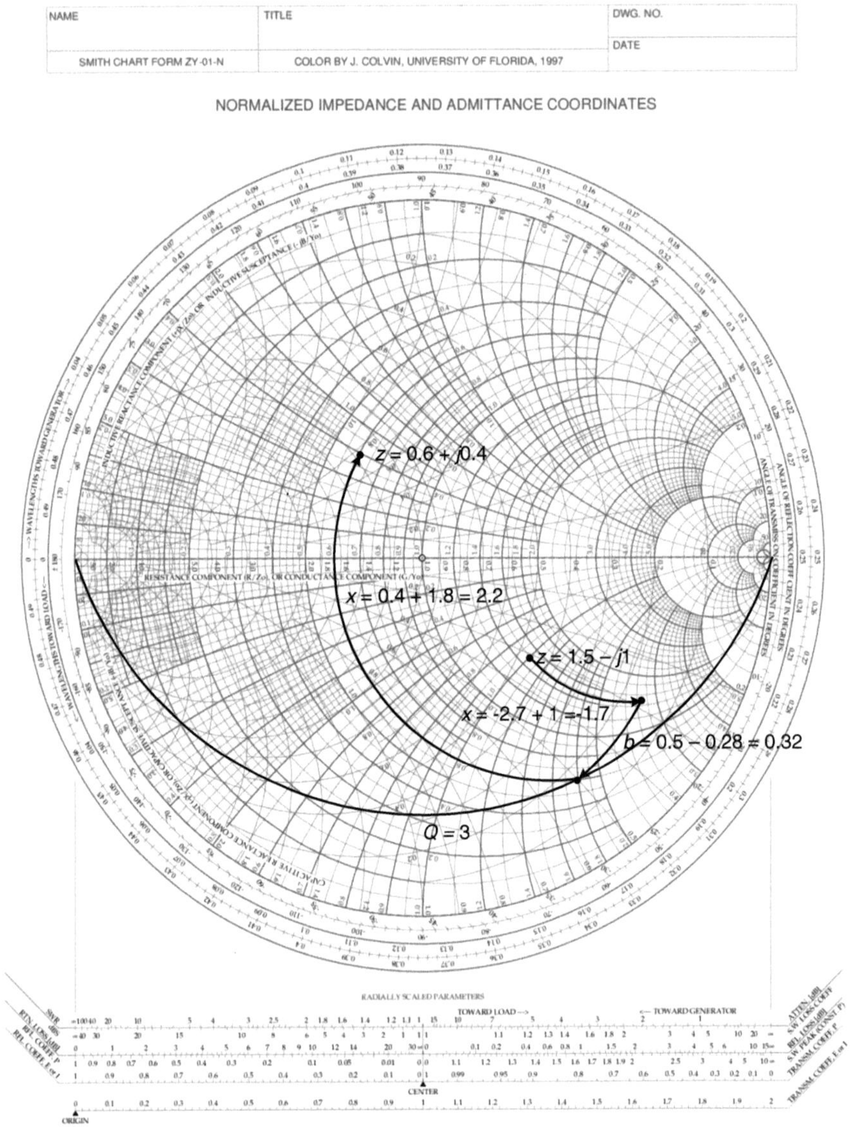

Fig. 6.16 Matching procedure of task 4

6.5 Transmission-Line Impedance Matching

Another way to match impedances in a power amplifier system is with the aid of transmission lines. As pointed out several times in this book, the size (length) of the transmission line is dependent on the line wavelength, so they only become practical (off-chip) at about 2 GHz. Around this frequency, a combination of lumped and distributive elements can have an advantage.

6.5.1 Variations of Transmission-Line Matching Networks

As with lumped-element matching networks, several variations of transmission-line matching networks are available. Typically, microstrip lines are used. These include:

- Matching networks with a combination of lumped elements and transmission lines;
- Matching networks based on single-stub microstrip lines with fixed characteristic impedances;
- Matching networks based on single-stub microstrip lines with different characteristic impedances;
- Matching networks based on double-stub microstrip lines; and
- Matching networks involving a quarter-wavelength transformer.

Figures 6.17, 6.18 and 6.19 show some examples of matching networks. However, when dealing with power amplifiers, the quarter-wavelength transformer, discussed in detail in Chap. 5, has the widest practical use and it will form the focus of our transmission-line tuning section.

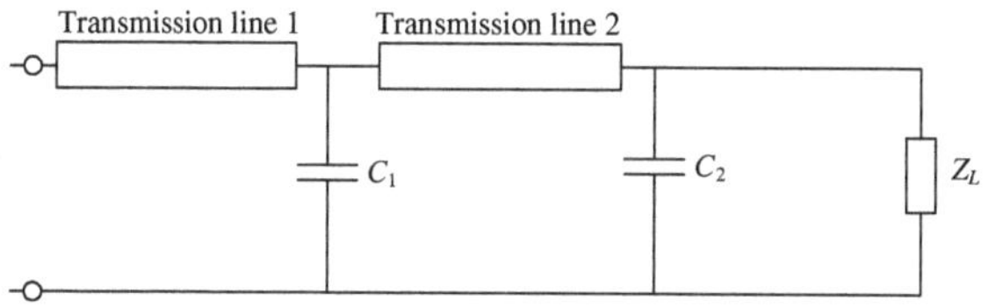

Fig. 6.17 Matching networks with a combination of transmission lines and lumped elements (in this case capacitors)

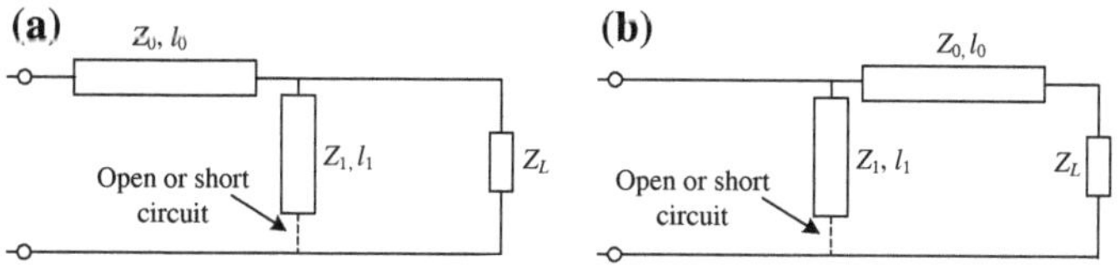

Fig. 6.18 Two variations of a single-stub matching networks

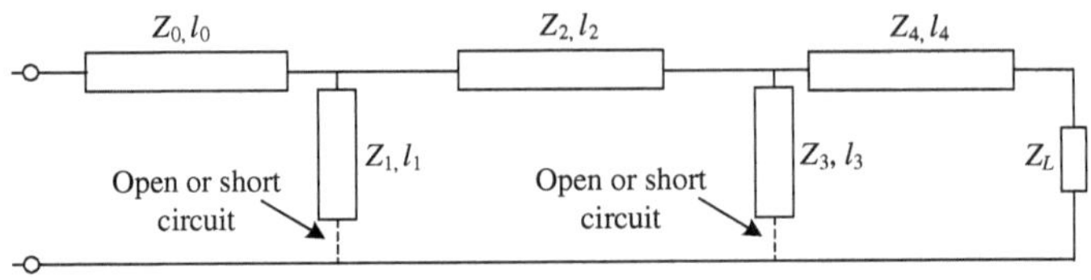

Fig. 6.19 A double-stub matching network

6.5.2 Quarter-Wave Transformer Impedance Matching

Note that the quarter-wave transformer can only be used to match real impedances. If complex impedance matching is required, this can be accomplished by adding short microstrip line sections of the required length to compensate for imaginary quantities.

The simplest form of the quarter-wave transformer was illustrated in Fig. 5.28 in Chap. 5. Recall that in this chapter we calculated its input impedance to be

$$Z_{in} = \frac{Z_1^2}{Z_L}, \tag{6.28}$$

where Z_1 was defined as is its characteristic impedance and Z_L was the impedance of the load connected to the other end of the transformer. This means that the simplest way to match the real source with impedance R_S to the load with real impedance R_L is by inserting a quarter-wave transformer with the characteristic impedance

$$Z_1 = \sqrt{R_S R_L}. \tag{6.29}$$

All that is now left to the designer is to find the line width w and the substrate height h, for a specific substrate material, using microstrip equations presented in Chap. 5, to reach the desired characteristic impedance. This can, however, be an iterative task. Furthermore, the height and width of the microstrip line can be unrealistic, depending on the impedances to be matched and the type of substrate used. In this case, it may prove prudent to introduce more than one quarter-wave transformer section. A two quarter-wave transformer configuration separated by the virtual impedance R_V is illustrated in Fig. 6.20. Then the virtual resistance is

$$R_V = \sqrt{R_S R_L}, \tag{6.30}$$

and the two characteristic impedances required are

$$Z_1 = \sqrt{R_S R_V}, \tag{6.31}$$

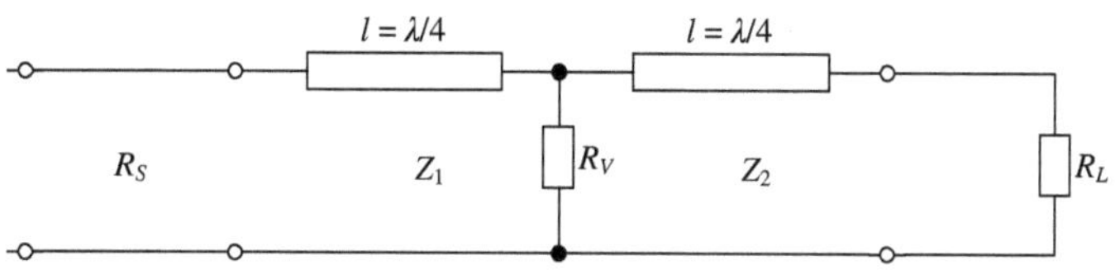

Fig. 6.20 Two quarter-wave transformers separated by a virtual impedance R_V

and

$$Z_2 = \sqrt{R_V R_L}. \tag{6.32}$$

Having multiple quarter-wave transformers also allows the designer to optimize the bandwidth of the matching network. In fact, any number of quarter-wave transformers can be used. With every next transformer, the microstrip line is narrower (or wider, depending on the impedances), thus forming a tapered transmission line. Unless used at very high frequencies (mm-wave and higher), this transformer does not really have much practical use in view of the required length of the line.

6.6 Impedance Matching Using MATLAB

By now it has become apparent that if we move from matching of real impedances to matching of complex impedances, and from two-element networks to multiple-element networks, the solution involving manual calculations becomes too complex and the graphical solution, although relatively easy, becomes inaccurate. This has resulted in the development of a great number of different EDA matching tools. Some of these tools are free and are readily available. A number of tools were reviewed by Bowick et al. [3].

The idea of this book, however, remains to present readers with enough information to develop their own algorithms for power amplifier design. This includes impedance matching. Later chapters deal with the topic of algorithm development using analytical equations presented in this chapter. However, this does not stop us from reviewing some MATLAB functions that are already available for RF and microwave designers to use.

Orfanidis [6] developed an "Electromagnetic Waves & Antennas" MATLAB toolbox[1] containing routines for accomplishing various RF and microwave tasks. Functions for the design of quarter-wave transformers and L-, Π- and T-networks are particularly in line with the topic of this text. Some useful functions are included in Table 6.1. Note that for the particular matching function to work, conditions

[1]The toolbox is available from http://www.ece.rutgers.edu/~orfanidi/ewa.

Table 6.1 MATLAB functions used for impedance matching

Function	Description	Arguments
`X12 = lmatch(ZG, ZL,type)`	L-section matching network	`ZG, ZL`: matching arguments `Type`: 'n' normal L-section; 'r' reverse L-section
`Zabc = pi2t (Z123)`	Π- to T-network transformation	`Z123`: 1x3 Π-network value matrix
`Z123 = t2pi (Zabc)`	T- to Π-network transformation	`Zabc`: 1x3 T-network value matrix
`X123 = pmatch (ZG,ZL,Z)`	Π-network matching design	`ZG, ZL`: matching arguments `Z`: reference impedance
`[Z1,Lm] = qwt1 (ZL,Z0,type)`	$\lambda/4$-transformer with series section matching network	`ZL, Z0`: matching arguments `Type`: 'min' or 'max'
`smith`	Draw a basic Smith chart	

discussed in previous sections of this chapter still have to be satisfied. Furthermore, as seen before, every type of matching will result in more than one solution, thus the output of each matching function is a matrix of values. However, not every solution will lead to practical inductor and capacitor values, so the designer still has to choose the most adequate solution for the particular application.

6.7 Summary

In this chapter, impedance matching was discussed. Both analytical and graphical approaches were used. Although analytical and graphical analysis are both important for understanding the matching theory, a third approach, which utilizes EDA, is the most practical because it is less tedious and less prone to errors. We described some EDA procedures found in other texts, but we also noted that in later chapters of this book we will be using analytical matching theory to develop our own procedures.

This is the last chapter of Part I of this book. From Chap. 7 onwards, we will use the theory described up to now to demonstrate the development and use of our own EDA algorithms, which can be used to speed up the power amplifier design process.

References

1. Pozar M. Microwave engineering. 4th ed. Hoboken: Wiley; 2012.
2. Ludwig Bretchko. RF circuit design: theory and applications. 1st ed. Upper Saddle River: Prentice Hall; 2000.
3. Bowick C, Blyler J, Ajluni C. RF circuit design. 2nd ed. Burlington: Elsevier; 2008.
4. Cripps SC. RF power amplifiers for wireless communications. Norwood: Artech House; 2006.

5. Kazimierczuk MK. RF power amplifiers. 2nd ed. Chiechester: Wiley; 2015.
6. Orfanidis JS. Electromagnetic waves and antennas. New Brunswick: Rutgers University; 2014.
7. Grebennikov A, Sokal NO. Switchmode RF power amplifiers. 1st ed. Burlington: Elsevier; 2007.

Part II
Developing EDA for Power Amplifier Design

Chapter 7
Intelligent Automated Design Ideas for Inductor Synthesis

In Chap. 5 of Part 1 of this book we explored the theory behind the behavior of various passive components at RF and microwave frequencies, together with the modeling necessary for accurate component design. In this chapter, we go one step further to show how the models can be used to automate the inductor design and at the same time optimize the inductor quality factor, seeing that for the inductors, this factor decreases abruptly with an increase in frequency, thus having the largest influence on the whole system. The chapter covers all inductors needed to design a stand-alone system, including filtering and matching inductors, RF chokes and transformers. With integrated inductors gaining ground owing to their smaller size and lower cost, a lot of effort in this chapter is put into integrated inductor design, but design of other inductors is not omitted. The Q-factor and its dependence on various integrated inductor design parameters are simulated and confirmed experimentally. The information presented in this chapter is not only beneficial for power amplifier design, but also for the design of other devices that require high-quality inductors, such as LNAs and DC-to-DC converters.

MATLAB scripts provided in this chapter were developed in MATLAB version 2014b.

7.1 Design of Integrated Inductors

Inductors have traditionally been difficult to integrate because of their inherent low Q-factors and their modeling complexity. Even though many different inductor configurations are available for an RF or a microwave designer to explore, support for integrated inductors in EDA tools and process design kits has been very limited in the past [1].

Some vendors provide libraries of several qualified integrated inductors for each RF-capable process. As seen in Chap. 5, each of these inductors operates at its peak Q-factor only in a small frequency range, making these libraries impractical for many applications. Other vendors provide p-cells of spiral inductors. Although the Q-factor of the specific geometry is calculated based on the technology parameters for a specific frequency, the technology-aware optimization is missing. It is more

M. Božanić and S. Sinha, *Power Amplifiers for the S-, C-, X- and Ku-bands*,
Signals and Communication Technology, DOI 10.1007/978-3-319-28376-0_7

practical, yet tedious, to use such p-cells, owing to the cut-and-try nature of this approach to inductor selection and the lack of automated design flow. In this section, we discuss how technology-aware integrated inductor design can be performed by introducing an inductor synthesis algorithm based on spiral inductor design equations and models presented in Chap. 5. If the algorithm is designed properly, it allows RF and microwave designers to optimize integrated inductors, given the inductor center frequency dictated by the device application and the geometry constraints. This not only lays down a foundation for system-level RF circuit performance optimization, but because inductors are often the largest parts of an RF system, it also allows for optimal usage of the chip floorplan.

7.1.1 Cut-and-Try Approach

When designing a capacitor, a designer may simply increase or decrease the area of the component until the required capacitance is obtained, as can be seen in capacitance equations presented in Chap. 5. A similar relationship between the length and total resistance holds for resistors. By modifying the length of a part of the process layer used for fabrication of the resistor or the number of segments, a designer can obtain the desired value of its resistance.

Inductance of a spiral inductor, however, is determined by several geometry parameters. This means that if one is to vary only one parameter, e.g. the number of turns, the other parameters need to meet some conditions. Thus the process of parameter sweep is not intuitive. This is best illustrated by varying some design parameters, as shown in Table 7.1. In this table, we use a data-fitted monomial

Table 7.1 Influence of varying parameters on the inductor DC inductance value, where each component varied is marked in bold letters

Inductor number	Outer diameter, d_{out} (μm)	Inner diameter, d_{in} (μm)	Turn width, w (μm)	Number of turns, n (μm)	Calculated turn spacing, s (μm)	Calculated inductance, L (nH)
1	**100**	50	10	2	5	0.46
2	**150**	50	10	2	30	0.53
3	**200**	50	10	2	55	0.62
4	**250**	**50**	10	2	80	0.73
5	250	**100**	10	2	55	1.07
6	250	**150**	**10**	2	30	1.50
7	250	**200**	10	**2**	5	2.09
8	250	150	**15**	2	20	1.43
9	250	150	**20**	2	10	1.40
10	250	150	**24**	2	2	1.44
11	250	150	10	**3**	10	3.22
12	250	150	10	**4**	3.33	5.65
13	255	150	10	**5**	0.63	9.20

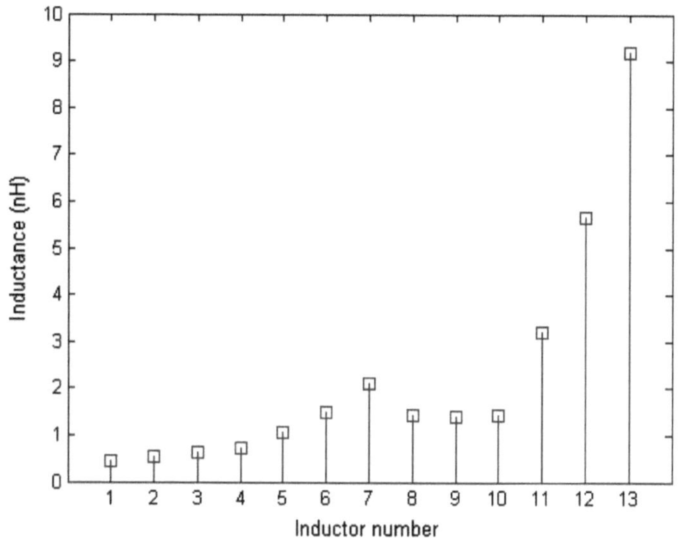

Fig. 7.1 Inductance value of thirteen inductors designed in Table 7.1

expression for square inductors to calculate inductance, and then we first vary the outer diameter, while keeping the inner diameter, width and number of turns constant. We then keep the outer diameter, width and number of turns constant, but vary the inner diameter. The third step is to vary the width of the spiral by keeping the rest constant, and the final step is to vary the number of turns. Resulting inductance values are also depicted in Fig. 7.1. We can observe that the number of turns has the highest influence on the inductance value, but the steps in inductance are discrete. Moreover, only a limited number of variation steps are possible (in this case four, where we increase the outer diameter slightly in the last step to accommodate five turns). This means that some other parameter needs to be varied (e.g. the outer diameter), if a bigger range of inductance values is needed, which could be followed by some tweaking of the width, and so on. Note that in this analysis, we completely ignored the quality factor, which changes entirely independently from the inductance. In addition, the geometry of these inductors differs dramatically, as can be seen in Fig. 7.2, which shows the layouts of these inductors.

The complexity of the inductance relationship is one of the reasons why various cut-and-try approaches are used in practice, such as the approach shown in Fig. 7.3.

7.1.2 Synthesis Approach

In this section, we show how an improvement to the iterative procedure described previously can be brought about. We introduce a search-algorithm-based routine that can find an inductor of the specified value, with the highest possible Q-factor,

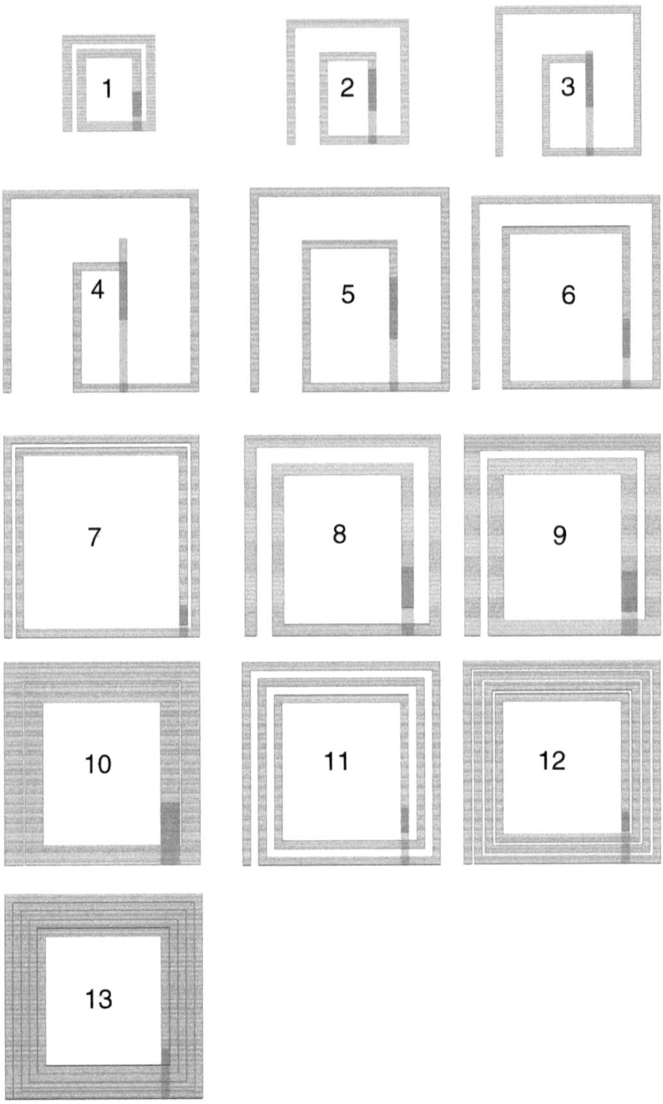

Fig. 7.2 Layouts of thirteen inductors designed in Table 7.1

occupying a limited area, and using predetermined technology layers (synthesis of the inductor structure). The nine-component single-π model of a single-layer spiral inductor, shown in Fig. 5.17, is used as the model for the calculations. The exact inductor shape (square, octagonal, circular) is not important, as long as a lookup table with correct coefficients for inductance calculations is used in the calculations.

We also deduce that for completeness, we need to make provision for the analysis of an inductor geometry in terms of inductance and the Q-factor. Now we

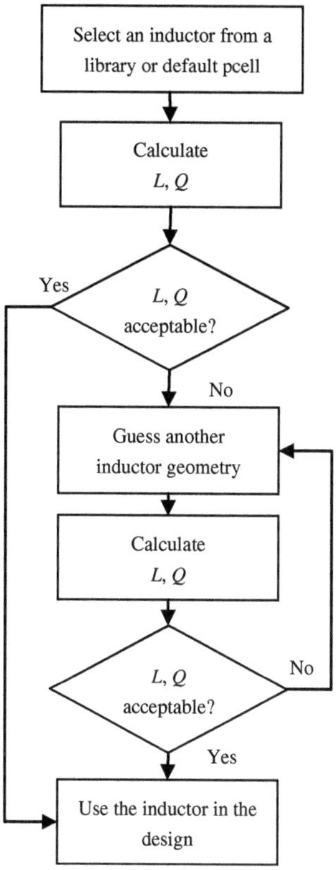

Fig. 7.3 Flow chart of conventional spiral inductor design procedure

can start by investigating which design equations should be used in the design algorithm.

7.1.2.1 Equations

In the nine-component model, it is necessary to calculate:

- One inductance, L_S,
- Two resistances, R_S and R_{sub}, and
- Three capacitances, C_S, C_{ox} and C_{sub}.

Thus, we use the equations tabulated in Table 7.2, where we have chosen to use the data-fitted monomial expression to calculate the inductance (with coefficients presented in Table 5.5). We also include the equations for the Q-factor and resonant frequency calculation, as well as for the calculation of the apparent inductance value.

The second step is to deduce the input parameters from the design equations.

Table 7.2 Summary of equations needed for the design

Calculation	Equation
Average diameter	$d_{avg} = \frac{d_{in} + d_{out}}{2}$
Length of the spiral	$l = 4(d_{in} + w) + 2n(2n - 1)(s + w)$
Spiral pitch	$s = \frac{\frac{d_{out} - d_{in}}{2} - nw}{n-1}$
Inductance	$L_{mon} = \beta d_{out}^{\alpha_1} w^{\alpha_2} d_{avg}^{\alpha_3} n^{\alpha_4} s^{\alpha_5}$
Parasitic resistance	$R_s = \frac{\rho l}{w t_{eff}}$
Effective spiral thickness	$t_{eff} = \delta(1 - e^{-t/\delta})$
Skin depth	$\delta = \sqrt{\frac{\rho}{\pi \mu f}}$
Parasitic capacitance	$C_s = nw^2 \frac{\varepsilon_{ox}}{t_{oxM1-M2}}$
Effective thickness (oxide or substrate)	$t_{eff} = \begin{cases} w\left[\frac{w}{t} + 2.42 - 0.44\frac{t}{w} + \left(1 - \frac{t}{w}\right)^6\right]^{-1}, & \text{for } \frac{t}{w} \le 1 \\ t_{eff} = \frac{w}{2\pi} \ln\left(\frac{8t}{w} + \frac{4w}{t}\right), & \text{for } \frac{t}{w} \ge 1 \end{cases}$
Effective oxide permittivity	$\varepsilon_{eff} = \frac{1+\varepsilon}{2} + \frac{\varepsilon - 1}{2}\left(1 + \frac{10t}{w}\right)^{-\frac{1}{2}}$
Oxide capacitance	$C_{ox} = \frac{wl\varepsilon_0 \varepsilon_{effox}}{t_{effox}}$
Substrate capacitance	$C_{sub} = \frac{wl\varepsilon_0 \varepsilon_{effsub}}{t_{effoxsub}}$
Substrate resistance	$R_{sub} = \frac{t_{effsub}}{\sigma_{eff} wl}$
Effective substrate conductivity	$\sigma_{eff} = \sigma\left[\frac{1}{2} + \frac{1}{2}\left(1 + \frac{10t}{w}\right)\right]^{-\frac{1}{2}}$
Quality factor	$Q = \frac{\omega L_S}{R_S} \cdot \frac{R_P}{R_P + \left[\left(\frac{\omega L_S}{R_S}\right)^2 + 1\right] R_S} \cdot \left[1 - (C_P + C_S) \cdot \left(\omega^2 L_S + \frac{R_S^2}{L_S}\right)\right]$
Quality factor resistance	$R_P = \frac{1}{\omega^2 C_{ox}^2 R_{sub}} + \frac{R_{sub}(C_{ox} + C_{sub})^2}{C_{ox}^2}$
Quality factor capacitance	$C_P = C_{ox} \cdot \frac{1 + \omega^2 (C_{ox} + C_{sub}) C_{sub} R_{sub}^2}{1 + \omega^2 (C_{ox} + C_{sub})^2 R_{sub}^2}$
Resonant frequency	$f_r = \frac{\omega_o}{2\pi} = \frac{1}{2\pi} \sqrt{\frac{1}{L_S \cdot (C_P + C_S)} - \left(\frac{R_S}{L_S}\right)^2}$
Effective inductance value close to resonant frequency	$L_{eff} = \frac{\mathrm{Im}(z)}{2\pi f_r}$

7.1.2.2 Input Parameters

By inspecting the equations, two groups of parameters are deduced: geometry and process parameters. The frequency of operation of the inductor also needs to be known for Q-factor calculation.

Geometry Parameters For the simple analysis of an inductor structure, the following input geometry parameters are sufficient:

- Outer diameter, d_{out};
- Inner diameter, d_{in};

- Turn width, w; and
- Number of turns, n.

For the main goal of the algorithm, which is the synthesis of the inductor structure, it is only necessary to specify the constraints on the inductor geometry to confine the search space. Note that the spiral inductor design guidelines listed in Chap. 5 also have to be taken into account, and none of the design rule checks (DRCs) of the specific process can be broken. The following parameters are required:

- Minimum value of inner diameter, $d_{in(min)}$,
- Maximum value of outer diameter, $d_{out(max)}$,
- Minimum value for turn spacing, s_{min}, and
- Minimum turn width, w_{min}.

Further, for synthesis, we need to specify the tolerance (in percentage) for the acceptable inductance values in order to aid the search, and the grid resolution (in micrometer) is also required, which determines the steps in the search. Table 7.3 summarizes the geometry input parameters, together with some typical values.

Process Parameters The inductance value of a high-Q structure is predominantly determined by its geometry. However, the silicon substrate introduces process-dependent parasitics, which are dependent on the process parameters. They decrease the Q-factor and add shift to the inductance value. The following substrate parameters need to be specified:

- Thickness of the metal used for the spiral, t,
- Resistivity of the metal used for the spiral, ρ,
- Permeability of the metal used for the spiral, μ,
- Thickness of the oxide between the spiral and underpass,
- Relative permittivity of the oxide between the spiral and underpass, ε_{rm},
- Thickness of the oxide between the substrate and the spiral, t_{sm},
- Relative permittivity of the oxide between the substrate and the spiral, ε_{rs},
- Thickness of the silicon substrate, t_{Si},

Table 7.3 Geometry parameters for the spiral inductor design

Parameter	Typical value(s)	Units	Analysis/synthesis
Outer diameter (d_{out})	500	μm	Analysis
Inner diameter (d_{in})	200	μm	Analysis
Turn width (w)	10	μm	Analysis
Number of turns (n)	2–5	–	Analysis
Minimum value of the inner diameter	200	μm	Synthesis
Maximum value of the outer diameter	500	μm	Synthesis
Minimum value for turn spacing (s)	0.5	μm	Synthesis
Minimum turn width	1	μm	Synthesis
Inductance value tolerance	5	%	Synthesis
Grid resolution	1	μm	Synthesis

Table 7.4 Process parameters for the spiral inductor design

Parameter	Typical value	Unit
Thickness of metal used for the spiral	1000	nm
Resistivity of metal used for the spiral (ρ)	2.82×10^{-8}	Ωm
Permeability of metal used for the spiral (μ)	1.257×10^{-6}	H/m
Thickness of oxide between the spiral and underpass (t_m)	1000	nm
Relative permittivity of oxide between the spiral and underpass (ε_{rm})	4	–
Thickness of oxide between substrate and the spiral (t_{sm})	5000	nm
Relative permittivity of oxide between substrate and the spiral (ε_{rm})	4	–
Thickness of the silicon substrate (t_{Si})	1000	μm
Relative permittivity of the silicon substrate ($\varepsilon_{r\mathrm{Si}}$)	11.7	–
Resistivity of the silicon substrate (ρ_{Si})	0.2	Ωm

- Relative permittivity of the silicon substrate, $\varepsilon_{r\mathrm{Si},}$, and
- Resistivity of the silicon substrate, ρ_{Si}.

The process parameters are usually obtained or calculated from parameters obtained in the datasheets supplied by the process foundry. Table 7.4 summarizes the technology input parameters and their typical values.

Operating frequency (f_0) The operating frequency may be understood as the frequency at which the Q-factor will be highest for a particular geometry. When inductors are designed for power amplifiers, as expected, the operating frequency is the center frequency used for power amplifier design (center frequency of the channel).

7.1.2.3 Development of the Inductor Design Routine

The main (synthesis) part of the inductor design routine is a search algorithm that looks into a range of possible geometries and identifies a geometry that will result in the required inductance within a certain tolerance. As a result, more than one geometry will result in the correct inductance at a given frequency, but with each of these geometries having different Q-factors. Thus, for the optimum design, it makes sense to choose the geometry that gives the highest Q-factor.

The accuracy of the algorithm depends on the tolerance for the required inductance values and on the search grid resolution. Higher tolerance of the inductance value will result in less accurate inductance values, but there will be a greater chance that inductor geometry resulting in the particular inductance will be found. The chance of finding a solution also depends on the grid resolution. With an increase in the grid resolution, there is a greater probability that the high-Q inductor will be found, but the time of execution and memory requirements of the search algorithm will also increase, resulting in a tradeoff that can be resolved for a particular application. The search algorithm is illustrated in Fig. 7.4. The algorithm

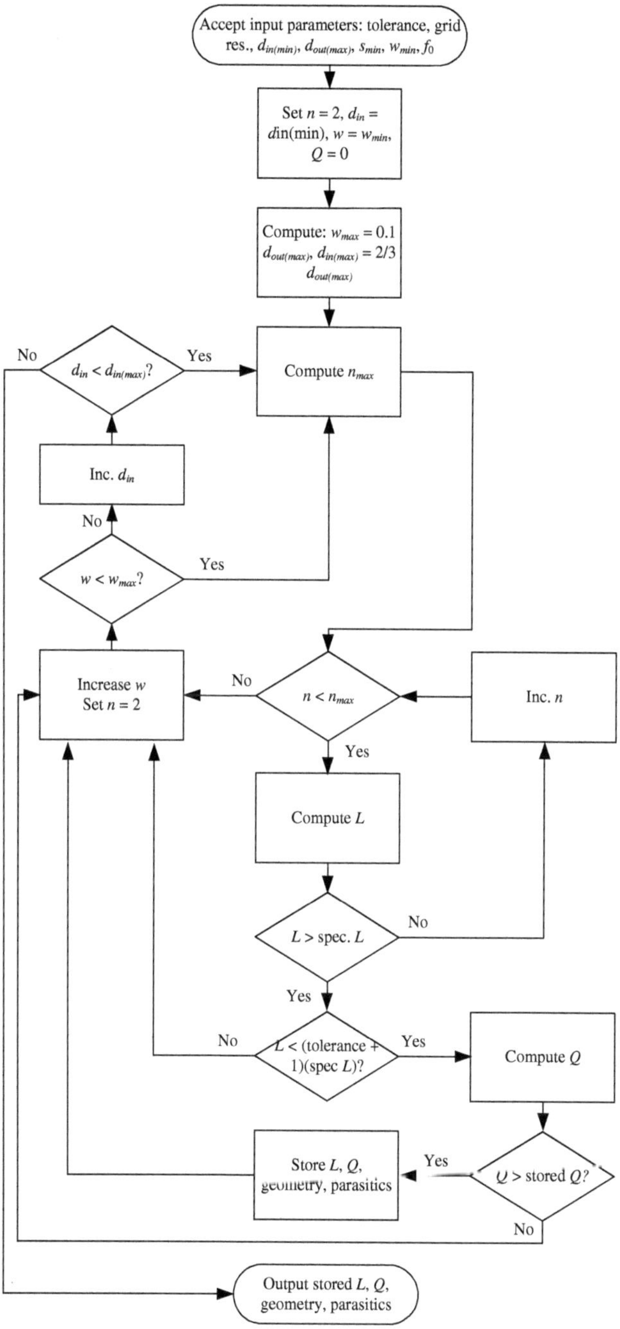

Fig. 7.4 Spiral inductor inductance search algorithm

Table 7.5 Comparison of computational efforts and trade-offs of different grid resolution and tolerance settings for the synthesis of a 2 nH inductor

Grid	Tolerance	0.1 %		0.5 %		1 %		5 %	
(μm)	System	Time (s)	Q-factor	Time (s)	Q-factor	Time (s)	Q-factor	Time (s)	Q-factor
0.1	Core2duo	147	6.82	147	6.82	147	6.82	151	6.82
	i7	55.6		55.7		55.5		56.7	
0.2	Core2duo	36.7	6.82	36.7	6.82	36.9	6.82	36.8	6.82
	i7	14.4		14.7		14.6		14.8	
0.5	Core2duo	6.01	6.82	5.98	6.82	5.99	6.82	5.97	6.82
	i7	3.62		3.08		2.81		2.79	
1	Core2duo	1.54	6.78	1.57	6.81	1.58	6.81	1.54	6.81
	i7	1.17		1.09		1.19		1.19	
2	Core2duo	–	None	0.435	6.81	0.435	6.81	0.438	6.81
	i7	–		0.483		0.478		0.521	
5	Core2duo	–	None	0.116	4.82	0.121	6.78	0.112	6.78
	i7	–		0.181		0.178		1.175	

consists of three nested loops: in the innermost loop, n is varied. In the middle loop, w is varied and in the outermost loop, d_{in} is varied, all within specified parameters.

To illustrate the computational effort of the search algorithm, a comparison of the execution time for the synthesis of a typical 2 nH inductor in the ams AG 0.35 μm BiCMOS S35 process, for various tolerances and grid resolutions on two different systems, is presented in Table 7.5. From this comparison the higher grid resolution (in this case a resolution higher than 1 μm) does not add to the quality of synthesized inductors and thus a lower grid resolution can be used, bringing the time consumed for the inductor synthesis to an acceptable level even for an older computer system.

The analysis part of the inductor design routine is a straightforward calculation, given the design parameters. If we consolidate both the synthesis and analysis parts of the inductor design routine, the flow chart may look like the one in Fig. 7.5.

7.1.2.4 MATLAB Code

Next we present MATLAB code for some of the important parts of the inductor design program:

- The code for the inductance search algorithm is provided in Fig. 7.6,
- Figure 7.7 lists the code for setup of the inductance search algorithm,
- Figure 7.8 lists the code for entering process parameters,
- Figure 7.9 lists the code for calculating the parasitics of the spiral inductor,
- Figure 7.10 lists the code for the analysis of spiral inductors with known geometry, and
- Figure 7.11 lists the code for the main spiral inductor design program, which also includes the procedure for netlist extraction and export of the inductor layout, procedures that will be discussed later.

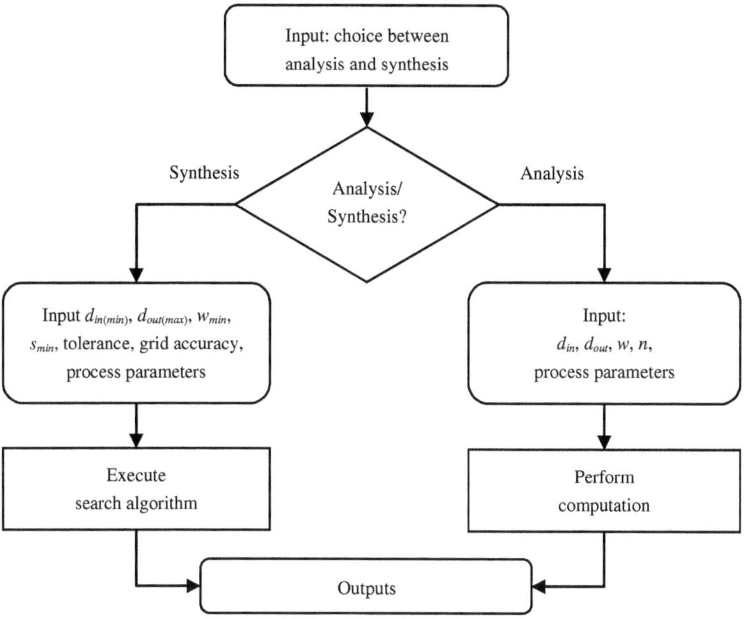

Fig. 7.5 Flow diagram of the inductor design routine complete with analysis and synthesis parts

7.1.2.5 Design Outputs

Finally, we can summarize the inductor design routine outputs that we find useful for power amplifier design. This is done in Table 7.6.

Simple programming techniques may be used to interpret numerical results to export the SPICE netlist. The SPICE netlist is complete with the inductance value and parasitics calculated for a single-π inductor model are saved with a .spc extension. This can be used in SPICE simulations without the need for a designer to draw the schematic of the inductor with its parasitics in the schematic editor. The netlist is exported in T-Spice format, which is recognized by most simulators, including Cadence Virtuoso. The exact spice format is used for illustration purposes, but the routine can easily be modified for another netlist format. The extension of the routine that handles SPICE netlist export is shown in Fig. 7.12.

Exporting the layout of the designed inductor structure in GDSII[1] format can also be beneficial and will be demonstrated in Chap. 9 when dealing with layout and packaging.

[1]GDS stands for Graphic Database System.

```
%This procedure searches for the inductance geometry with the
%highest quality factor given the inductance
%Initialize all storage variables to zero
Qstored = 0; fostored = 0; Lcstored =0; Rsstored = 0; RSistored =
0; CSistored = 0; Coxstored = 0; Csstored = 0; wstored = 0;
sstored = 0; dinstored = 0; doutstored = 0; nstored = 0;
fprintf('\nLooking for geometry with highest Q-factor...\n\n');
%Initialize geometry parameters to default minimum/maximum values
Lc = 0;
dout = 0;
s = smin;
din = dinmin;
w = wmin;
n = 2;
%Inductance search algorithm
while (din < 2*doutmax/3)
  s = smin;
  w = wmin;
  while (w <= doutmax/10)
    n = 2;
    dout = 0;
    while (dout < doutmax)
      dout = din + 2*n*w + 2*(n-1)*s;
      if (dout > doutmax)
        break
      end%if
      davg = (din + dout) / 2;
      Lc = b * dout^a1 * w^a2 * davg^a3 * n^a4 * s^a5;
      calcParasitics; %Procedure to calculate parasitics
      Lcc = Lc/1e9;
      Lzz = Lz*1e9;
      if (Lzz > Ls)
         if (Lzz < (1 + tolerance) * Ls)
           %Calculate Q-factor
           Rp = 1/(omega^2*Cox^2*RSi) + RSi*(Cox + CSi)^2/Cox^2;
           Cp = Cox*(1 + omega^2*(Cox + CSi)*CSi*RSi^2)/(1 +
omega^2*(Cox + CSi)^2*RSi^2);
           Q = omega*Lcc/Rs*Rp/(Rp + ((omega*Lcc/Rs)^2 + 1)*Rs)*(1
- (Cp + Cs)*(omega^2*Lcc + Rs^2/Lcc));
           fo = 1/(2*pi)*sqrt(1/(Lcc*(Cp + Cs)) - (Rs/Lcc)^2);
           if (Q > Qstored)
             Qstored = Q; fostored = fo; Lclfstored = Lc;
             Lcstored = Lzz; Rsstored = Rs; RSistored = RSi;
             CSistored = CSi; Coxstored = Cox; Csstored = Cs;
             wstored = w; sstored = s; dinstored = din;
             doutstored = dout; nstored = n;
           end%if
         end%if
         Lc = 0;
         n = 1;
         break
       end%if
       n = n + 1;
     end%while
     w = w + resolution;
   end%while
  din = din + resolution;
end%while
%==== OUTPUT PARAMS ====%
if Qstored < 1
    fprintf('Could not find a geometry for %.2f nH\nlimited to
dinmin and doutmax with Q greater than 1 at %.2f MHz.\n', Ls,
f/1e6)
end%if
if Qstored >=1
  fprintf('Ls = %.2f nH \n', Lcstored);
  fprintf('Lslf = %.2f nH \n', Lclfstored);
  fprintf('Q = %.2f \n', floor(100 * Qstored + 0.5) / 100);
  fprintf('fo = %.2f GHz\n', floor(fostored/1e7 + 0.5) / 100);
  fprintf('w = %.2f um\n', floor(100*wstored + 0.5) / 100);
  fprintf('s = %.2f um\n', floor(100*sstored + 0.5) / 100);
  fprintf('din = %.2f um\n', floor(100*dinstored + 0.5) / 100);
  fprintf('dout = %.2f um\n', floor(100*doutstored + 0.5) / 100);
  fprintf('n = %d\n', floor(100*nstored + 0.5) / 100);
end%if
```

Fig. 7.6 MATLAB code for the inductance search algorithm (indSearch.m)

```
%Setup the inductance search algorithm search parameters
tolerance = input('Please enter the tolerance for the inductance
value (%) (-1 to skip): ');
if (tolerance <= 0)
    tolerance = 0.01;
else
    tolerance = tolerance/100;
end
resolution = input('Please enter the search grid resolution (um)
(-1 to skip): ');
if (resolution <= 0)
    resolution = 1;
end
%Setup the geometry parameters
cminpar = input('Do you want to change any of the default geometry
parameters (y/n)? ', 's');
if cminpar == 'y' || cminpar == 'Y'
  dinmin = input('\nPlease enter the minimum value for din (um) (-
1 to skip): ');
  if (dinmin < 30)
    dinmin = 30;
    fprintf('Using the default value of %.0f um for minimum
din.\n', dinmin);
  end%if
  doutmax = input('Please enter the maximum value for dout (um) (-
1 to skip): ');
  if (doutmax <= 0)
    doutmax = 500;
    fprintf('Using the default value of %.0f um for maximum
dout.\n', doutmax);
  end%if
  smin = input('Please enter the minimum value for turn spacing s
(um) (-1 to skip): ');
  if (smin <= 0)
    smin = 2;
    fprintf('Using the default value of %.0f um for s.\n', smin);
  end%if
  wmin = input('Please enter the minimum turn width w (um) (-1 to
skip): ');
  if (wmin <= 0)
    wmin = 2;
    fprintf('Using the default value of %.0f um for w.\n', wmin);
  end%if
else
    %Default parameters
  dinmin = 30;
  doutmax = 500;
  smin = 2;
  wmin = 2;
  fprintf('Using the default values of %.1f, %.1f, %.1f and %.1f
for dinmin, doutmax, s and wmin respectively.\n', dinmin, doutmax,
smin, wmin);
end%if
```

Fig. 7.7 MATLAB code for the setup of inductance search algorithm parameters (indSetup.m)

```
%Enter process parameters
%Change default process parameters - any process can be used
%Enter default process parameters here
TM3default = 1000; rhodefault = 2.82e-8;
MAldefault = 1.257e-6; ToxM2M3default = 1000;
E = 8.85e-12; EroxM2M3default = 4;
ToxM3Sdefault = 5000; EroxM3Sdefault = 4;
TSidefault = 1000; ErSidefault = 11.7;
rhoSidefault = 0.2;

ctech = input('Do you want to change any of the default process
parameters (y/n)? ', 's');
if ctech == 'y' || ctech == 'Y'
    fprintf('\nPlease enter the thickness of the top metal (nm) (-
1 to skip): \nOld value: %.0f  -->  ', TM3default);
    TM3 = input('New value: ');
    if (TM3 <= 0)
      TM3 = TM3default * 1e-9;
    else
      TM3 = TM3 * 1e-9;
    end%if
    fprintf('Please enter rho (ohm.m) (-1 to skip): \nOld value:
%.2e  -->  ', rhodefault);
    rho = input('New value: ');
    if (rho <=0)
      rho = rhodefault;
    end%if
    fprintf('Please enter MAl (H/m) (-1 to skip): \nOld value:
%.2e -->  ', MAldefault);
    MAl = input('New value: ');
    if (MAl <= 0)
      MAl = MAldefault;
    end%if
    delta = sqrt(rho / (pi * MAl * f));
    teff = delta*(1-exp(-TM3/delta));
    fprintf('Please enter thickness of the oxide between top two
metals (nm) (-1 to skip): \nOld value: %.0f  -->  ',
ToxM2M3default);
    ToxM2M3 = input('New value: ');
    if (ToxM2M3 <= 0)
      ToxM2M3 = ToxM2M3default * 1e-9;
    else
      ToxM2M3 = ToxM2M3 * 1e-9;
    end%if
    fprintf('Please enter Er of the oxide between top two metals
(-1 to skip): \nOld value: %.2f  -->  ', EroxM2M3default);
    EroxM2M3 = input('New value: ');
    if (EroxM2M3 <= 0)
      EroxM2M3 = EroxM2M3default;
    end%if
    EoxM2M3 = E * EroxM2M3;
     fprintf('Please enter thickness of the oxide between the sub-
strate and the top metal (nm) (-1 to skip): \nOld value: %.0f  -->
', ToxM3Sdefault);
    ToxM3S = input('New value: ');
```

Fig. 7.8 MATLAB code for entering process parameters (entProcParam.m)

```
    if (ToxM3S <= 0)
        ToxM3S = ToxM3Sdefault * 1e-9;
    else
      ToxM3S = ToxM3S * 1e-9;
    end%if
    fprintf('Please enter Er of the oxide between the substrate
and the top metal (-1 to skip): \nOld value: %.2f  -->  ', Er-
oxM3Sdefault);
    EroxM3S = input('New value: ');
    if (EroxM3S <= 0)
      EroxM3S = EroxM3Sdefault;
    end%if
    fprintf('Please enter TSi (um) (-1 to skip): \nOld value: %.0f
-->  ',TSidefault);
    TSi = input('New value: ');
    if (TSi <= 0)
      TSi = TSidefault * 1e-6;
    else
      TSi = TSi*1e-6;
    end%if
    fprintf('Please enter ErSi (-1 to skip): \nOld value: %.2f  --
>  ', ErSidefault);
    ErSi = input('New value: ');
    if (ErSi <= 0)
      ErSi = ErSidefault;
    end%if
    %Rsi
    fprintf('Please enter rhoSi (ohm.m) (-1 to skip): \nOld value:
%.2f  -->  ', rhoSidefault);
    rhoSi = input('New value: ');
    if (rhoSi <= 0)
      rhoSi = rhoSidefault;
    end%if
else
    %If the parameters are not changed, use default parameters
    TM3 = TM3default * 1e-9;
    ToxM3S = ToxM3Sdefault * 1e-9;
    rho = rhodefault;
    MAl = MAldefault;
    delta = sqrt(rho / (pi * MAl * f));
    teff = delta*(1-exp(-TM3/delta));
    ToxM2M3 = ToxM2M3default * 1e-9;
    EroxM2M3 = EroxM2M3default;
    EoxM2M3 = E * EroxM2M3;
    EroxM3S = EroxM3Sdefault;
    TSi = TSidefault * 1e-6;
    ErSi = ErSidefault;
    rhoSi = rhoSidefault;
end%if
fprintf('\n');
```

Fig. 7.8 (continued)

```
%This procedure calculates the parasitics of spiral inductors
ell = 4*n*(din+w) + (2*n)*(2*n-1) * (w+s);
Rs = rho * ell / (w * teff);
ws = w / 1e6;
ells = ell / 1e6;
Cs = n * ws * ws * EoxM2M3 / ToxM2M3;
if (ToxM3S <= ws)
   ToxM1Seff = ws * (ws/ToxM3S + 2.42 - 0.44*ToxM3S/ws + (1 -
ToxM3S/ws)^6)^(-1);
else
   ToxM1Seff = ws/(2*pi) * log(8*ToxM3S/ws + 4*ws/ToxM3S);
end%if
EoxM1Seff = E *((ExroxM3S + 1)/2 + (ExroxM3S - 1)/2*(1 +
10*ToxM3S/ws)^(-1/2));
Cox = ells * ws * EoxM1Seff / ToxM1Seff / 2;
TSieffC = ws/(2*pi) * log(8*TSi/ws + 4*ws/TSi);
ESieff = E *((ErSi + 1)/2 + (ErSi - 1)/2*(1 + 10*TSi/ws)^(-1/2));
CSi = ells * ws * ESieff / TSieffC / 2;
sigmaeff = 1/rhoSi * (1/2 + 1/2*(1 + 10*TSi/ws)^(-1/2));
RSi = 2 * TSieffC / (ells * ws * sigmaeff);

%Effective Ls
%Complex number calculations
% i = sqrt(-1)
ZSi = 1/(i*omega*CSi);
Zox = 1/(i*omega*Cox);
Zs = 1/(i*omega*Cs);
Lss = Lc / 1e9;
ZLs = i*omega*Lss;
ZA = Zox + ZSi * RSi / (ZSi + RSi);
ZB = Zs * ( ZLs + Rs) / (Zs + ZLs + Rs);
Z = ZA * ZB / (ZA + ZB);
R = real(Z);
Lz = imag(Z) / omega;
```

Fig. 7.9 MATLAB code for calculation of the parasitics of any designed spiral inductor (calcParasitics.m)

7.1.3 MEMS Inductors

As discussed previously, the idea behind MEMS is that the fabrication process is modified in some way to decrease the influence of the process parasitics and thus allow the inductor to mimic one of the off-chip inductors. If the MEMS inductor is still a spiral inductor, then the same design procedure listed in the previous section can be applied, but taking care to analyze which parasitics will have decreased influence in response to the process change.

For example, consider the case where the oxide below the spiral is removed. This means that the formula for C_{ox} will change to resemble lower capacitance because of the relative permittivity changes (silicon-dioxide is effectively replaced by air). The capacitance between the spiral and the underpass is likely to change as well. In this case, all it takes is change in one process parameter.

```
%This procedure calculates the inductance and quality factor values
for any
%given square inductor geometry
%Input geometry parameters
dout = input('Please enter the dout diameter (dout) (um): ');
din = input('Please enter the din diameter (din) (um): ');
w = input('Please enter the turn width (w) (um): ');
n = input('Please enter the number of turns: ');
ratio = floor(100*(din / dout) + 0.5)/100;
%Perform basic checks
nmin = (dout/2 - din/2 + 1)/(w + 1);
while (nmin < 2) || (n > nmin) || (n < 2)
    if (nmin <2)
        fprintf('Given geometry results in n smaller than 2. Please
re-enter the geometry: \n');
    end%if
    if (n > nmin)
        fprintf('The number of turns (n) is too small for the chosen
geometry. Please re-enter the geometry: \n');
    end%if
    if (n < 2)
        fprintf('Cannot use less than two turns. Please choose n as 2
or greater. Please re-enter the geometry: \n');
    end%if
    dout = input('Please enter the dout diameter (dout) (um): ');
    din = input('Please enter the din diameter (din) (um): ');
    w = input('Please enter the turn width (w) (um): ');
    n = input('Please enter the number of turns: ');
    ratio = floor(100*(din / dout) + 0.5)/100;
    nmin = (dout/2 - din/2 + 1)/(w + 1);
end%while
%Enter operating frequency
f = input('Please enter the operating frequency (MHz): ')*1e6;
omega = (2*pi*f); %Radian frequency
s = (dout / 2 - din / 2 - n * w) / (n - 1); %Pitch between the turns
of the spiral
davg = (din + dout) / 2; %Average diameter
Lc = b * dout^a1 * w^a2 * davg^a3 * n^a4 * s^a5; %Inductance
entProcParam; %Call the procedure for entering process parameters
calcParasitics; %Call the procedure that calculates parasitics and Q
factor
Lcc = Lc/1e9;
Lzz = Lz*1e9;
%Calculate Q-factor
Rp = 1/(omega^2*Cox^2*RSi) + RSi*(Cox + CSi)^2/Cox^2;
Cp  =  Cox*(1  +  omega^2*(Cox  +  CSi)*CSi*RSi^2)/(1  +  omega^2*(Cox  +
CSi)^2*RSi^2);
Q  =  omega*Lcc/Rs*Rp/(Rp  +  ((omega*Lcc/Rs)^2  +  1)*Rs)*(1  -  (Cp  +
Cs)*(omega^2*Lcc + Rs^2/Lcc));
fo = 1/(2*pi)*sqrt(1/(Lcc*(Cp + Cs)) - (Rs/Lcc)^2);
%Outputs
fprintf('Ls = %.2f nH \n', Lzz);
fprintf('Lslf = %.2f nH \n', Lc);
fprintf('Q = %.2f \n', floor(100 * Q + 0.5) / 100);
fprintf('fo = %.2f GHz\n', floor(fo/1e7 + 0.5) / 100);
fprintf('w = %.2f um\n', floor(100*w + 0.5) / 100);
fprintf('s = %.2f um\n', floor(100*s + 0.5) / 100);
fprintf('din = %.2f um\n', floor(100*din + 0.5) / 100);
fprintf('dout = %.2f um\n', floor(100*dout + 0.5) / 100);
fprintf('n = %d\n', floor(100*n + 0.5) / 100);
%Store the parasitic parameters for netlist extraction
Lcstored = Lzz; Lclfstored = Lc; Rsstored = Rs; RSistored = RSi;
CSistored = CSi; Coxstored = Cox; Csstored = Cs; nstored = n;
doutstored = dout; wstored = w; sstored = s;
```

Fig. 7.10 MATLAB code for analysis of spiral inductors with known geometry (indGeom.m)

```
%This program finds a geometry for given inductance or calculates
the inductance for given geometry. GDS and CIF file extraction is
included as well as netlist extraction.
try
    clear all; %Clear all variables from the workspace
    fprintf('\nWelcome to Inductor Calculator v2.2 for Matlab.
Please follow the prompts.\n');    %Define constants
    b = 1.62e-3; a1 = -1.21; a2 = -0.147; a3 = 2.40; a4 = 1.78;
    a5 = -0.030;
    choice = input('\nPlease enter 1 to specify geometry or 2 to
specify inductance: ');
    if choice == 1 || choice == 2
      if choice == 2 %If choice is 2, program finds a geometry
for given inductance
          Ls = input('Please enter the wanted inductance value
(Ls) (nH): '); %Wanted inductance
          indSetup;
          fprintf('\n');
          f = input('Please enter the operating frequency (MHz):
')*1e6; %Frequency
          omega = (2*pi*f); %Radian frequency
          entProcParam; %Call the procedure for entering process
parameters
          indSearch; %Inductance search algorithm
      end%if
      if choice == 1
          indGeom;
      end%if
      %if choice == 2 && Qstored > 1;
          %Export the netlist
          cnet = input('\nDo you want to export the netlist of
this inductor into a SPC file (y/n)? ', 's');
          if cnet == 'y' || cnet == 'Y';
            exportNetlist;
          end%if
          %Export the layout
          cwcif = input('\nDo you want to export the layout of
this inductor into a CIF or GDS file (c - CIF, g - GDS, n -
none)? ', 's');
          if cwcif == 'c' || cwcif == 'C';
            writecif; %Export layout into a text (CIF) file
          elseif cwcif == 'g' || cwcif == 'G';
            convertGDS; %Export layout into a stream (GDS) file
          end
          fprintf('\n');
      %end
    else
      fprintf('Wrong choice. Please run the program again.\n\n');
    end%if
catch
    fprintf('MATLAB returns: \n%s\nError occurred while running
the program. Please try running the program again.\n\n', las-
terr);
end%try
```

Fig. 7.11 MATLAB code for the main spiral inductor design program (indcalc.m)

Table 7.6 Summary of inductor design routine outputs

Parameter	Unit
Effective inductance value of the inductor at the operating frequency, L_S	nH
Nominal inductance value of the inductor ($Q \to \infty$), L_{inf}	nH
Q-factor of the inductor at the operating frequency, Q	–
Resonant frequency of the inductor, f_r	GHz
Width of the spiral	μm
Spacing between the turns of the spiral	μm
Input diameter of the spiral	μm
Output diameter of the spiral	μm
Number of turns of the spiral	–

```
%This procedure exports the netlist of a spiral inductor
try
    inductorName = input('Please enter the inductor name (press
"s" to skip): ', 's');
    if inductorName(1) == 's' || inductorName(1) == 'S';
        inductorName = 'MyInductor';
    end%if
    fileName = strcat(inductorName, '_net.spc');
    fid = fopen(fileName, 'w')    ;
    fprintf(fid, '*SUBCKT gen. by Inductor Calculator v2.2\n');
    fprintf(fid, '.SUBCKT %s L1 L2 GND\n', inductorName);
    fprintf(fid, 'CS L1 L2 %.2ffF\n', floor(1e17 * Csstored + 0.5)
/ 100);
    fprintf(fid, 'LS N4 L1 %.2fn\n', floor(100 * Lclfstored + 0.5)
/ 100);
    fprintf(fid, 'RS N4 L2 %.2f\n', floor(100 * Rsstored + 0.5) /
100);
    fprintf(fid, 'Csi1 N2 GND %.2ffF\n', floor(1e17 * CSistored +
0.5) / 100);
    fprintf(fid, 'Csi2 N3 GND %.2ffF\n', floor(1e17 * CSistored +
0.5) / 100);
    fprintf(fid, 'Cox1 L1 N2 %.2ffF\n', floor(1e17 * Coxstored +
0.5) / 100);
    fprintf(fid, 'Cox2 L2 N3 %.2ffF\n', floor(1e17 * Coxstored +
0.5) / 100);
    fprintf(fid, 'Rsi1 N2 GND %.2f\n', floor(100 * RSistored +
0.5) / 100);
    fprintf(fid, 'Rsi2 N3 GND %.2f\n', floor(100 * RSistored +
0.5) / 100);
    fprintf(fid, '.ENDS\n\n');
    fclose(fid);
    fprintf('File %s created successfully.\n', fileName);
catch
    fprintf('MATLAB reports:\n%s\n', lasterr);
    fprintf('Netlist export failed.\n');
end%try
```

Fig. 7.12 MATLAB code for inductor netlist export (exportNetlist.m)

Now consider the case where substrate removal is implemented. This could potentially result in a simpler model for an inductor because R_{sub} and C_{sub} components can be ignored. This would allow simplification of the inductor design program by removing adequate calculations.

The expected result in either case is an increased Q-factor of the inductor.

7.1.4 Integrated Transformers

The inductor design program can be expanded to include spiral transformer design. The extent of the required modification of the program will depend on whether the one- or two-layer configurations from Chap. 5 are used.

In the case of the interleaved one-layer transformer configuration (Fig. 5.24a), the capacitance for each spiral will remain unchanged, but the inductance of each spiral will change. However, since this configuration is symmetric, both spirals will have the same inductance and Q-factor, making this configuration practical for implementing a 1:1 transformer.

In the case of two-layer configuration (Fig. 5.24b), four metal layers are needed to include the two underpasses. The prominent capacitance in the bottom spiral will be between the spiral itself and the substrate, and in the top spiral it will be between the top spiral and bottom spirals, so the capacitance equations will need to be changed accordingly. Typically, the metals used for each spiral could also differ in thickness, so the resistance of the two spirals will be different as well. Since all the parasitics of each spiral differ, the apparent value of inductance will differ as well. Thus, this configuration is effectively a 1:*n* transformer.

7.1.5 Verification of the Inductor Model and the Search Algorithm

To gain some confidence in the inductor design program and the inductance search algorithm, we have designed some inductors in ams AG 0.35 μm BiCMOS S35 process in different metal layers. Ten third-metal (3M) and ten thick-metal (TM) inductors fabricated over a standard resistivity substrate at common frequencies of 1, 2, 2.4 and 5 GHz were designed. The starting inductor value designed for was 0.5 nH, followed by nine inductors in increments of 0.5 nH. Tables 7.7 and 7.8 show geometric parameters, low-frequency inductance values and the Q-factor of each designed 3M and TM inductor respectively. To verify the predicted values, electromagnetic (EM) simulation on the designed inductors was performed in Virtuoso Spiral Inductor Modeler [2]. This modeler invokes electro-static and magneto-static EM solvers separately to extract the capacitive and inductive parameters of the spiral inductor structure, if a process file with information on metal

Table 7.7 Third-metal inductors designed with inductance search algorithm

f_0 (GHz)	L (nH)	Calc. L (nH)	Calc. Q	EM L (nH)	EM Q	d_{out} (μm)	d_{in} (μm)	w (μm)	s (μm)	n
1	0.5	0.50	3.37	0.50	3.00	220	30	47	1	2
1	1	1.00	4.63	0.94	4.15	291	93	49	1	2
1	1.5	1.48	5.22	1.40	4.75	347	149	49	1	2
1	2	1.95	5.47	1.90	5.11	397	199	49	1	2
1	2.5	2.39	5.50	2.33	2.33	441	243	49	1	2
1	3	2.83	5.39	2.75	5.27	483	285	49	1	2
1	3.5	3.29	5.10	3.22	3.22	500	326	43	1	2
1	4	3.67	4.80	3.52	4.52	426	164	43	1	3
1	4.5	4.12	4.66	3.97	4.56	427	189	39	1	3
1	5	4.55	4.60	4.52	4.41	431	211	36	1	3
2	0.5	0.49	5.79	0.45	4.29	210	32	55	1	2
2	1	0.97	7.16	0.91	5.92	288	90	49	1	2
2	1.5	1.41	6.89	1.32	6.11	339	141	49	1	2
2	2	1.86	6.24	1.73	6.23	349	191	39	1	2
2	2.5	2.32	5.64	2.26	5.68	363	233	32	1	2
2	3	2.76	5.12	2.72	5.27	384	270	28	1	2
2	3.5	3.21	4.81	3.01	5.33	276	152	20	1	3
2	4	3.65	4.53	3.45	5.19	283	171	18	1	3
2	4.5	4.06	4.27	3.86	4.98	294	188	17	1	3
2	5	4.51	4.05	4.35	4.76	241	123	14	1	4
2.4	0.5	0.49	6.72	0.45	4.94	216	30	46	1	2
2.4	1	0.95	7.63	0.90	6.38	286	88	49	1	2
2.4	1.5	1.39	6.95	1.34	6.34	316	142	43	1	2
2.4	2	1.86	6.18	1.82	6.25	320	190	32	1	2
2.4	2.5	2.30	5.51	2.26	5.47	339	229	27	1	2
2.4	3	2.76	5.04	2.71	5.32	249	131	19	1	3
2.4	3.5	3.17	4.69	3.12	4.89	262	150	18	1	3
2.4	4	3.61	4.39	3.59	4.74	262	168	15	1	3
2.4	4.5	4.03	4.12	4.03	4.60	220	110	13	1	4
2.4	5	4.47	3.91	4.50	4.55	216	122	11	1	4
5	0.5	0.47	9.48	0.41	6.05	200	30	42	1	2
5	1	0.92	8.22	0.88	6.93	209	95	28	1	2
5	1.5	1.36	6.76	1.34	5.88	222	140	20	1	2
5	2	1.81	5.71	1.80	5.48	169	87	13	1	3
5	2.5	2.22	5.03	2.20	5.15	176	106	11	1	3
5	3	2.60	4.46	2.59	4.45	186	122	10	1	3
5	3.5	2.97	4.02	2.96	4.03	160	82	9	1	4
5	4	3.46	3.70	3.52	4.43	148	94	6	1	4
5	4.5	3.98	3.43	4.07	4.30	141	103	4	1	4
5	5	4.29	3.27	4.42	4.24	106	38	4	1	7

Table 7.8 Thick-metal inductors designed with inductance search algorithm

f_0 (GHz)	L (nH)	Calc. L (nH)	Calc. Q	EM L (nH)	EM Q	d_{out} (μm)	d_{in} (μm)	w (μm)	s (μm)	n
1	0.5	0.50	7.43	0.38	4.71	216	30	48	2	2
1	1	0.99	10.1	0.93	8.13	299	95	50	2	2
1	1.5	1.47	11.1	1.39	9.96	355	151	50	2	2
1	2	1.95	11.5	1.84	11.1	406	202	50	2	2
1	2.5	2.39	11.4	2.30	11.7	451	247	50	2	2
1	3	2.81	10.9	2.70	12.0	493	289	50	2	2
1	3.5	3.29	10.3	3.20	11.6	499	331	41	2	2
1	4	3.71	9.78	3.49	9.62	403	173	37	2	3
1	4.5	4.17	9.49	3.95	9.89	409	197	34	2	3
1	5	4.60	9.21	4.40	9.99	418	218	32	2	3
2	0.5	0.49	11.4	0.43	9.17	222	30	47	2	2
2	1	0.96	13.0	0.88	13.4	295	91	50	2	2
2	1.5	1.42	11.9	1.35	14.6	316	148	41	2	2
2	2	1.88	10.8	1.82	14.2	324	196	31	2	2
2	2.5	2.33	9.80	2.27	13.8	348	236	37	2	2
2	3	2.79	9.08	2.66	11.6	268	134	21	2	3
2	3.5	3.21	8.55	3.10	11.6	276	154	19	2	3
2	4	3.64	8.05	3.54	11.6	283	173	17	2	3
2	4.5	4.09	7.59	3.96	11.4	294	190	16	2	3
2	5	4.49	7.25	4.39	10.3	240	124	13	2	4
2.4	0.5	0.49	12.4	0.43	10.7	219	31	46	2	2
2.4	1	0.94	13.0	0.86	14.4	286	90	48	2	2
2.4	1.5	1.41	11.6	1.36	15.3	289	149	34	2	2
2.4	2	1.88	10.3	1.83	14.7	302	194	26	2	2
2.4	2.5	2.34	9.25	2.24	11.8	229	113	18	2	3
2.4	3	2.78	8.61	2.69	12.0	238	134	16	2	3
2.4	3.5	3.18	8.01	3.08	12.0	256	152	16	2	3
2.4	4	3.63	7.47	3.56	11.7	256	170	13	2	3
2.4	4.5	4.54	7.04	4.00	10.6	213	113	11	2	4
2.4	5	4.46	6.67	4.36	10.6	223	123	11	2	4
5	0.5	0.47	14.4	0.41	17.3	200	32	41	2	2
5	1	0.93	12.1	0.89	18.9	199	99	24	2	2
5	1.5	1.38	9.83	1.35	17.0	215	143	17	2	2
5	2	1.82	9.37	1.74	14.0	163	89	11	2	3
5	2.5	2.26	7.31	2.18	13.5	170	108	9	2	3
5	3	2.70	6.74	2.62	10.6	123	47	6	2	5
5	3.5	3.09	6.23	3.04	12.0	182	138	6	2	3
5	4	3.42	5.88	3.32	10.4	137	61	6	2	5
5	4.5	3.79	5.44	3.70	10.8	163	103	6	2	4
5	5	4.07	5.02	3.98	9.65	149	73	6	2	5

```
processType CMOS
numLayers 11

layer 0
      name tab
      type ground
      thickness 0.1e-6
      epsr 1
      rho 1.7857e-8
layer 1
      name substrate
      type dielectric
      thickness 725e-6
      epsr 11.7
      rho 0.19
layer 2
      name sio2
      type dielectric
      thickness 1e-6
      epsr 4.0
      rho 1e10
layer 3
      name metal1
      type metal
      thickness 1e-6
      epsr 4
      rho 1e10
      rhom 2.82e-8
layer 4
      name sio2
      type dielectric
      thickness 1e-6
      epsr 4
      rho 1e10
layer 5
      name metal2
      type metal
      thickness 1e-6
      epsr 4
      rho 1e10
      rhom 2.82e-8
layer 6
      name sio2
      type dielectric
      thickness 1e-6
      epsr 4
      rho 1e10
layer 7
      name metal3
      type metal
      thickness 1e-6
      epsr 4
      rho 1e10
      rhom 2.82e-8
layer 8
      name sio2
      type dielectric
      thickness 1e-6
      epsr 4
      rho 1e10
layer 9
      name metal4
      type metal
      thickness 1e-6
      epsr 4
      rho 1e10
      rhom 2.82e-8
layer 10
      name sio2
      type dielectric
      thickness 1e-6
      epsr 3.9
      rho 1e10
```

Fig. 7.13 An example process file for EM simulation

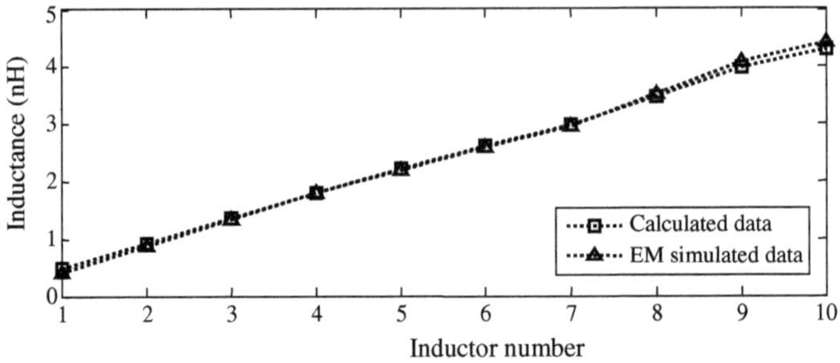

Fig. 7.14 Inductance of 5 GHz 3M inductors designed with inductance search algorithm and simulated by means of EM simulation

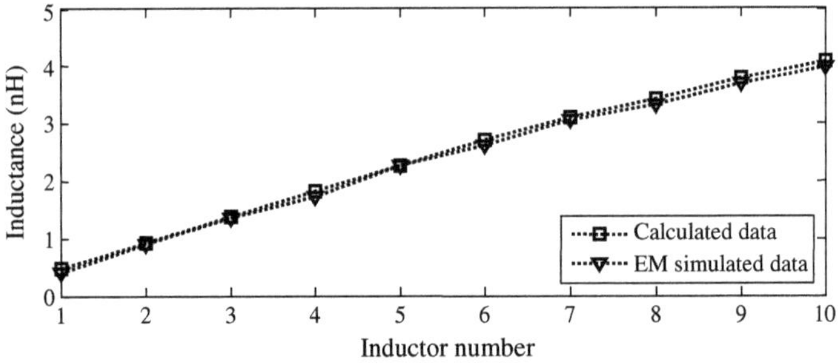

Fig. 7.15 Inductance of 5 GHz TM inductors designed with inductance search algorithm and simulated by means of EM simulation

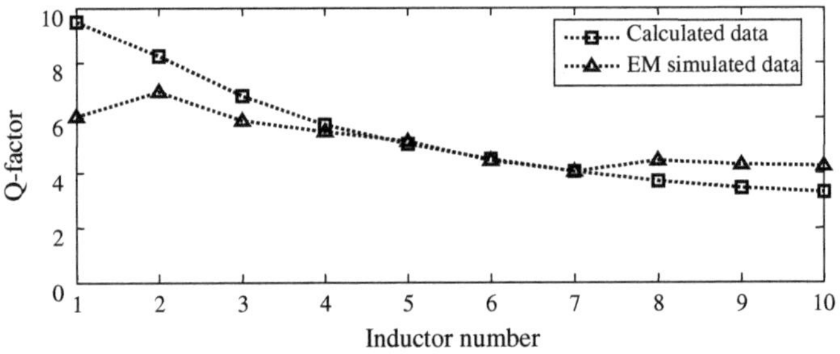

Fig. 7.16 Q-factor of 5 GHz 3M inductors designed with inductance search algorithm and simulated by means of EM simulation

and dielectric layers is provided (an example shown in Fig. 7.13). Figures 7.14, 7.15, 7.16 and 7.17 show a graphical comparison of designed and simulated inductances and Q-factors for 5 GHz inductors.

Figures show that inductance and Q-factor values obtained using the inductor design routine correspond well with simulated inductance values, with somewhat

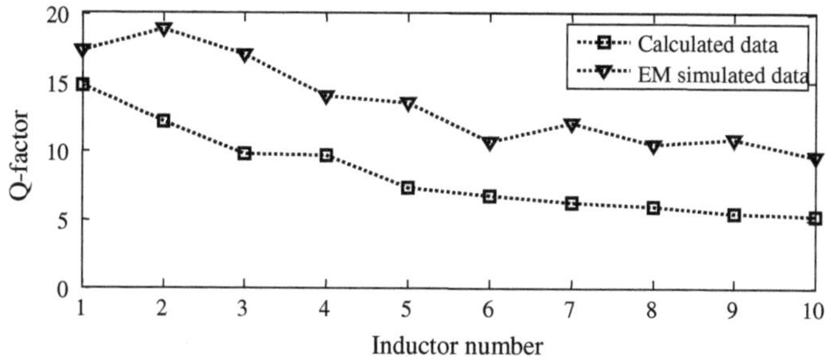

Fig. 7.17 Q-factor of 5 GHz TM inductors designed with inductance search algorithm and simulated by means of EM simulation

poorer correspondence of the Q-factor for TM inductors. This is explained by the fact that in the case of TM inductors, which have lower spiral resistance and less prominent influence of the substrate, as the impedance of parasitic elements in the *RL* model of the spiral (where we allowed oxide and substrate effects to be ignored) approaches that of the inductive reactance near the peak frequency, the nine-component model yields a pessimistic estimate of the actual Q-factor of the spiral [3]. The fact that the estimate of the Q-factor is pessimistic is advantageous, but where higher accuracy is needed, the use of one of the more detailed models presented in Chap. 5 may be explored.

Furthermore, the inductance routine was used to predict inductances and Q-factors of several spiral inductor geometries provided and measured by ams AG. The measurement results showed that inductance values are correctly predicted (within 3.7 %) by the inductor models used for the inductance search algorithm with Q-factors exhibiting the same behavior as shown by EM simulations. Details of this study can be found in [4].

7.2 Bond Wires

We can now use the same methodology that we used for spiral inductors to develop a design routine for the design of bond wires, which present a good alternative to spiral inductors when working with integrated circuits. The principle will remain the same, where we will first identify the determining equations, tabulate the required input parameters and develop analysis and synthesis routines for the specified inductance and maximum Q-factor. It is important to note here that the model of the bondwire will be much simpler than the nine-component model of the spiral inductor used previously, so the task of identifying proper equations will be much simpler.

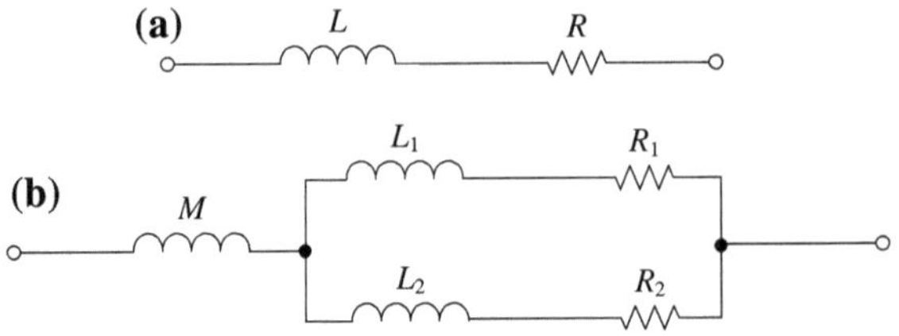

Fig. 7.18 Models of bond wire inductors: **a** single bond wire and **b** double bond wire

7.2.1 Model

As seen in Chap. 5, it is normally assumed that the bond wire is placed well above any conducting planes to avoid generation of any parasitic capacitances. Thus the bond wire inductor can typically be modeled by the inductor with a resistor in series, as shown in Fig. 7.18a. If two bond wires are used, then one can expand the model to include the mutual inductance, as shown in Fig. 7.18b. Note that in this case, two inductors, L_1 and L_2, and two parasitic resistors, R_1 and R_2, are in parallel, which means that although the total parasitic resistance for two wires will decrease, the same will happen with inductance. Thus if we were to use the model of Fig. 7.18a for a two-wire inductor the total inductance will be $L_{total} = M + L_1||L_2$ and $R_{total} = R_1||R_2$. Using this simplification over and over again, the same model can be used for any number of bond wires in parallel.

7.2.2 Equations

To demonstrate the concept, we will develop a program that analyzes and synthesizes bond wire inductors made up of one or two bond wires, where in the latter case the shapes of the two wires are identical. Thus, the following quantities need calculating:

- Inductances L and M, and
- Resistance R.

Required equations are tabulated in Table 7.9.

7.2.3 Input Parameters

Again, we can create two branches of the program, one for synthesis and the other one for analysis. For analysis of a bond wire inductor, the following geometry parameters are needed:

Table 7.9 Summary of equations needed for the design

Calculation	Equation
Bond wire inductance	$L = \frac{\mu_0 l}{2\pi}\left(\ln\frac{2l}{a} - 0.75\right)$
Bond wire mutual inductance (two wires)	$M = \frac{\mu_0 l}{2\pi}\left(\ln\frac{2l}{s} - \frac{s}{l} - 1\right)$
Wire resistance	$R = \frac{1}{2\pi a\delta\sigma}$
Skin depth	$\delta = \sqrt{\frac{\rho}{\pi\mu f}}$
Quality factor	$Q = \frac{X_L}{R}$
Reactance	$X_L = 2\pi f$

- Length of the wire, l,
- Radius of the wire, a, and
- Pitch between wires (for two-wire inductor), s.

For synthesis, we note that the radius of the wire is fixed, and that in the case of double bond wire, the spacing between bond wires can be fixed to the minimum allowed for maximum mutual coupling. So in this case, there is no need to search for the best inductor; straightforward calculations can be performed, given the wanted inductance value. Furthermore, the following process parameters are required:

- Permeability of air,
- Permeability of the wire metal, μ ($\mu_r \approx 1$), and
- Conductivity/resistivity of the wire metal, σ/ρ.

Finally, the frequency of operation is needed for skin depth calculation and Q-factor calculation. All required parameters are tabulated in Table 7.10.

Table 7.10 Input parameters for bond wire inductor design

Parameter	Typical value(s)	Units	Analysis/synthesis
Length of the wire	–	mm	Analysis
Radius of the wire	–	mm	Both
Pitch between wires	–	mm	Both
Wanted inductance	–	nH	Synthesis
Permeability of air	1.257×10^{-6}	H/m	Both
Permeability of metal	1.257×10^{-6}	H/m	Both
Resistivity of the metal	2.82×10^{-8}	Ωm	Both
Frequency of operation	–	GHz	Both

7.2.4 Development of the Inductor Design Routine

We can now use the same approach for the development of the routine as for the spiral inductor. The main program allows the user to choose between analysis and synthesis and is executed in a manner similar to Fig. 7.5. The analysis option is a straightforward calculation, given input parameters. Here the user can also be allowed to choose whether to use one wire or two wires placed closely together.

For synthesis, we calculate each one-wire and two-wire option. Note that in either case, we initially face a problem where we need to solve the inductance equation for l in terms of L, μ_0 and a, or L, M, μ_0, a and s, but the solution in either case is tedious to calculate manually. Thus we first use MATLAB `solve` command in command line to solve for l. For single bondwire, the command is:

```
>> solve('L = M0*ell/(2*pi)*(log(2*ell/a) - 0.75)*1e6', ell)
```

where `M0` is permeability and `log` is the natural logarithm. The solution is:

```
ell = 0.5*exp(lambertw(0.594e-5/M0/a*L) + 0.75)*a;
```

where `w = lambertw(x)` function is the solution to `w*exp(w) = x`.

For the double bond wire we have to solve $L/2 + M$ equation for l in a similar manner using MATLAB, to obtain

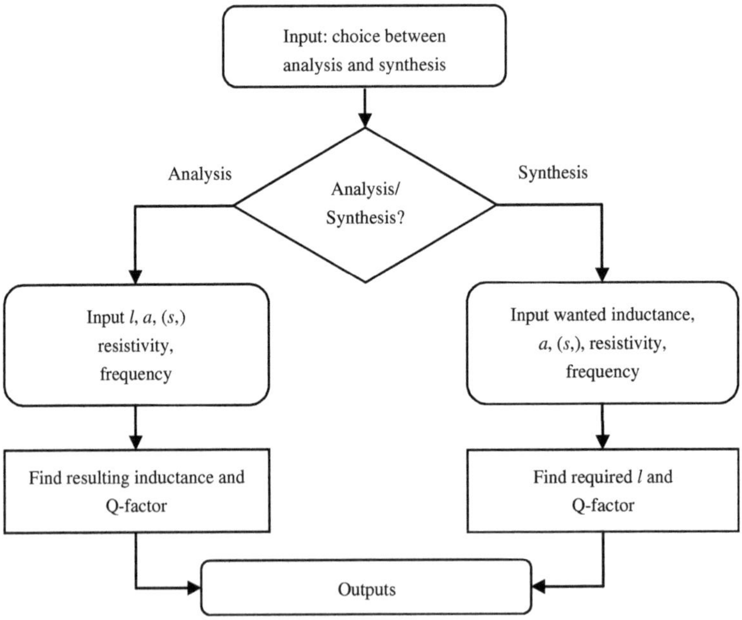

Fig. 7.19 Bond wire inductance design algorithm

```
%Choice between permeability or inductance index formula
%Setup constants
M0 = 1.257e-6;
rhodefault = 2.82e-8;

choice = input('\nPlease enter 1 for analysis and 2 for synthesis:
');
if choice == 1 || choice == 2
    if choice == 1
        ell = input('Please enter bond wire length (mm): ');
    end%if;
    a = input('Please enter bond wire radius (mm): ');
    s = input('Please enter spacing between two wires (mm) (-1 to
skip): ');
    rho = input('Please enter wire resistivity (ohm.m) (-1 to
skip): ');
    if rho < 0
        rho  = rhodefault;
        fprintf('Using the default value of %.2e ohm.m for resis-
tivity.\n', rho);
    end%if
    if choice == 1
        Lc = M0*ell/(2*pi)*(log(2*ell/a) - 0.75)*1e6;
        Lc2 = 0;
        if s > 0
            M = M0*ell/(2*pi)*(log(2*ell/s) - s/ell - 1)*1e6;
            Lc2 = Lc/2 + M;
            ell2 = ell;
        end%if
    else %if choice == 2
        L = input('Please enter the wanted inductance value (nH):
'); %Wanted inductance
        ell=0.5*exp(lambertw(0.594e-5/M0/a*L)+0.75)*a;
        Lc = L;
        if s > 0
            ell2 = 0.5*exp(lambertw(0.267e-
5/M0/s*(3.142*L+50.*M0*s)*exp(0.333*log(s/a)-0.917))-
0.333*log(s/a)+0.917)*s;
            Lc2 = L;
        end%if
    end%if;
    f0 = input('Please enter the operating frequency (GHz): ');
    f0 = f0*1e9;
    delta = sqrt(rho/(pi*M0*f0)); %skin depth
    R = rho*ell / (2*pi*a*delta); %resistance of one-wire option
    R2 = rho*ell2 / (2*pi*a*delta)/2; % Resitance of two-wire op-
tion
    Q = 2 * pi * f0 * Lc / R / 1e9; % 1e9 conversion from GHz to
Hz
    Q2 = 2 * pi * f0 * Lc2 / R2 / 1e9; % 1e9 conversion from GHz
to Hz

    fprintf('\n');
    fprintf('One-wire option: \n----------------------\n');
    fprintf('\n');
    fprintf('ell = %.2f mm\n', floor(100*ell + 0.5) / 100);
    fprintf('Lc = %.2f nH\n', floor(100*Lc + 0.5) / 100);
    fprintf('Rc = %.2f Ohm\n', floor(100*R + 0.5) / 100);
    fprintf('Q = %.0f\n', floor(100*Q + 0.5) / 100);
    fprintf('\n');
    if s > 0
        fprintf('Two-wire option: \n----------------------\n');
        fprintf('\n');
        fprintf('ell = %.2f mm\n', floor(100*ell2 + 0.5) / 100);
        fprintf('Lc = %.2f nH\n', floor(100*Lc2 + 0.5) / 100);
        fprintf('Rc = %.2f Ohm\n', floor(100*R2 + 0.5) / 100);
        fprintf('Q = %.0f\n', floor(100*Q2 + 0.5) / 100);
        fprintf('\n');
    end%if
else
   fprintf('Wrong choice. Please run the program again.\n\n');
end%if
```

Fig. 7.20 MATLAB code for bond wire search inductance (bondwire.m)

Table 7.11 Summary of inductor design routine outputs

Parameter	Unit
Nominal inductance value of the inductor ($Q \rightarrow \infty$), L_{inf}	nH
Q-factor of the inductor at the operating frequency, Q	–
Final length of the wire (one-wire and two-wire options)	mm
Final radius of the wire	mm
Spacing between two wires (two-wire option)	mm

```
ell2 = 0.5*exp(lambertw(0.267e-
5/M0/s*(3.142*L + 50.*M0*s)*exp(0.333*log(s/a) - 0.917))-
0.333*log(s/a) + 0.917)*s;
```

In this case, note that the fact that we are using MATLAB comes in very handy for complicated equation solving. If MATLAB is not available to the reader, the simplest way would be to solve the equation for l inside the inductor design program using one of the numerical methods described in [5]. Such an approach will be used in Chap. 8 to design quarter-wave transmission lines.

We then incorporate the two solutions for l in the code. The bond wire inductor design algorithm is illustrated in Fig. 7.19, and the sample code is shown Fig. 7.20.

7.2.5 *Design Outputs*

Once again, we summarize the inductor design routine outputs that we find useful for power amplifier design. This is done in Table 7.11.

7.3 Discrete Inductors

The design process of the discrete inductor is typically much simpler than that of the integrated inductors and bond wire inductors. As these inductors typically have higher Q-factors than integrated inductors, the simplest approach is to choose an

Table 7.12 Summary of equations for discrete inductor design

Calculation	Equation
Solenoidal inductor (inductance)	$L = \frac{0.394 r^2 N^2}{9r + 10l}$
Toroid inductor (inductance)	$L = \frac{0.004 \pi N^2 \mu_i A_c}{l_e}$
Toroid inductor (inductance in terms of inductance index)	$L = N^2 A_L$

```
%Choice between permeability or inductance index formula
choice = input('\nPlease enter 1 to use permeability formula and 2
to use inductance-index formula: ');
if choice == 1 || choice == 2
    if choice == 1
        Mi = input('\nPlease enter the initial permeability of the
core (-1 to skip): ');
        if (Mi < 0)
            Mi = 500; % Ferrite
            fprintf('Using the default value of 500 for initial
permeability .\n')
        end%if;
        le = input('\nPlease enter the effective length of the
core in cm (-1 to skip): ');
        if (le < 0)
            le = 1.5;
            fprintf('Using the default value of %.2f cm for effec-
tive length.\n', le);
        end%if;
        Ac = input('\nPlease enter the effective area of the core
in cm2 (-1 to skip): ');
        if (Ac < 0)
            Ac = 1;
            fprintf('Using the default value of %.2f cm2 for ef-
fective area.\n', le);
        end%if;
    else
        AL = input('\nPlease enter inductance index (nH/turn2) (-1
to skip): ');
        if (AL < 0)
            AL = 495;
            fprintf('Using the default value of %.2f nH/turn2 for
inductance index.\n', AL);
        end%if;
    end%if;
    L = input('Please enter the wanted inductance value (nH): ');
%Wanted inductance
    if choice == 1
        N = sqrt(L*le/(0.004*pi*Mi*Ac));
    else
        N = sqrt(L/AL);
    end%if;
    fprintf('\n');
    fprintf('N = %.0f\n', floor(100*N + 0.5) / 100);
else
   fprintf('Wrong choice. Please run the program again.\n\n');
end%if
```

Fig. 7.21 MATLAB routine for calculating the number of turns of a toroid (toroid.m)

inductor from the manufacturer with the correct value of inductance for a desired frequency and application, as described in Chap. 5, with several inductor examples given. However, a situation may arise where a manually wire-wound inductor of a carefully controlled inductance value is required.

In this case, several steps in the design process can be identified:

1. Determine the shape of the inductor (i.e. solenoid or toroid),
2. Determine whether the air gap or ferrite inductor is sufficient, and finally
3. Determine the exact geometry of the inductor (radius of the wire, number of turns, length/effective length).

To assist the process, the discrete inductor equations from Chap. 5 will be tabulated in Table 7.12. Typically, even for an in-house wire-wound inductor, it is unlikely that the designer will have much choice of type and size of core available. Therefore, geometry parameters, such as radius r and length l of the solenoid, the effective length l_e and the cross sectional A_c area of the toroid, can be treated as fixed parameters, as is the case of the initial permittivity of the material used for the core. The only remaining degree of freedom is the number of turns N. Thus the MATLAB routine can be made to calculate the number of turns given the required inductance (and pre-programmed geometry parameters), as illustrated for a toroid in Fig. 7.21.

If knowing the Q-factor is paramount, the program could be extended to include the resistance of the wire calculation and subsequently the Q-factor calculation. Once again, the wire resistance equation would be used for this purpose.

7.4 Design Examples

In this section, we briefly demonstrate how to use MATLAB programs developed in this chapter. In fact, we only demonstrate the use of the spiral inductor subroutine through three tasks, whereas the subroutines for the design of bond wires and discrete inductors are demonstrated through various tasks in Chap. 8.

Task 1 Design a square spiral inductor required to have an inductance of 3 nH and operate at 5 GHz. Use the default process parameters.

Solution: Clearly, a synthesis of an inductor is needed. Thus we run `indcalc` from the MATLAB main window and choose option 2. Then we enter 3 nH for the wanted inductance value and 5000 MHz for the operating frequency. We choose not to export the netlist or the layout because netlist export will be demonstrated in Chap. 8 and layout export will be demonstrated in Chap. 9. We also choose not to modify any process parameters, grid resolution or tolerance. The designed inductor has an output diameter of 198 μm, inner diameter of 124 μm, width of 11 μm and spacing of 2 μm. Its Q-factor is 4.76 and its resonant frequency 12.25 GHz.

The interaction with the program is shown in Fig. 7.22.

Task 2 Design the same inductor as in Task 1, but allow for the minimum spacing between the turns to be 1 μm.

```
>> indcalc

Welcome to Inductor Calculator v2.2 for Matlab. Please follow the
prompts.

Please enter 1 to specify geometry or 2 to specify inductance: 2
Please enter the wanted inductance value (Ls) (nH): 3
Please enter the tolerance for the inductance value (%) (-1 to
skip): -1
Please enter the search grid resolution (um) (-1 to skip): -1
Do you want to change any of the default geometry parameters
(y/n)? n
Using the default values of 30.0, 500.0, 2.0 and 2.0 for dinmin,
doutmax, s and wmin respectively.

Please enter the operating frequency (MHz): 5000
Do you want to change any of the default process parameters
(y/n)? n

Looking for the geometry for highest quality factor Q...

Ls = 3.01 nH
Lslf = 2.59 nH
Q = 4.76
fo = 12.25 GHz
w = 11.00 um
s = 2.00 um
din = 124.00 um
dout = 198.00 um
n = 3

Do you want to export the netlist of this inductor into a SPC
file (y/n)? n

Do you want to export the layout of this inductor into a CIF or
GDS file (c - CIF, q - GDS, n - none)? n
```

Fig. 7.22 Interaction with the spiral inductor design program of task 1

Solution: We follow the same procedure as in Task 1, but we choose to change geometry parameters. We now notice that the new inductor using a smaller turn spacing can achieve the same inductance with a smaller floorplan (d_{out} = 187 μm), while also achieving Q-factor improvements, as expected. The interaction with the program is shown in Fig. 7.23.

Task 3 Now, what would the Q-factor of the inductor designed in Task 2 be, if the metal used to lay out the spiral should be 20 % thicker?

Solution: We can now run the program again, but choose option 1 for analysis. Then we can enter the geometry we obtained in Task 2, but need to remember to modify the process parameters to increase the thickness of the top metal by 20 %.

```
>> indcalc

Welcome to Inductor Calculator v2.2 for Matlab. Please follow the
prompts.

Please enter 1 to specify geometry or 2 to specify inductance: 2
Please enter the wanted inductance value (Ls) (nH): 3
Please enter the tolerance for the inductance value (%) (-1 to
skip): -1
Please enter the search grid resolution (um) (-1 to skip): -1
Do you want to change any of the default geometry parameters
(y/n)? y

Please enter the minimum value for din (um) (-1 to skip): -1
Using the default value of 30 um for minimum din.
Please enter the maximum value for dout (um) (-1 to skip): -1
Using the default value of 500 um for maximum dout.
Please enter the minimum value for turn spacing s (um) (-1 to
skip): 1
Please enter the minimum turn width w (um) (-1 to skip): -1
Using the default value of 2 um for w.

Please enter the operating frequency (MHz): 5000
Do you want to change any of the default process parameters
(y/n)? n

Looking for the geometry for highest quality factor Q...

Ls = 3.01 nH
Lslf = 2.63 nH
Q = 4.91
fo = 12.75 GHz
w = 10.00 um
s = 1.00 um
din = 123.00 um
dout = 187.00 um
n = 3

Do you want to export the netlist of this inductor into a SPC
file (y/n)? n

Do you want to export the layout of this inductor into a CIF or
GDS file (c - CIF, g - GDS, n - none)? n
```

Fig. 7.23 Interaction with the spiral inductor design program of task 2

The new thickness is 1200 nm and the Q-factor now reaches 5.23, which is about 0.3 higher than with the standard metal thickness. The interaction with the program is shown in Fig. 7.24.

```
>> indcalc

Welcome to Inductor Calculator v2.2 for Matlab. Please follow the
prompts.

Please enter 1 to specify geometry or 2 to specify inductance: 1
Please enter the dout diameter (dout) (um): 187
Please enter the din diameter (din) (um): 123
Please enter the turn width (w) (um): 10
Please enter the number of turns: 3
Please enter the operating frequency (MHz): 5000
Do you want to change any of the default process parameters
(y/n)? y
Please enter the thickness of the top metal (nm) (-1 to skip):
Old value: 1000  --> New value: 1200
Please enter rho (ohm.m) (-1 to skip):
Old value: 2.82e-008  --> New value: -1
Please enter MAl (H/m) (-1 to skip):
Old value: 1.26e-006 --> New value: -1
Please enter thickness of the oxide between top two metals (nm)
(-1 to skip):
Old value: 1000  --> New value: -1
Please enter Er of the oxide between top two metals (-1 to skip):
Old value: 4.00  --> New value: -1
Please enter thickness of the oxide between the substrate and the
top metal (nm) (-1 to skip):
Old value: 5000  --> New value: -1
Please enter Er of the oxide between the substrate and the top
metal (-1 to skip):
Old value: 4.00  --> New value: -1
Please enter TSi (um) (-1 to skip):
Old value: 1000  --> New value: -1
Please enter ErSi (-1 to skip):
Old value: 11.70  --> New value: -1
Please enter rhoSi (ohm.m) (-1 to skip):
Old value: 0.20  --> New value: -1

Ls = 3.02 nH
Lslf = 2.63 nH
Q = 5.23
fo = 12.75 GHz
w = 10.00 um
s = 1.00 um
din = 123.00 um
dout = 187.00 um
n = 3

Do you want to export the netlist of this inductor into a SPC
file (y/n)? n

Do you want to export the layout of this inductor into a CIF or
GDS file (c - CIF, g - GDS, n - none)? n
```

Fig. 7.24 Interaction with the spiral inductor design program of task 3

7.5 Summary

In this chapter we started sharing our ideas on developing custom EDA tools with the readers. We decided to start with the inductor, which is the passive with the lowest Q-factor, particularly in integrated designs. We developed subroutines for the design of integrated spiral inductors, bond wires and discrete toroidal inductors. Each subroutine used equations applicable to the particular model. Procedures for toroid and bond wire design turned out to be very simple to implement in MATLAB, while for spiral inductor design, we needed to implement a search algorithm alongside regular calculations, leading to a somewhat more complex flow of the final subroutine. Finally, we demonstrated the use of the spiral inductor procedures with several examples, where the type of problem demanded that both synthesis and analysis parts needed to be used.

In Chap. 8, we will go on to develop procedures for the design of various power amplifier output stages, after which we will demonstrate how different subroutines from Chaps. 7 and 8 can be brought together to form a fully comprehensive power amplifier design program.

References

1. Fakhfakh M, Tlelo-Cuautle E, Siarry P. Computational intelligence in analog and mixed-signal (AMS) and radio-frequency (RF) circuit design. 1st ed. Cham: Springer; 2015.
2. Zhu X. Modeling and simulation of on-chip spiral inductors and transformers. San Jose: Cadence Design Systems; 2000.
3. IBM Corporation. BiCMOS-7WL Model Reference Guide. Armonk 2008.
4. Božanić M, Sinha S. Design methodology for SiGe-based class-E power amplifier. In: Proceedings of the 1st South African Conference on Semi and Superconductor Technology; 2009; Stellenbosch.
5. Mathews JH, Fink KK. Numerical methods using MATLAB. Upper Saddle River: Pearson; 2004.

Chapter 8
Full Power Amplifier System Design

In Chaps. 3 and 4, the theory of operation of most commonly used power amplifier stages was presented. In this chapter, this theory is used to develop several power amplifier design subroutines (procedures) that can be used in the power amplifier design cycle to simplify the calculations and the process. This chapter builds on Chap. 7 where routines for inductor design were introduced. These procedures will now be incorporated with newly developed power amplifier procedures to allow the development of a program that will enable the rapid design of fully functional power amplifier systems. The process of program development is illustrated step by step, and as before, it is explained in detail which equations from Part 1 are applicable, which input parameters need to be identified and where to find their typical values, and finally, what the output design parameters and files are. Readers of this book can use the program developed in this chapter as is, or they can use it as an example or a starting point to develop their own EDA programs.

This chapter is organized as follows: First, the procedures for Class-E, Class-F, Class-A/AB/B/C and Class-J power amplifier design are developed. This is followed with the development of impedance matching algorithms based on the theory described in Chap. 6. Finally, the development of the above-mentioned single power amplifier design program is attempted. The envisioned comprehensive program should allow for stage selection, include the impedance matching subroutine and allow for different inductor types to be selected, depending on the application. As a result, this program will become suitable for integrated, discrete and hybrid solutions. The stage selection capability allows for numerous applications, different frequency bands and different modulation schemes.

As in the previous chapter, it was confirmed that MATLAB scripts run in MATLAB version 2014b.

M. Božanić and S. Sinha, *Power Amplifiers for the S-, C-, X- and Ku-bands*,
Signals and Communication Technology, DOI 10.1007/978-3-319-28376-0_8

8.1 Subroutine for Design of Class-E Zero-Voltage-Switching Power Amplifiers

The chapter commences with the development of the ZVS Class-E power amplifier design subroutine, because of the simplicity of the output stage. As stated before, the basic principles of the Class-E power design process amplifier (to be precise, the design process of any power amplifier) remain the same for discrete and integrated solutions, and in the case of the latter, technology-independent. In this section, we consider only the variation with the large feed inductor (RFC). The development of the extension of the program, in which finite DC feed inductance is used, is left as an exercise to the reader.

A simplified Class-E ZVS power amplifier schematic from Chap. 4 is repeated for convenience in Fig. 8.1, but with an HBT transistor used instead of the MOSFET, and all information that does not appear in the final design equation removed from the schematic. The sudden shift in the type of active device used is to demonstrate the concept of technology independence.

8.1.1 Equations

The process of developing the subroutine starts with inspecting the equations listed in Chap. 4 in order to identify which equations can be used for the design, so that a table of useful equations can be made and that all other equations (such as equations used for derivation or analysis) can be omitted. Some equations are rewritten and some approximations are performed so that the parameters that need calculating are expressed in terms of design parameters, listed in the following section. These equations are tabulated in Table 8.1.

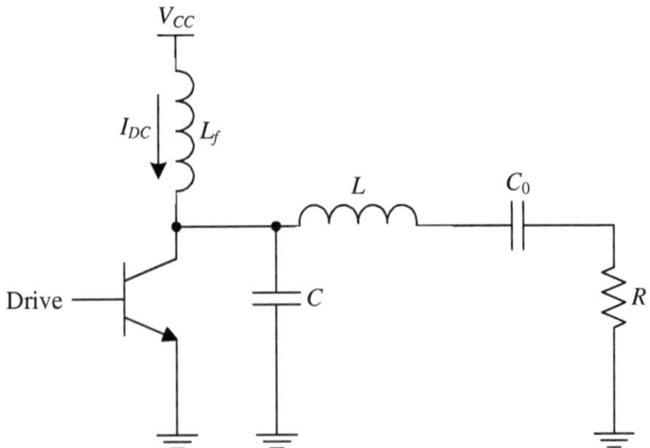

Fig. 8.1 Circuit diagram of a single-ended Class-E power amplifier

Table 8.1 Summary of equations needed for the design of Class-E ZVS power amplifier

Calculation	Equation
Maximum collector voltage	$V_{\max} = 3.562V_{CC}$
Load resistance	$R = 0.577\frac{V_{CC}^2}{P_O}$
Total inductance (L + L_0)	$L = \frac{Q_L R_L}{\omega_o}$
Shunt capacitance	$C = \frac{1}{5.447\omega R}$
Series capacitance	$C_0 = \frac{5.447C}{Q_L}\left(1 + \frac{1.42}{Q_L - 2.08}\right)$
Feed resistor	$L_f = 8.669\frac{R}{f}$
DC current	$I_{DC} = \frac{V_{CC}}{1.734R_L}$
Maximum collector current	$I_{\max} = 2.8621I_{DC}$
Angular frequency	$\omega_o = 2\pi f_o$

8.1.2 Input Parameters

For the ZVS Class-E power amplifier design, the following input parameters are required:

- *Center frequency of the channel* (f_o). This quantity is determined by the specifications of the transmitter system of which the power amplifier is part.
- *The loaded quality factor* (Q_L). This is the quality factor of the series resonator. It can be chosen freely by the power amplifier designer, but a trade-off must be considered between high efficiency and power (low Q_L) on the one side and THD of the output signal on the other side (high Q_L). A plausible Q-factor is in the range of 5–10 [1]. If efficiency is important, a lower Q_L can be chosen and harmonics removed by additional filters at the output of the amplifier [2]. The narrowband output matching network can serve this purpose.
- *Output power* (P_O). This quantity is also determined by the specifications of the transmitter system. A higher output power is achieved with a lower load resistance (R_L), which places more stringent requirements on the output impedance matching. High output power also demands a higher supply voltage, described next.
- *Supply voltage* (V_{CC}). This quantity can be chosen by the designer and is specific for a particular application, but it must be ensured that it does not exceed the value of $BV_{CEO}/3.56$ where BV_{CEO} is defined as collector-emitter breakdown voltage.

The subroutine can be expanded with some useful functionality if some optional parameters are specified:

- *Collector-emitter saturation voltage* (V_{CEsat}). This parameter, also specified in volts, is a process-dependent quantity, usually equal to 0.1 or 0.2 V. Since it is approximately equal to 0, its omission will result in minimal discrepancies between the actual and predicted power amplifier values.

Table 8.2 Input parameters required for the Class-E ZVS power amplifier subroutine

Parameter	Unit	Mandatory/optional
Center frequency of the channel (f_o)	MHz	Mandatory
Loaded quality factor (Q_L)	–	Mandatory
Output power (P_O)	mW	Mandatory
Supply voltage (V_{CC})	V	Mandatory
Collector-emitter saturation voltage (V_{CEsat})	V	Optional
Maximum and minimum inductance values ($L_{\max}$ and $L_{\min}$)	nH	Optional
Collector-emitter breakdown voltage (BV_{CEO})	V	Optional
Collector-emitter capacitance, C_{ce}	pF	Optional

- *Maximum and minimum inductance values* ($L_{\max}$ *and* $L_{\min}$). These two quantities can be included by the designer to limit extremely high or low values of inductors.
- *Collector-emitter breakdown voltage* (BV_{CEO}). This quantity warns the user if the V_{CC} specified is high enough to result in the transistor entering breakdown.
- *Collector-emitter parasitic capacitance* (C_{ce}). This parameter allows for automatic calculation of the actual capacitance value to be placed in parallel to the transistor parasitic collector-emitter capacitance.

Parameters needed for the Class-E power amplifier subroutine are summarized in Table 8.2.

8.1.3 Description of the Subroutine

Because of the simplicity of Class-E ZVS power amplifier, this subroutine is carried out in a linear fashion, where all required Class-E parameters are calculated using applicable equations. Note that from Class-E ZVS theory, the capacitor C must include the output (parasitic) capacitance of the transistor used for amplification if the output stage is to have any practical application. If this condition is satisfied one can subtract the capacitance values to obtain the actual capacitor value. By introducing limits on the inductance, the Q-factor can be optimized to avoid excessive inductance values.

A flow diagram of the Class-E ZVS design subroutine is given in Fig. 8.2. Complete MATLAB code listing for the subroutine is given in Fig. 8.3.

8.1.4 Subroutine Outputs

The subroutine outputs are summarized in Table 8.3.

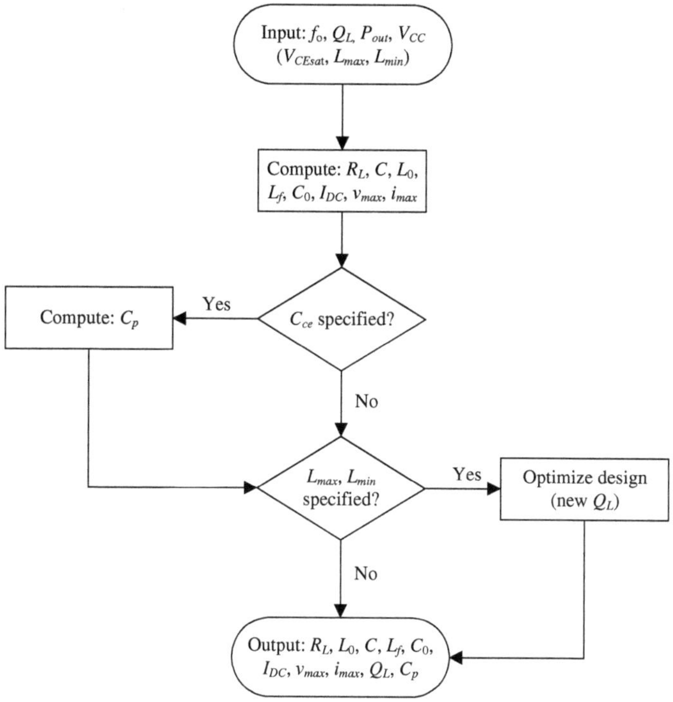

Fig. 8.2 Flow chart of the Class-E ZVS power amplifier design subroutine

8.2 Subroutine for Design of the Class-F Power Amplifiers

In this section, we proceed to deal with a somewhat more complex power amplifier class: Class F. The complexity comes from the fact that a number of Class-F power amplifier variations are possible. In Chap. 4 it was discussed that the efficiency of integrated Class-F power amplifiers depends on the number of harmonic resonators used. It is therefore necessary first to evaluate which actual Class-F configuration will be the best fit across a range of practical applications. Here, to illustrate the concept and complexity of the Class-F subroutine development, it is sufficient to consider two different Class-F power amplifier configurations: the Class-F power amplifier with resonators up to third harmonic (Class F_3), shown in Fig. 8.4a for convenience, and the Class-F power amplifier with resonators up to the fifth harmonic (Class F_{35}) shown in Fig. 8.4b. The theoretical maximum efficiencies for the two circuits are 81.7 and 90.5 % respectively. The former is included because of its particularly simple output filter, while the latter is included because it offers fairly high efficiency while still keeping the output filter reasonably simple. Including a greater number of resonators further increases the complexity of the circuit and is not deemed feasible here. A subroutine for the design of the most efficient Class-F power amplifier circuit, the Class-F power amplifier with quarter-wave transmission

```
%Enter PA design parameters
fo = input('Please enter the frequency of operation, fo (MHz):
');
if (fo <= 0)
    fo = 1e9; %Frequency
else
    fo = fo * 1e6; end
w = 2*pi*fo; %Radian frequency
QL = input('Please enter the loaded quality factor, QL: ');
if (QL < 1)
    QL = 10; %Loaded quality factor
end
Po = input('Please enter the required output power, Po (mW): ');
if (Po <= 0)
    Po = 0.01; %Output power
else
    Po = Po/1000; end
Vcc = input('Please enter the value of the regulated power supply
(Vcc/Vdd): ');
if (Vcc <= 0)
    Vcc = 1; %Power supply
end
VCEsat = input('Please enter VCEsat/VDSsat (-1 to skip): ');
if (VCEsat <= 0)
    VCEsat = 0; %Saturation VCE voltage
end
Lmax = input('Please enter the highest allowed inductance (nH) (-
1 to skip): ');
if (Lmax <= 0)
    Lmax = 0; %Maximum inductance
else
    Lmax = Lmax * 1e-9; end
Lmin = input('Please enter the lowest allowed inductance (nH) (-1
to skip): ');
if (Lmin <= 0)
    Lmin = 0; %Minimum inductance
else
    Lmin = Lmin * 1e-9; end

BVCE = input('Please enter collector-emitter breakdown voltage
(V) (-1 to skip): ');
if (BVCE <= 0)
    BVCE = 0; %Breakdown voltage
end
%Calculate
vcp = 3.56 * Vcc;
if (vcp > BVCE && BVCE > 0)
    NVcc = BVCE / 3.56;
    fprintf('Warning: Vcc should not exceed %.2f or the transis-
tor will go into breakdown!\n', NVcc); end
RL = 0.577*(Vcc - VCEsat)^2/Po; %Load resistance
L2 = QL*RL/w; %Series inductance
Lf = 8.669*RL/fo; %Feed inductor
C1 = 1/(w*RL*5.447); %Parallel capacitance
C2 = C1*(5.447/QL)*(1+1.42/(QL - 2.08)); %Series capacitance
```

Fig. 8.3 MATLAB code of the Class-E ZVS power amplifier design subroutine (ClassE.m)

```
Idc = Vcc / (1.734*RL); %DC current consumption
isp = 2.86 * Idc; %Maximum collector current dissipation
%Q-factor adjustment
if (Lmax ~= 0)
    if (L2 > Lmax)
        QLB = w*Lmax/RL;
        if (QLB > 1)
            L2 = Lmax;
            C2 = C1*(5.447/QLB)*(1+1.42/(QLB - 2.08));
            fprintf('New QL = %.2f \n', floor(100 * QLB + 0.5) /
100);
        else
            fprintf('Cannot find QL greater than 1 to keep L2 <=
Lmax\n');
        end
    end
end
if (Lmin ~= 0)
    if (L2 < Lmin)
        QLB = w*Lmin/RL;
        L2 = Lmin;
        C2 = C1*(5.447/QLB)*(1+1.42/(QLB - 2.08));
        fprintf('New QL = %.2f \n', floor(100 * QLB + 0.5) /
100);
    end
end
Cce = input('Please enter the value of Cce/Cds (pF) (-1 to skip):
');
if (Cce > 0)
    Cce = Cce * 1e-12;
    if (C1 >= Cce)
        Cp = C1 - Cce;
    else
        Cp = 0;
    end
else
    Cp = C1; %Ideal transistor
end
%Display calculated values
fprintf('\nRESULTS:\n');
fprintf('R = %.2f Ohm \n', floor(100 * RL + 0.5) / 100);
fprintf('L = %.2f nH \n', floor(1e11 * L2 + 0.5) / 100);
fprintf('Lf = %.2f nH \n', floor(1e11 * Lf + 0.5) / 100);
fprintf('C0 = %.2f pF \n', floor(1e14 * C2 + 0.5) / 100);
fprintf('C = %.2f pF \n', floor(1e14 * C1 + 0.5) / 100);
fprintf('Cp = %.2f pF \n', floor(1e14 * Cp + 0.5) / 100);
fprintf('Idc = %.2f mA \n', floor(1e5 * Idc + 0.5) /100);
fprintf('vcem = %.2f V \n', floor(1e2 * vcp + 0.5) / 100);
fprintf('imax = %.2f mA \n', floor(1e5 * isp + 0.5) / 100);
```

Fig. 8.3 (continued)

line, will be developed in a separate section, where the procedure will cater for the design of the quarter-wave line as well. The development of routines of other Class-F power amplifier variations is again left as an exercise to readers.

Table 8.3 Output parameters of the Class-E ZVS power amplifier design subroutine

Parameter	Unit
Optimum load resistance (R_L)	Ω
Shunt capacitor (C)	pF
Series capacitor (C_0)	pF
Series inductor (L)	nH
Feed inductor (RFC) (L_f)	nH
DC and maximum currents (I_{DC}, $I_{\max}$)	mA
Maximum collector voltage (v_{CEM})	V
Actual shunt capacitor required (C_p)	pF
Loaded quality factor for maximum allowed inductance	–

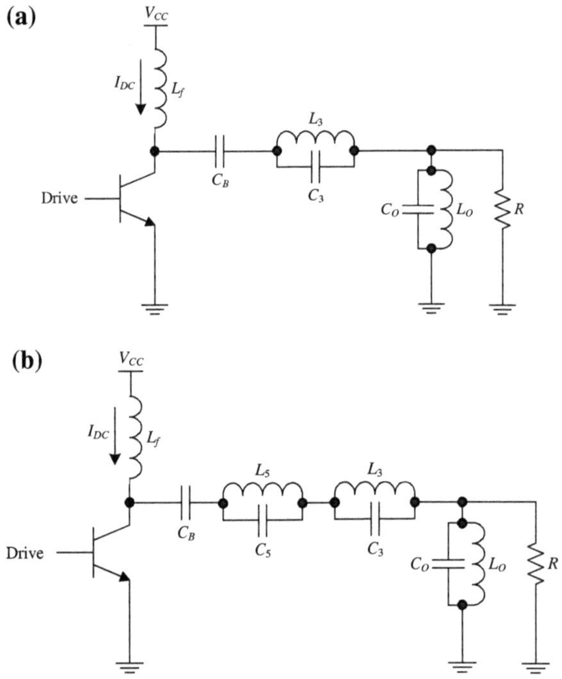

Fig. 8.4 Class-F power amplifier circuits: **a** Class F_3 and **b** Class F_{35}

8.2.1 Equations

As always, the design equations for power amplifiers need to be identified and tabulated first. This is done in Table 8.4. For both power amplifier variations, configurations with maximally flat waveforms are used.

Table 8.4 Summary of equations needed for the design of Class-F_3 and Class-F_{35} power amplifiers with maximally flat waveforms

Calculation	Equation (Class F_3)	Equation (Class F_5)
Optimum load resistance	$R = 0.6328\frac{V_{CC}^2}{P_O}$	$R = 0.6866\frac{V_{CC}^2}{P_O}$
Base capacitance (given filter inductance)	$C_O = \frac{1}{(\omega_o)^2}L_O$	$C_O = \frac{1}{(\omega_o)^2}L_O$
Third harmonic capacitance	$C_3 = \frac{1}{(3\omega_o)^2}L_3$	$C_3 = \frac{1}{(3\omega_o)^2}L_3$
Fifth harmonic capacitance	–	$C_5 = \frac{1}{(5\omega_o)^2}L_5$
Peak output voltage	$V_m = \frac{9}{8}V_{CC}$	$V_m = \frac{75}{64}V_{CC}$
Peak collector voltage	$v_{CEM} = 2V_{CC}$	$v_{CEM} = 2V_{CC}$
DC current	$I_{DC} = \frac{9}{8}\frac{V_{CC}}{\gamma_1 R}$	$I_{DC} = \frac{75}{64}\frac{V_{CC}}{\gamma_1 R}$
Gamma	$\gamma_1 = \frac{\theta - \sin\theta\cos\theta}{\sin\theta - \theta\cos\theta}$	$\gamma_1 = \frac{\theta - \sin\theta\cos\theta}{\sin\theta - \theta\cos\theta}$
RFC	$L_{RFC} \geq \frac{10R}{\omega_0}$	$L_{RFC} \geq \frac{10R}{\omega_0}$
Blocking capacitance	$C_B \geq \frac{R}{10\omega_0}$	$C_B \geq \frac{R}{10\omega_0}$
Angular frequency	$\omega_o = 2\pi f_o$	$\omega_o = 2\pi f_o$

8.2.2 *Input Parameters*

As in the case of the Class-E design, several input parameters are needed for the Class-F subroutine to execute:

- *Center frequency of the channel* (f_o). Again, this is determined by the specifications of the transmitter system of which the power amplifier is part of.
- *Output power* (P_O). This quantity is also determined by the specifications of the transmitter system.
- *Supply voltage* (V_{CC}). A quantity specific to the particular application as described before.
- *Filtering inductances* (L_O, L_3 *and* L_5). The inductances used to build Class-F resonant tanks. They can be calculated from each tank's loaded Q-factor, or if the quality is not important, they can be specified to match as standard (or previously designed) values.
- *Half of conduction angle* (θ). Parameter describing the biasing of the driving transistor, defined in the same fashion as in Class-A/AB/B/C power amplifier theory.
- *Collector-emitter breakdown voltage* (BV_{CEO}). This quantity is an optional parameter and is used by the program to warn the user if the V_{CC} specified is high enough to send the transistor into breakdown.

Parameters needed for the Class-F power amplifier routine are summarized in Table 8.5.

Table 8.5 Input parameters required for the Class-F power amplifier software routine

Parameter	Unit	Mandatory/optional
Center frequency of the channel (f_O)	MHz	Mandatory
Output power (P_O)	mW	Mandatory
Supply voltage (V_{CC})	V	Mandatory
Filtering inductances (L_O, L_3 and L_5)	nH	Mandatory
Half of conduction angle (θ)	o	Optional
Collector-emitter breakdown voltage (BV_{CEO})	V	Optional

8.2.3 Description of the Subroutine

Similar to the case of the Class-E power amplifier, for each variation the routine is developed such that the design is performed for optimum load resistance. The only design quantities are the filtering and blocking capacitors and inductors, but the subroutine also calculates the voltage and current levels that have to be maintained for designed operation (in this case, maximum flatness) by adjusting the input signal levels. The procedure will also warn if the transistor used enters breakdown, if the breakdown voltage is specified.

The MATLAB code of the subroutine is listed in Fig. 8.5.

8.2.4 Subroutine Outputs

The outputs of the Class-F software subroutine are tabulated in Table 8.6.

8.3 Subroutine for Design of Class-F Power Amplifiers with Quarter-Wave Transmission Line

If the wavelength of the system allows for the use of the quarter-wave transmission line, a Class-F power amplifier with the quarter-wave transmission line or Class-F_∞ power amplifier can allow for high efficiency (100 % in theory). From a programming perspective, the main difference between developing a routine for the design of, say a Class-F_{35} amplifier and a Class-F_∞ amplifier, is that it is necessary to design the transmission line as well. In principle, a transmission line with any characteristic impedance can be used, but this calls for non-standard load impedances and impedance matching. However, if the impedance matching is incorporated in the design of the stage, a subroutine for possibly the simplest power amplifier stage can be created. In this section, it will be assumed that a lossless microstrip line can be designed.

```
%Enter PA design parameters
harmonics = 0;
theta_deg = 120; % Default theta
%Choose between the 3 abd 5 harmonic designs
while (harmonics ~= 1 && harmonics ~= 2)
    harmonics = input('Please enter 1 for 3-harmonic design or 2
for 5-harmonic design: ');
    if (harmonics ~= 1 && harmonics ~= 2)
       fprintf('Wrong entry. Please try again.\n');
    end
end
fo = input('Please enter the frequency of operation, fo (MHz):
');
if (fo <= 0)
    fo = 1e9; %Frequency of operation
else
    fo = fo * 1e6; end
w = 2*pi*fo;
Po = input('Please enter the required output power, Po (mW): ');
if (Po <= 0)
    Po = 0.01; %Output power
else
    Po = Po/1000; end
Vcc = input('Please enter the value of the regulated power supply
(Vcc/Vdd): ');
if (Vcc <= 0)
    Vcc = 1; %Voltage supply
end
BVCE = input('Please enter collector-emitter breakdown voltage
(V) (-1 to skip): ');
if (BVCE <= 0)
    BVCE = 0; %Collector-emitter breakdown voltage
end
L0 = input('Please enter the inductance for the base filter (nH):
');
if (L0 <= 0)
    L0 = 1e-9; %Base filter inductance
else
    L0 = L0 * 1e-9; end
L3 = input('Please enter the inductance for the third harmonic
filter (nH): ');
if (L3 <= 0)
    L3 = 1e-9; %Third harmonic filter inductance
else
    L3 = L3 * 1e-9; end
if harmonics == 2;
    L5 = input('Please enter the inductance for the fifth har-
monic filter (nH): ');
    if (L5 <= 0)
        L5 = 1e-9; %Fifth harmonic filter inductance
    else
        L5 = L5 * 1e-9; end
else
    L5 = 0;
end
theta = input('Please enter the theta (deg) (-1 to skip): ');
if (theta <= 0)
    theta = theta_deg/180*pi;
else
    theta = theta/180*pi;
end
```

Fig. 8.5 MATLAB code of the Class-F power amplifier design subroutine (ClassF.m)

```
%Calculate optimum resistance and waveforms
%Common calculations
g1 = (theta-sin(theta)*cos(theta))/(sin(theta)-
theta*cos(theta));%gamma
VCEM = 2*Vcc;
if harmonics == 1  %Class F3
    RL = 0.6328*Vcc^2/Po;
    Vm = 9/8*Vcc;
    Idc = 9/8*Vcc/(g1*RL);
else %Class F35
    RL = 0.6866*Vcc^2/Po;
    Vm = 75/64*Vcc;
    Idc = 75/64*Vcc/(g1*RL);
end
Lf = 10*RL/w;
CB = RL/(10*w);
%Calculate Filters
%base
C0 = 1/((2*pi*fo)^2*L0);
%3rd harmonic
C3 = 1/((2*pi*3*fo)^2*L3);
if harmonics == 2;
    C5 = 1/((2*pi*5*fo)^2*L3);
end
%Display calculated values
fprintf('\nRESULTS:\n');
fprintf('R = %.2f Ohm \n', floor(100 * RL + 0.5) / 100);
fprintf('L0 = %.2f nH \n', floor(1e11 * L0 + 0.5) / 100);
fprintf('C0 = %.2f pF \n', floor(1e14 * C0 + 0.5) / 100);
fprintf('L3 = %.2f nH \n', floor(1e11 * L3 + 0.5) / 100);
fprintf('C3 = %.2f pF \n', floor(1e14 * C3 + 0.5) / 100);
fprintf('Lf = %.2f nH \n', floor(1e11 * Lf + 0.5) / 100);
fprintf('CB = %.2f pF \n', floor(1e14 * CB + 0.5) / 100);
if harmonics == 2;
    fprintf('L5 = %.2f nH \n', floor(1e11 * L5 + 0.5) / 100);
    fprintf('C5 = %.3f pF \n', floor(1e14 * C5 + 0.5) / 100);
end
fprintf('Idc = %.2f mA \n', floor(1e5 * Idc + 0.5) /100);
fprintf('vm = %.2f V \n', floor(1e2 * Vm + 0.5) / 100);
fprintf('VCEM = %.2f V \n', floor(1e2 * VCEM + 0.5) / 100);

if (VCEM > BVCE && BVCE > 0)
    NVcc = BVCE / 2;
    fprintf('Warning: Vcc should not exceed %.2f or the transis-
tor will go into breakdown!\n', NVcc);
end
```

Fig. 8.5 (continued)

A simplified schematic of a Class-F_∞ power amplifier with the parallel resonant tank is repeated in Fig. 8.6, but with an HBT as a driving transistor.

8.3.1 Equations

The required equations for Class-F_∞ power amplifier design are tabulated in Table 8.7.

Table 8.6 Output parameters of the Class-F power amplifier design subroutine

Parameter	Unit
Optimum load resistance (R_L)	Ω
Nominal frequency resonant capacitor (C_O)	pF
Third- and fifth-harmonic frequency resonant capacitors (C_3 and C_5)	pF
Feed inductor (RFC) (L_f)	nH
Blocking capacitor (C_B)	pF
DC current (I_{DC})	mA
Peak collector voltage and peak output voltage (v_{CEM} and V_m)	V

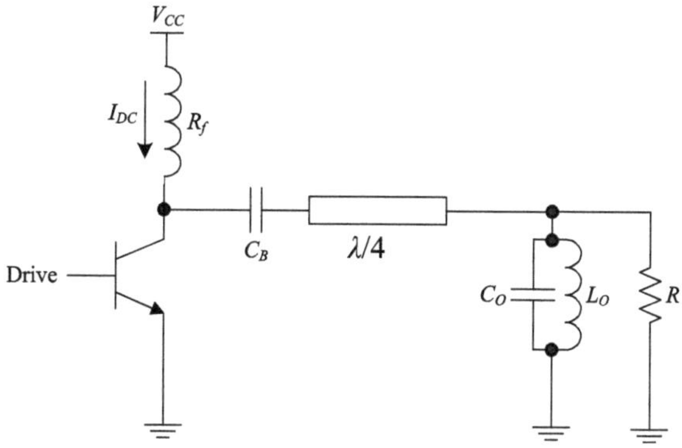

Fig. 8.6 Class-F$_\infty$ power amplifier circuit

8.3.2 Input Parameters

In addition to the center frequency of the channel (f_o), output power (P_O), supply voltage (V_{CC}) and the inductance values for a nominal resonant tank (L_O), which have all been described previously, the following parameters are required to complete the design of the class Class-F$_\infty$ power amplifier:

- *Default resistance of the antenna.* The presence of this parameter allows for the transmission line to be designed in such a way as to include impedance matching. If this parameter is not specified, a typical impedance of the antenna of 50 Ω will be used.
- *The relative dielectric constant of the transmission line substrate* (ε_r). This parameter is needed to calculate the effective dielectric constant of the substrate, which is then used in characteristic impedance calculations.
- *Limits on the transmission line height and width values* (w_{min}, w_{max}, h_{min}, h_{max}). Because of the complexity of the characteristic impedance calculations, a search algorithm for the transmission line will be developed. These parameters put constraints on the search space.

Parameters needed for the Class-F$_\infty$ part of the routine are summarized in Table 8.8.

Table 8.7 Summary of equations needed for the design of a Class-F$_\infty$ power amplifier

Calculation	Equation
Peak collector voltage	$V_m = \frac{4}{\pi} V_{CC}$
Optimum load resistance (input of the transmission line	$R_i = \frac{V_m^2}{2P_O}$
Base capacitance (given filter inductance)	$C_O = \frac{1}{(\omega_o)^2} L_o$
Peak collector voltage	$v_{CEM} = 2V_{CC}$
DC current	$I_{DC} = \frac{8}{\pi^2} \frac{V_{CC}}{R_i}$
RFC	$L_{RFC} \geq \frac{10R_i}{\omega_0}$
Blocking capacitance	$C_B \geq \frac{R_i}{10\omega_0}$
Angular frequency	$\omega_o = 2\pi f_o$
Wavelength	$\lambda = \frac{v}{f}$
Length of the transmission line	$l = \lambda/4$
Required characteristic impedance	$Z_0 = \sqrt{R_i R}$
Z_f	$Z_f = \sqrt{\frac{\mu_0}{\varepsilon_0}} \approx 376.8\,\Omega$
Width-to-height ratio	$r = w/h$
Characteristic impedance of the narrow line	$Z_0 = \frac{Z_f}{2\pi\sqrt{\varepsilon_{eff}}} \ln\left(\frac{8}{r} + \frac{r}{4}\right)$
Effective dielectric constant (narrow line)	$\varepsilon_{eff} = \frac{\varepsilon_r+1}{2} + \frac{\varepsilon_r-1}{2}\left[\left(1+\frac{12}{r}\right)^{-\frac{1}{2}} + 0.04(1-r)^2\right]$
Characteristic impedance of the wide line	$Z_0 = \frac{Z_f}{\sqrt{\varepsilon_{eff}}\left[1.393 + r + \frac{2}{3}\ln(r+1.444)\right]}$
Effective dielectric constant (narrow line)	$\varepsilon_{eff} = \frac{\varepsilon_r+1}{2} + \frac{\varepsilon_r-1}{2}\left(1+\frac{12}{r}\right)^{-\frac{1}{2}}$

Table 8.8 Input parameters required for the Class-F$_\infty$ power amplifier software routine

Parameter	Unit	Mandatory/optional
Center frequency of the channel (f_O)	MHz	Mandatory
Output power (P_O)	mW	Mandatory
Supply voltage (V_{CC})	V	Mandatory
Filtering inductance (L_O)	nH	Mandatory
Antenna resistance (Z_A)	Ω	Optional
Relative dielectric constant of the substrate (ε_r)	–	Mandatory

8.3.3 *Description of the Subroutine*

The routine follows the same linear flow as in the case of the routine for Class-F$_3$ and Class-F$_{35}$ design, with the exception that a separate procedure that performs the quarter-wave design is used. This will allow the later use of this procedure when devising a routine for impedance matching.

Before attempting to design the procedure for the quarter-wave transmission line design, we note that the transmission line width and the substrate height always

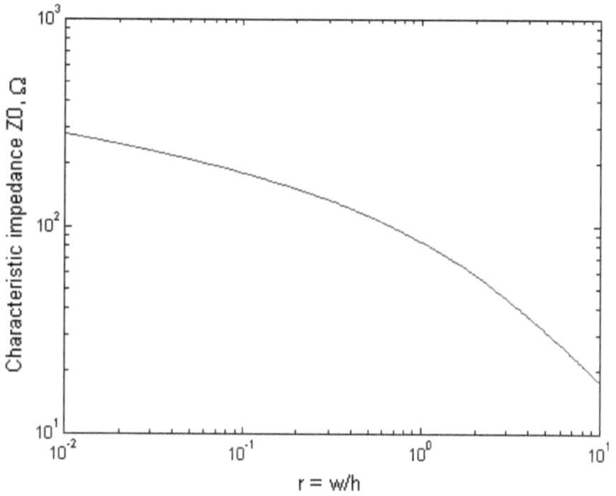

Fig. 8.7 Z_0 as a function of $r = w/h$

appear as a ratio, which we have defined as ratio r in Table 8.7. Thus, the quarter-wave microstrip line will be fully described if this ratio is calculated. Since the ratio is defined by a piecewise function (for $w/h < 1$ and $w/h > 1$), we make a decision here to solve the characteristic impedance equation iteratively. To confirm that there is only one solution to this equation, one can first plot the characteristic impedance equation, as shown in Fig. 8.7. From this plot it is evident that the impedance as a function of r is monotonic, i.e. it has no maxima or minima; there is indeed only one solution, and one can proceed with the iterative solution scenario. The flow chart of the resulting iterative quarter-wave line design procedure is shown in Fig. 8.8 and its MATLAB code is shown in Fig. 8.9.

The MATLAB code of the complete Class-F_∞ power amplifier design subroutine is given in Fig. 8.10.

8.3.4 Subroutine Outputs

The outputs of the Class-F_∞ design subroutine are tabulated in Table 8.9.

8.4 Subroutine for Design of Class-A/AB/B/C Power Amplifiers

The next exercise is to develop a design subroutine for the design of the combined A/AB/B/C power amplifier class. Although simplified equations for Class-A and Class-B power amplifiers exist, as seen in Chap. 3, the equations for Class-AB and Class-C power amplifiers also cover the design of the other two classes.

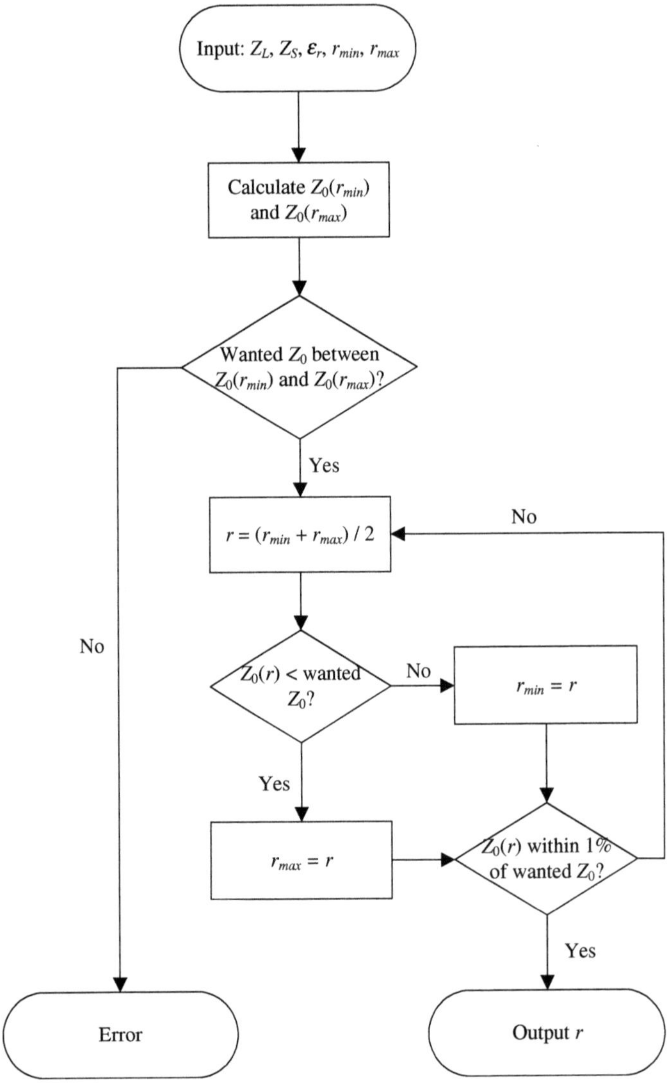

Fig. 8.8 Flow chart of the quarter-wave transmission line design procedure

As was the case with Class-F power amplifiers, the circuit diagram and biasing (conduction angle) fully define the class, and all that is left for the subroutine to do is to calculate the component values, voltages and currents. Seeing that the efficiency changes with the conduction angle, it was decided to include the efficiency calculation as well. The circuit diagram of the combined class is repeated in Fig. 8.11.

```
%Quarter-wave transmission line design program
%Initial values
rmin = 0.1; %minimum w/h ratio
rmax = 10; %maximum w/h ratio
er = input('Please enter relative dielectric permittivity (-1 to
skip): ');
if (er < 0)
    er = 3;
end;
tolerance = 0.01; %1%
Z0d = sqrt(ZS*ZL); %characteristic impedance (design value)
%Check if there is a solution within initial conditions
eff = (er+1)/2+(er-1)/2*((1+12/rmin)^(-1/2)+0.04*(1-rmin)^2);
Z0 = 376.8/(2*pi*sqrt(eff))*log(8/rmin+rmin/4);
eff2 = (er+1)/2+(er-1)/2*((1+12/rmax)^(-1/2)+0.04*(1-rmax)^2);
Z02 = 376.8/(sqrt(eff2)*(1.393+rmax+2/3*log(rmax+1.444)));
if(Z02 < Z0d && Z0> Z0d) %There is a solution, find it
    r = rmin;
    while(Z0 < Z0d*(1-tolerance) || Z0 > Z0d*(1+tolerance))
      r = (rmin + rmax) /2;
      if (r < 1)
        eff = (er+1)/2+(er-1)/2*((1+12/r)^(-1/2)+0.04*(1-r)^2);
        Z0 = 376.8/(2*pi*sqrt(eff))*log(8/r+r/4);
      else
        eff = (er+1)/2+(er-1)/2*((1+12/r)^(-1/2)+0.04*(1-r)^2);
        Z0 = 376.8/(sqrt(eff)*(1.393+r+2/3*log(r+1.444)));
      end
      if (Z0 < Z0d)
          rmax = r;
      else
          rmin = r;
      end
    end
    %Outputs
    fprintf('\nRESULTS:\n');
    fprintf('Z0 = %.2f Ohm \n', floor(100 * Z0d + 0.5) / 100);
    fprintf('r = w/h = %.2f \n', floor(1e2 * r + 0.5) / 100);
else
    fprintf('There is no solution within specified initial condi-
tions!\n');
end;
```

Fig. 8.9 MATLAB code of the quarter-wave transmission line design procedure (qwt.m)

8.4.1 Equations

When designing Class-A/AB/B/C power amplifiers or developing a routine for their design, the challenge is to find appropriate equations and arrange them in the order appropriate for the task. This is done in Table 8.10.

```
%Enter PA design parameters
fo = input('Please enter the frequency of operation, fo (MHz): ');
if (fo <= 0)
    fo = 1e9; %Frequency of operation
else
    fo = fo * 1e6; end
w = 2*pi*fo;
Po = input('Please enter the required output power, Po (mW): ');
if (Po <= 0)
    Po = 0.01; %Output power
else
    Po = Po/1000; end
Vcc = input('Please enter the value of the regulated power supply
(Vcc/Vdd): ');
if (Vcc <= 0) %Voltage supply
    Vcc = 1; end
BVCE = input('Please enter collector-emiiter breakdown voltage (V)
(-1 to skip): ');
if (BVCE <= 0)
    BVCE = 0; %Collector-emitter breakdown voltage
end
L0 = input('Please enter the inductance for the base filter (nH):
');
if (L0 <= 0)
    L0 = 1e-9; %Base filter inductance
else
    L0 = L0 * 1e-9; end
RL = input('Please enter antenna impedance (Ohms) (-1 to skip):
');
if (RL <= 0)
    RL = 50; end
%Calculate optimum resistance and waveforms
VCEM = 2*Vcc; Vm = 4/pi*Vcc; Ri = 0.5*Vm^2/Po;
Idc = 8/pi^2*Vcc/Ri; Lf = 10*Ri/w; CB = Ri/(10*w);
%length
vp = 299792458; %phase velocity
ell = vp/fo/4*1000; %length in mm
C0 = 1/((2*pi*fo)^2*L0); %Calculate base filters
ZS = Ri; %For impedance matching / quarter-wave transformer
ZL = RL; qwt; %Look for quarter-wave transformer w/h ratio
%Display calculated values
fprintf('length of QWT = %.2f mm \n', floor(1e2 * ell + 0.5) /
100);
fprintf('Ri = %.2f Ohm \n', floor(100 * Ri + 0.5) / 100);
fprintf('L0 = %.2f nH \n', floor(1e11 * L0 + 0.5) / 100);
fprintf('C0 = %.2f pF \n', floor(1e14 * C0 + 0.5) / 100);
fprintf('Lf = %.2f nH \n', floor(1e11 * Lf + 0.5) / 100);
fprintf('CB = %.2f pF \n', floor(1e14 * CB + 0.5) / 100);
fprintf('Idc = %.2f mA \n', floor(1e5 * Idc + 0.5) /100);
fprintf('vm = %.2f V \n', floor(1e2 * Vm + 0.5) / 100);
fprintf('VCEM = %.2f V \n', floor(1e2 * VCEM + 0.5) / 100);
 if (VCEM > BVCE && BVCE > 0)
    NVcc = BVCE / 2;
    fprintf('Warning: Vcc should not exceed %.2f or the transistor
will go into breakdown!\n', NVcc); end
```

Fig. 8.10 MATLAB code of the Class-F_∞ power amplifier design subroutine (ClassFQWT.m)

Table 8.9 Output parameters of the Class-F$_\infty$ power amplifier design subroutine

Parameter	Unit
Optimum load resistance (R)	Ω
Nominal frequency resonant capacitor (C_O)	pF
Feed inductor (RFC) (L_f)	nH
Blocking capacitor (C_B)	pF
DC current (I_{DC})	mA
Peak collector voltage and peak output voltage (V_{CEM} and V_m)	V
Wavelength (λ)	μm
Transmission line length	μm
Transmission line *w/h*	–

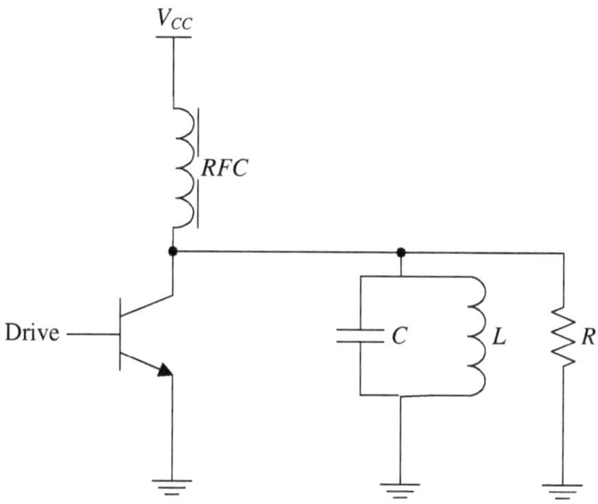

Fig. 8.11 Circuit diagram of Class-A/AB/B/C power amplifier

8.4.2 Input Parameters

Input parameters are summarized in Table 8.11; all parameters have been described in previous sections of this chapter. In this case, we decide not to choose the filtering inductance, but rather to calculate it in terms of the bandwidth (or the loaded Q-factor).

8.4.3 Description of the Subroutine

The Class-A/AB/B/C power amplifier design subroutine executes in a linear fashion, where all required calculations are performed using previously defined equations. The MATLAB code of the subroutine is given in Fig. 8.12.

Table 8.10 Summary of equations needed for design of Class-A/AB/B/C power amplifiers

Calculation	Equation (Class-A/AB/B/C)
Maximum output voltage	$V_m = V_{CC}$
Optimum load resistance	$R = \frac{V_{CC}^2}{2P_O}$
Maximum load current	$I_m = \frac{V_m}{R}$
Gamma	$\gamma_1 = \frac{\theta - \sin\theta\cos\theta}{\sin\theta - \theta\cos\theta}$
DC current	$I_{DC} = \frac{I_m}{\gamma_1}$
Peak collector voltage	$v_{CEM} = 2V_{CC}$
DC power	$P_{DC} = I_{DC}V_{CC}$
Collector efficiency	$\eta = \frac{P_O}{P_{DC}}$
RFC	$L_{RFC} \geq \frac{10R}{\omega_0}$
Blocking capacitance	$C_B \geq \frac{R}{10\omega_0}$
Angular frequency	$\omega_o = 2\pi f_o$
Loaded Q-factor (given bandwidth	$Q_L = \frac{f_o}{BW}$
Resonant tank reactance	$X = \frac{R}{Q_L}$
Resonant tank inductance	$L = \frac{X}{\omega_o}$
Resonant tank capacitance	$C = \frac{1}{\omega_o X}$

Table 8.11 Input parameters required for the Class-A/AB/B/C power amplifier software routine

Parameter	Unit	Mandatory/optional
Center frequency of the channel (f_O)	MHz	Mandatory
Output power (P_O)	mW	Mandatory
Supply voltage (V_{CC})	V	Mandatory
Bandwidth of the resonant tank	MHz	Mandatory
Half of conduction angle (θ)	°	Optional
Collector-emitter breakdown voltage (BV_{CEO})	V	Optional

8.4.4 Subroutine Outputs

The outputs of the Class-A/AB/B/C design subroutine are tabulated in Table 8.12.

8.5 Subroutine for Design of the Class-J Power Amplifiers

The schematic of the Class-J power amplifier is repeated in Fig. 8.13, but with an HBT as a driving transistor and all parameters not appearing in design equations omitted, as was done with previously analyzed power amplifier classes. The subroutine for the development of Class-J power amplifiers is perhaps the most tedious to develop, because for every set of specifications there are numerous combinations

```
%Enter PA design parameters
theta_deg = 120; % Default theta
fo = input('Please enter the frequency of operation, fo (MHz):
');
if (fo <= 0)
    fo = 1e9; %Frequency of operation
else
    fo = fo * 1e6;
end
w = 2*pi*fo;
Po = input('Please enter the required output power, Po (mW): ');
if (Po <= 0)
    Po = 0.01; %Output power
else
    Po = Po/1000;
end
Vcc = input('Please enter the value of the regulated power supply
(Vcc/Vdd): ');
if (Vcc <= 0)
    Vcc = 1; %Voltage supply
end
BVCE = input('Please enter collector-emitter breakdown voltage
(V) (-1 to skip): ');
if (BVCE <= 0)
    BVCE = 0; %Collector-emitter breakdown voltage
end
BW = input('Please enter the bandwidth of the resonant tank
(MHz): ');
if (BW <= 0)
    BW = 1e8; %Bandwidth
else
    BW = BW * 1e6;
end
theta = input('Please enter the theta (deg) (-1 to skip): ');
if (theta <= 0)
    theta = theta_deg/180*pi;
else
    theta = theta/180*pi;
end
```

Fig. 8.12 MATLAB code of the Class-A/AB/B/C power amplifier design subroutine (ClassABC.m)

```
%Calculate optimum resistance and waveforms + efficiency
g1 = (theta-sin(theta)*cos(theta))/(sin(theta)-
theta*cos(theta));%gamma
RL = 0.5*Vcc^2/Po;
VCEM = 2*Vcc;
Vm = Vcc;
Im = Vm/RL;
Idc = Im/g1;
Pdc = Idc*Vcc;
eff = Po/Pdc*100;
Lf = 10*RL/w;
CB = RL/(10*w);
%Calculate filter
QL = fo/BW;
X = RL/QL;
L = X/w;
C = 1 / (w*X);

%Display calculated values
fprintf('\nRESULTS:\n');
fprintf('R = %.2f Ohm \n', floor(100 * RL + 0.5) / 100);
fprintf('L = %.2f nH \n', floor(1e11 * L + 0.5) / 100);
fprintf('C = %.2f pF \n', floor(1e14 * C + 0.5) / 100);
fprintf('Lf = %.2f nH \n', floor(1e11 * Lf + 0.5) / 100);
fprintf('CB = %.2f pF \n', floor(1e14 * CB + 0.5) / 100);
fprintf('Idc = %.2f mA \n', floor(1e5 * Idc + 0.5) /100);
fprintf('Vm = %.2f V \n', floor(1e2 * Vm + 0.5) / 100);
fprintf('VCEM = %.2f V \n', floor(1e2 * VCEM + 0.5) / 100);
fprintf('Efficiency = %.2f % \n', floor(1e2 * eff + 0.5) / 100);
if (VCEM > BVCE && BVCE > 0)
    NVcc = BVCE / 2;
    fprintf('Warning: Vcc should not exceed %.2f or the transis-
tor will go into breakdown!\n', NVcc);
end
```

Fig. 8.12 (continued)

of design parameters that can yield a solution. Thus, to simplify the development challenge, we will make some initial assumptions:

1. In addition to the supply voltage (V_{CC}) and output power (P_O) specifications, the user of the subroutine will be required to specify the load resistance. This is done in order to fix the value of I_1 in equations that will be summarized in the next section. If solutions for different I_1 values are required (such as for the case where a solution cannot be found for the required load impedance), a wrapper subroutine that invokes this subroutine with different values of R_L can be written to increase the chances of finding a solution. The output impedance network needs to match real parts of impedances in addition to the imaginary part that needs to be matched by definition.
2. The design will be performed for the specified efficiency. Users can specify maximum theoretical efficiency or a lower figure. This will allow for the values of I_{CEM} and I_{DC} to be fixed in calculations, given V_{CC}, P_O and R_L.

Table 8.12 Output parameters of the Class-A/AB/B/C power amplifier design subroutine

Parameter	Unit
Optimum load resistance (R)	Ω
Resonant capacitor (C)	pF
Resonant inductor (L)	nH
Feed inductor (RFC) (L_f)	nH
Blocking capacitor (C_B)	pF
DC current (I_{DC})	mA
Peak collector voltage and peak output voltage (V_{CEM} and V_m)	V
Efficiency	%

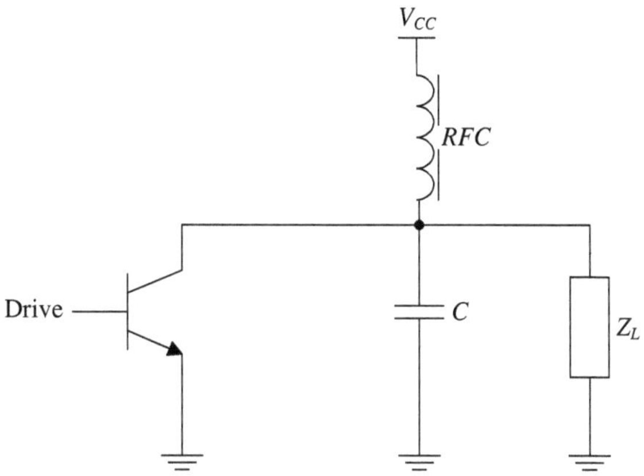

Fig. 8.13 Circuit of the Class-J power amplifier

3. The only remaining design parameters are the capacitor C, design angle ϕ and V_{off}. We elect to allow the user to choose the value ϕ and develop the procedure to find V_{off}. Finally, C is found as a part of the scaling factor that makes the mean value of V_o a unity value.

8.5.1 Equations

The required equations are tabulated in accordance with presented assumptions in Table 8.13. Note that we have used the relation $R_L = \mathrm{Re}(Z_L)$.

8.5.2 Input Parameters

Input parameters to the Class-J power amplifier design subroutine are summarized in Table 8.14; all parameters have been described previously.

Table 8.13 Summary of equations needed for the design of a Class-J power amplifier

Calculation	Equation
Required I_1	$I_1 = \sqrt{\frac{2P_O}{R_L}}$
DC current	$I_{DC} = \frac{P_O}{V_{CC}\eta}$
Peak collector current	$I_{CEM} = \pi I_{DC}$
Output voltage ($0 \le \omega t \le \pi$)	$v_o = \frac{1}{\omega C}\left[I_{CM}\left(1 - \frac{\theta}{\pi} - \cos\theta\right) + I_1(\cos(\theta + \phi) - \cos\phi)\right] + V_{OFF}$
Output voltage ($\pi \le \omega t \le 2\pi$)	$v_o = \frac{1}{\omega C}\left[I_{CM}\left(2 - \frac{\theta}{\pi}\right) + I_1(\cos(\theta + \phi) - \cos\phi)\right] + V_{OFF}$
Load reactance	$X_L = \frac{\tan\phi + \omega C R_L}{\omega C \tan\phi}$
Angular frequency	$\omega_o = 2\pi f_o$
RFC	$L_{RFC} \ge \frac{10R_i}{\omega_0}$

Table 8.14 Input parameters required for the Class-J power amplifier software routine

Parameter	Unit	Mandatory/optional
Center frequency of the channel (f_O)	MHz	Mandatory
Output power (P_O)	mW	Mandatory
Supply voltage (V_{CC})	V	Mandatory
Wanted load resistance (R_L)	Ω	Mandatory
Phase parameter ϕ	°	Mandatory

8.5.3 Description of the Subroutine

The subroutine is developed according to the assumptions made in the introductory section. Once all the parameters that can be calculated have been calculated and the value of ϕ has been chosen, we use MATLAB to find the minimum of the V_o function and decide on the value of V_{off} to make the function touch zero. In this step, a numerical solution approach is taken. Finally, we find the value of C which, at the specified angular frequency, averages the voltage function to unity. Recall that the phase parameter ϕ is achieved by the presence of the complex load, thus, the reactance of the load also needs to be calculated. Impedance matching in the main program can then be used to match the required complex load to the typical real load of the antenna.

The MATLAB code for the subroutine is listed in Fig. 8.14.

8.5.4 Subroutine Outputs

Outputs of the Class-J design subroutine are listed in Table 8.15.

```
%Enter PA design parameters
fo = input('Please enter the frequency of operation, fo (MHz):
');
if (fo <= 0)
    fo = 1e9; %Frequency of operation
else
    fo = fo * 1e6;
end
w = 2*pi*fo;
Po = input('Please enter the required output power, Po (mW): ');
if (Po <= 0)
    Po = 0.01; %Output power
else
    Po = Po/1000;
end
Vcc = input('Please enter the value of the regulated power supply
(Vcc/Vdd): ');
if (Vcc <= 0)
    Vcc = 1; %Voltage supply
end
RL = input('Please enter the required impedance (Ohm): ');
if (RL <= 0)
    RL = 50;
end
eff = input('Please enter the required efficiency (%): ');
if (eff <= 0) || (eff > 78.5);
    fprintf('\nSetting to default efficiency of 70 percent.\n');
    eff = 0.7;
else
    eff = eff/100;
end
phi = input('Please choose a value of phi (deg): ');
phi = phi/180*pi; %radians

%Calculations
I1 = sqrt(2*Po/RL);
Idc = Po / (Vcc*eff);
ICEM = pi*Idc;
%Find the minimum value of the waveform to calculate Voff
%Scale to get C (numeric solution)
theta1 = [0:180]*pi/180; %vector for 0 < theta < pi
theta2 = [181:360]*pi/180; % vector for pi < theta < 2pi
vo1 = ICEM*(1-theta1/pi-cos(theta1))+I1*(cos(theta1+phi)-
cos(phi)); % vo for 0 < theta < pi
vo2 = ICEM*(2-theta2/pi)+I1*(cos(theta2+phi)-cos(phi)); % vo for
pi < theta < 2pi
vo = horzcat(vo1, vo2); % concatenate to get waveform for 0 <
theta < 2pi
voff = -min(vo); % find the offset
voo = vo + voff; % vo with offset incorporated
scale_factor = 1/mean(voo); % caling factor for waveform to mean
to be 1
C = 1/(w*scale_factor); % capacitor value
XL = (tan(phi)+w*C*RL)/(w*C*tan(phi)); %Imaginary part of induc-
tance
Lf = 10*RL/w; %Feed inductor
```

Fig. 8.14 MATLAB code of the Class-J power amplifier design subroutine (ClassJ.m)

```
%Display calculated values
% fprintf('\nRESULTS:\n');
fprintf('RL = %.2f Ohm \n', floor(100 * RL + 0.5) / 100);
fprintf('XL = %.2f Ohm \n', floor(100 * XL + 0.5) / 100);
fprintf('C = %.2f pF \n', floor(1e14 * C + 0.5) / 100);
fprintf('Lf = %.2f nH \n', floor(1e11 * Lf + 0.5) / 100);
fprintf('voff = %.2f V \n', floor(100 * voff + 0.5) / 100);
fprintf('I1 = %.2f mA \n', floor(1e5 * I1 + 0.5) /100);
fprintf('Idc = %.2f mA \n', floor(1e5 * Idc + 0.5) /100);
fprintf('ICEM = %.2f mA \n', floor(1e5 * ICEM + 0.5) /100);
```

Fig. 8.14 (continued)

Table 8.15 Output parameters of the Class-J power amplifier design subroutine

Parameter	Unit
Design current (I_1)	mA
DC current (I_{DC})	mA
Peak collector current (I_{CEM})	mA
Offset voltage (V_{off})	V
Capacitor (C)	pF
Feed inductor (L_f)	nH
Load reactance (X_L)	Ω

8.6 Subroutine for Impedance Matching

As seen earlier in this chapter, it rarely occurs that the optimum impedance for a particular power amplifier stage is equal to a typical antenna impedance (50 Ω). Thus, before one can move on to develop a program that will encompass the procedures for the design of different power amplifier output stages, the subroutine for output impedance matching needs to be developed as well.

Power amplifiers can use both wideband and narrowband matching and the choice of an output network depends on the specific application. Therefore, for completeness, the subroutine needs to incorporate L-, T- and Π-networks, as well as quarter-wave matching, and to cater for both real and complex matching. Figures of different variations of L-, T- and Π-networks have all been shown in Chap. 6 and will not be repeated here.

8.6.1 Input Parameters

Even before inspecting the equation, in this case it is clear which input parameters are needed for the output impedance matching subroutine:

- *Source impedance of the matching network* (R_S). This impedance is equivalent to the optimum load resistance of the power amplifier and is specified in ohms. For some power amplifiers, such as Class-J, this quantity is complex ($Z_S = R_S + jX_S$).
- *Load impedance of the matching network* (R_L). This impedance is equivalent to the impedance of the antenna. It is specified in ohms and is usually equal to 50 Ω.
- *Center frequency of the matching network* (f_O). This is the center frequency of the channel. The matching network will only operate correctly at the center frequency. This quantity is specified in megahertz.
- *Bandwidth of the matching network* (*BW*). This parameter, also specified in megahertz, is needed for narrowband matching networks (T- and Π-networks).

8.6.2 Subroutine Outputs

For lumped impedance matching, inductance (L_i) and capacitance (C_i) parameters of different matching networks, where $i = 1$ or 2, are the outputs of the impedance matching subroutine. For matching using the quarter-wave transformer, the required characteristic impedance and the length, width and height of the microstrip line are the outputs.

8.6.3 Equations

The equations required for each matching network are tabulated in Table 8.16. When devising a routine for L-network matching, equations for matching of complex impedances are used, to cater for the Class-J amplifier. Where applicable, X_S and/or X_L can be set to zero.

8.6.4 Description of the Subroutine

The subroutine is developed in such a way as to calculate the matching components for all possible matching network variations using previously identified equations. This includes the lumped component matching and matching using the quarter-wave transformer. The above approach allows the user to select the best variation based on the specific implementation and the feasibility of the implementation of passives (inductors) or the transmission line.

Furthermore, when dealing with lumped matching, different matching network variations will result in different component sizes, some of which may be

Table 8.16 Summary of equations required for impedance matching subroutine

Calculation	Network	Equation
Shunt matching component	L (shunt component next to R_L)	$(X_p)_{1,2} = \frac{X_L \pm R_L Q}{\frac{R_L}{R_S} - 1}$
Series matching component	L (shunt component next to R_L)	$(X_s)_{1,2} = -(X_S \pm R_S Q)$
Shunt matching component	L (shunt component next to R_S)	$(X_p)_{1,2} = \frac{X_S \pm R_S Q}{\frac{R_S}{R_L} - 1}$
Series matching component	L (shunt component next to R_S)	$(X_s)_{1,2} = -(X_L \pm R_L Q)$
Smallest of the source and load resistance	T	$R_{small} = \min(R_S, R_L)$
Virtual resistance	T	$R_V = R_{small}(Q^2 + 1)$
Primary series reactance	T	$X_{S1} = QR_S$
Primary parallel reactance	T	$X_{P1} = \frac{R_V}{Q}$
Secondary Q-factor	T	$Q_2 = \sqrt{\frac{R_V}{R_L} - 1}$
Secondary series reactance	T	$X_{S2} = Q_2 R_L$
Secondary parallel reactance	T	$X_{P2} = \frac{R_V}{Q_2}$
Largest of the source and load resistance	Π	$R_{large} = \max(X_S, X_L)$
Virtual resistance	Π	$R_V = R_{small}(Q^2 + 1)$
Primary series reactance	Π	$X_{S2} = QR_V$
Primary parallel reactance	Π	$X_{P2} = \frac{R_L}{Q}$
Secondary Q-factor	Π	$Q_1 = \sqrt{\frac{R_S}{R_V} - 1}$
Secondary series reactance	Π	$X_{S1} = Q_1 R_V$
Secondary parallel reactance	Π	$X_{P1} = \frac{R_P}{Q_1}$
Primary Q-factor	T or Π	$Q = f_0/BW$
Wavelength	Quarter-wave	$\omega_o = 2\pi f_o$
Microstrip line length	Quarter-wave	$\lambda = \frac{v}{f}$
Required characteristic impedance	Quarter-wave	$Z_0 = \sqrt{R_S R_L}$
Z_f	Quarter-wave	$Z_f = \sqrt{\frac{\mu_0}{\varepsilon_0}} \approx 376.8\,\Omega$
Width-to-height ratio	Quarter-wave	$r = w/h$
Characteristic impedance of the narrow line	Quarter-wave	$Z_0 = \frac{Z_f}{2\pi\sqrt{\varepsilon_{eff}}} \ln\left(\frac{8}{r} + \frac{r}{4}\right)$
Effective dielectric constant (narrow line)	Quarter-wave	$\varepsilon_{eff} = \frac{\varepsilon_r + 1}{2} + \frac{\varepsilon_r - 1}{2}\left[\left(1 + \frac{12}{r}\right)^{-\frac{1}{2}} + 0.04(1 - r)^2\right]$
Characteristic impedance of the wide line	Quarter-wave	$Z_0 = \frac{Z_f}{\sqrt{\varepsilon_{eff}}\left[1.393 + r + \frac{2}{3}\ln(r + 1.444)\right]}$
Effective dielectric constant (narrow line)	Quarter-wave	$\varepsilon_{eff} = \frac{\varepsilon_r + 1}{2} + \frac{\varepsilon_r - 1}{2}\left(1 + \frac{12}{r}\right)^{-\frac{1}{2}}$
Effective dielectric constant	Quarter-wave	$\varepsilon_{eff} = \frac{\varepsilon_r + 1}{2} + \frac{\varepsilon_r - 1}{2}\left[\left(1 + \frac{12}{r}\right)^{-\frac{1}{2}} + 0.04(1 - r)^2\right]$

impractical, whether owing to difficulty with integration (when dealing with ICs) or owing to values difficult to reach with standard component values (discrete implementations). Another concern might be a necessity to eliminate separate coupling (blocking) capacitors, in which case a designer is likely to choose a network with a capacitor in series. In case of L-matching networks, the program will analyze the sizes of source and load resistances in order to determine whether the configuration with the shunt component next to the load or next to the source is applicable.

The complete MATLAB listing of the impedance matching procedure is given in Fig. 8.15. There is no separate procedure for the design of the quarter-wave transformer, and the same procedure developed for Class-F_∞ power amplifiers is used. Secondly, because the parameters at the input of this procedure are meant to be inherited from the main program, no inputs are required from the user. To make the routine standalone, a wrapper such as the one in Fig. 8.16 can be introduced.

8.7 Complete System Integration

Subroutines for designing different power amplifier blocks described in Sects. 8.1–8.6 are combined into one stand-alone comprehensive program. This program can be used to perform the complete output stage design, including the output impedance matching and design of all inductors needed in the power amplifier for integrated solutions. Subroutines are combined in such a way that certain parameters, such as operating frequency, needed for more than one sub-routine, are only entered once, which is a distinct advantage of using one complete routine for the design instead of using each of the subroutines separately. Matching impedances are deduced during program execution. For integrated power amplifier design utilizing spiral inductors, the required inductor values are obtained from stage design parameter calculations or from matching network calculations.

Thus the first step is to tabulate input and output parameters of the the main program.

8.7.1 Input Parameters

All input parameters of the full program are summarized in Table 8.17. This table also includes the section number in which the detailed description of each parameter is given. The table is complete with the parameters for spiral inductor design parameters, which is useful for integrated power amplifier design.

```
%L network
fprintf('\nL network: \n');
if RL > RS
    Q = sqrt(RL/RS - 1 + XL^2/(RS*RL));
    Xp1 = (XL + RL*Q)/(RL/RS - 1);
    Xp2 = (XL - RL*Q)/(RL/RS - 1);
    Xs1 = -(XS + RS*Q);
    Xs2 = -(XS - RS*Q);
    LL1 = Xp1/wo;
    CL1 = 1 / (Xs1*wo);
    LL2 = Xp2/wo;
    CL2 = 1 / (Xs2*wo);
    CL3 = 1 / (Xp1*wo);
    LL3 = Xs1/wo;
    CL4 = 1 / (Xp2*wo);
    LL4 = Xs2/wo;
    fprintf('First component in series with source and second in
shunt with load \n');
    fprintf('Option 1:\n');
    fprintf('CL1 = %.3f pF\n', floor(1e15 * CL1 + 0.5) / 1000);
    fprintf('LL1 = %.3f nH\n', floor(1e12 * LL1 + 0.5) / 1000);
    fprintf('Option 2:\n');
    fprintf('CL2 = %.3f pF\n', floor(1e15 * CL2 + 0.5) / 1000);
    fprintf('LL2 = %.3f nH\n', floor(1e12 * LL2 + 0.5) / 1000);
    fprintf('Option 3:\n');
    fprintf('LL3 = %.3f nH\n', floor(1e12 * LL3 + 0.5) / 1000);
    fprintf('CL3 = %.3f pF\n', floor(1e15 * CL3 + 0.5) / 1000);
    fprintf('Option 4:\n');
    fprintf('LL4 = %.3f nH\n', floor(1e12 * LL4 + 0.5) / 1000);
    fprintf('CL4 = %.3f pF\n', floor(1e15 * CL4 + 0.5) / 1000);
else
    Q = sqrt(RS/RL - 1 + XS^2/(RS*RL));
    Xp1 = (XS + RS*Q)/(RS/RL - 1);
    Xp2 = (SL - RS*Q)/(RS/RL - 1);
    Xs1 = -(XL + RL*Q);
    Xs2 = -(XL - RL*Q);
    LL5 = Xp1/wo;
    CL5 = 1 / (Xs1*wo);
    LL6 = Xp2/wo;
    CL6 = 1 / (Xs2*wo);
    CL7 = 1 / (Xp1*wo);
    LL7 = Xs1/wo;
    CL8 = 1 / (Xp2*wo);
    LL8 = Xs2/wo;
    fprintf('\nFirst component in shunt with source and second in
series with load \n');
    fprintf('Option 1:\n');
    fprintf('LL5 = %.3f nH\n', floor(1e12 * LL5 + 0.5) / 1000);
    fprintf('CL5 = %.3f pF\n', floor(1e15 * CL5 + 0.5) / 1000);
    fprintf('Option 2:\n');
    fprintf('LL6 = %.3f nH\n', floor(1e12 * LL6 + 0.5) / 1000);
    fprintf('CL6 = %.3f pF\n', floor(1e15 * CL6 + 0.5) / 1000);
    fprintf('Option 3:\n');
    fprintf('CL7 = %.3f pF\n', floor(1e15 * CL7 + 0.5) / 1000);
    fprintf('LL7 = %.3f nH\n', floor(1e12 * LL7 + 0.5) / 1000);
```

Fig. 8.15 MATLAB code for the impedance matching subroutine (calcMatch.m)

```
    fprintf('Option 4:\n');
    fprintf('CL8 = %.3f pF\n', floor(1e15 * CL8 + 0.5) / 1000);
    fprintf('LL8 = %.3f nH\n', floor(1e12 * LL8 + 0.5) / 1000);
end;%if

%For all other network types, ignore reactive parts (XS and XL)
%Only symmetric networks
%T networks
Rsmall = RS;
if RL < Rsmall
    Rsmall = RL;
end
R = Rsmall*(QL^2 + 1);
XS1 = QL*RS;
XP1 = R/QL;
Q2 = (R/RL - 1)^(1/2);
XP2 = R/Q2;
XS2 = Q2*RL;
CT1 = XP1*XP2/(XP1+XP2);
CT2 = XP1*XP2/abs(XP1-XP2);
LM1 = XS1/wo;
LM2 = XS2/wo;
LM3 = CT1/wo;
CM1 = 1/(CT1*wo);
CM2 = 1/(XS1*wo);
CM3 = 1/(XS2*wo);
%Results for T network
fprintf('\nind-cap-ind T network: \n');
fprintf('L1 = %.3f nH (side of the smaller resistance)\n',
floor(1e12 * LM1 + 0.5) / 1000);
fprintf('C1 = %.3f pF\n', floor(1e15 * CM1 + 0.5) / 1000);
fprintf('L2 = %.3f nH (side of the larger resistance)\n',
floor(1e12 * LM2 + 0.5) / 1000);
fprintf('cap-ind-cap T network: \n');
fprintf('C1 = %.3f pF (side of the smaller resistance)\n',
floor(1e15 * CM2 + 0.5) / 1000);
fprintf('L1 = %.3f nH\n', floor(1e12 * LM3 + 0.5) / 1000);
fprintf('C2 = %.3f pF (side of the larger resistance)\n',
floor(1e15 * CM3 + 0.5) / 1000);
%Pi network
RH = RS;
if RL > RH
    RH = RL;
end
R = RH/(QL^2 + 1);
XS2 = QL*R;
XP2 = RL/QL;
Qi = (RS/R - 1)^(1/2);
XP1 = RS/Q2;
XS1 = Qi*R;
CT1 = XS1 + XS2;
CT2 = abs(XS1-XS2);
LM4 = XP1/wo;
LM5 = XP2/wo;
LM6 = CT1/wo;
```

Fig. 8.15 (continued)

```
CM4 = 1/(CT1*wo);
CM5 = 1/(XP1*wo);
CM6 = 1/(XP2*wo);
%Results for Pi network
fprintf('\nind-cap-ind Pi network: \n');
fprintf('L1 = %.3f nH (side of the smaller resistance)\n',
floor(1e12 * LM4 + 0.5) / 1000);
fprintf('C1 = %.3f pF\n', floor(1e15 * CM4 + 0.5) / 1000);
fprintf('L2 = %.3f nH (side of the larger resistance)\n',
floor(1e12 * LM5 + 0.5) / 1000);
fprintf('cap-ind-cap Pi network: \n');
fprintf('C1 = %.3f pF (side of the smaller resistance)\n',
floor(1e15 * CM5 + 0.5) / 1000);
fprintf('L1 = %.3f nH\n', floor(1e12 * LM6 + 0.5) / 1000);
fprintf('C2 = %.3f pF (side of the larger resistance)\n',
floor(1e15 * CM6 + 0.5) / 1000);
%Quarter-wave transformer
fprintf('\nQuarter-wave transformer:\n \n');
ZS = RS;
ZL = RL;
qwt; %Invoke quarter-wave transformer procedure
```

Fig. 8.15 (continued)

```
%Calculates the match between real source and load given the fre-
quency of
%operation and bandwidth (for narrowband networks)
clear all;
RS = input('Please enter the value for the source resistance RS
(ohm): ');
XS = input('Please enter the value for the source reactance XS
(ohm) (0 to skip): ');
RL = input('Please enter the value for the load resistance RL
(ohm): ');
XL = input('Please enter the value for the load reactance XL
(ohm) (0 to skip): ');
fo = input('Please enter the operating frequency (MHz): ');
BW = input('Please enter the bandwidth (MHz): ');
fo = fo * 1e6;
BW = BW * 1e6;
wo = 2*pi*fo;
QL = fo/BW;
calcMatch; %Invoke matching subroutine
```

Fig. 8.16 MATLAB code for the matching subroutine wrapper that makes the program standalone (match.m)

8.7.2 *Routine Outputs*

Output parameters combine most of the output parameters calculated by separate subroutines. They are summarized in Table 8.18. As in the case of input parameters, the number of the section in which the parameter is described is included in the

Table 8.17 Summary of input parameters for the complete power amplifier design program

Parameter	Unit	Mandatory/ optional	Section(s)
Center frequency of the channel (f_o)	MHz	Mandatory	8.1, 8.2, 8.3, 8.4, 8.5
Loaded quality factor (Q_L)	–	Mandatory (Class E)	8.1
Output power (P_O)	mW	Mandatory	8.1, 8.2, 8.3, 8.4, 8.5
Supply voltage (V_{CC})	V	Mandatory	8.1, 8.2, 8.3, 8.4, 8.5
Collector-emitter saturation voltage (V_{CEsat})	V	Optional (Class E)	8.1
Maximum and minimum inductance values ($L_{\max}$ and $L_{\min}$)	nH	Optional (Class E)	8.1
Collector-emitter breakdown voltage (BV_{CEO})	V	Optional	8.1, 8.2, 8.3, 8.4
Collector-emitter capacitance, C_{ce}	pF	Optional (Class E)	8.1
Filtering inductances (L_0, L_3 and L_5)	nH	Mandatory (Class F)	8.2, 8.3
Half of conduction angle (θ)	o	Optional (Class F, Cls. A/AB/B/C)	8.2, 8.4
Antenna resistance (Z_A)	Ω	Optional (Class F_∞)	8.3
Relative dielectric constant of the substrate (ε_r)	–	Mandatory (Transmission line)	8.3, 8.6
Bandwidth of the resonant tank	MHz	Mandatory	8.4
Wanted load resistance (R_L)	Ω	Mandatory (Class J)	8.5
Parameter ϕ	o	Mandatory (Class J)	8.5
Bandwidth of the matching network (BW)	MHz	Mandatory	8.6
Minimum value of the inner diameter	μm	Optional (IC)	7.1
Maximum value of the outer diameter	μm	Optional (IC)	7.1
Minimum value for turn spacing	μm	Optional (IC)	7.1
Minimum turn width	μm	Optional (IC)	7.1
Inductance value tolerance	%	Optional (IC)	7.1
Grid resolution	μm	Optional (IC)	7.1
Thickness of metal in which the inductor spiral is laid out	nm	Optional (IC)	7.1
Resistivity of metal used for the spiral (ρ)	Ω m	Optional (IC)	7.1
Permeability of metal used for the spiral (μ)	H/m	Optional (IC)	7.1
Thickness of oxide between the spiral and underpass (t_m)	nm	Optional (IC)	7.1
Relative permittivity of oxide between the spiral and underpass (ε_{rm})	–	Optional (IC)	7.1
Thickness of oxide between substrate and the spiral (t_{sm})	nm	Optional (IC)	7.1
Relative permittivity of oxide between substrate and the spiral (ε_{rm})	–	Optional (IC)	7.1

(continued)

Table 8.17 (continued)

Parameter	Unit	Mandatory/ optional	Section(s)
Thickness of the silicon substrate (t_{Si})	μm	Optional (IC)	7.1
Relative permittivity of the silicon substrate (ε_{rSi})	–	Optional (IC)	7.1
Resistivity of the silicon substrate (ρ_{Si})	Ω m	Optional (IC)	7.1

table. For integrated implementations, all relevant spiral inductor parameters are also displayed.

8.7.3 Description and Flow Diagram of the Power Amplifier Design Routine

Upon starting the program, the user can choose the output stage in which to perform the power amplifier design. Depending on the choice, the applicable subroutine developed in Sects. 8.1–8.6 is executed. Following the chosen output stage design, the user may choose to perform the output impedance matching to the standard antenna impedance. If this option is chosen, the matching subroutine developed in Sect. 8.6 is executed.

The next choice presented to the user of the program is to export the SPICE netlist of the complete power amplifier system. For integrated circuit solutions the likely choice here is to accept the netlist export option; however, this is not applicable to discrete designs. The generated netlist is exported in T-spice format, already discussed in Chap. 7. It includes the subcircuit of the chosen matching network and the subcircuits of all spiral inductors, provided that particular options have been selected. The program is prohibited from exporting the netlist for the Class-F_∞ amplifier, because the SPICE simulator might not be able to handle the transmission line, and for the Class-J amplifier, because one cannot be certain how the load is implemented.

The next step is also applicable only to integrated design: the user can choose to design spiral inductors for all inductors (including matching inductors), needed for full power amplifier design. As seen in Chap. 7, the design of spiral inductors is process-specific, so by this time the designer should have chosen the process for fabrication. For illustration purposes, we use the same process as in Chap. 7 (ams AG S35 μm process), and we include the capability for both three-metal and thick-metal inductor design. If spiral inductor design is indeed selected at this step, the user then has a choice to override any of the inductor values (which may be practical if the user wants to correct any of the inductor values based on previous experience). The main program then invokes the inductance search algorithm separately for each required inductor, and if the netlist export option has been selected, exports the SPICE netlist subcircuit of each inductor into the netlist file.

Table 8.18 Summary of output parameters for the complete power amplifier design program

Parameter	Unit	Remark	Section(s)
Optimum load resistance (R_L)	Ω	All classes except Class-J	8.1, 8.2, 8.3, 8.4
DC current (I_{DC})	mA	–	8.1, 8.2, 8.3, 8.4, 8.5
Maximum current (I_{CEM})	mA	–	8.5
Maximum collector voltage (V_{CEM})	V	–	8.1, 8.2, 8.3, 8.4, 8.5
Shunt capacitor (C)	pF	Class E, Class J	8.1
Series capacitor (C_0)	pF	Class E	8.1
Inductor (L)	nH	Class E	8.1
Feed inductor (L_f)	nH	–	8.1, 8.2, 8.3, 8.4, 8.5
Actual shunt capacitor required	pF	Class E (Optional)	8.1
Loaded quality factor for maximum allowed inductance	–	Class E (Optional)	8.1
Nominal frequency resonant capacitor (C_0)	pF	Class F, Class-A/AB/B/C	8.2, 8.3, 8.5
Resonant inductor, L_0	nH	Class-A/AB/B/C	8.5
Third-harmonic frequency resonant capacitor (C_3)	pF	Class F	8.2
Fifth-harmonic frequency resonant capacitor (C_5)	pF	Class F	8.2
Peak output voltage (V_{om})	V	Class F	8.1, 8.2, 8.3
Blocking capacitor (C_B)	pF	Class F	8.2, 8.3
Wavelength (λ)	μm	Transmission line	8.3, 8.6
Transmission line length	μm	Transmission line	8.3, 8.6
Transmission line w/h	–	Transmission line	8.3, 8.6
Design current (I_1)	mA	Class-J	8.5
Offset voltage (V_{off})	V	Class-J	8.5
Load reactance (X_L)	Ω	Class-J	8.5
Matching network inductors and capacitors	nH or pF	L, T and Π networks	8.6
Effective inductance value of inductor at operating frequency (L_S)	nH	Each ind (opt.)	7.1
Nominal inductance value of the inductor (L_{inf})	nH	Each ind (opt.)	7.1
Q-factor of the inductor at the operating frequency (Q)	–	Each ind (opt.)	7.1
Resonant frequency of the inductor (f_r)	GHz	Each ind (opt.)	7.1
Width of the spiral	mm	Each ind (opt.)	7.1
Spacing between the turns of the spiral	mm	Each ind (opt.)	7.1
Input diameter of the spiral	mm	Each ind (opt.)	7.1
Output diameter of the spiral	mm	Each ind (opt.)	7.1
Number of turns of the spiral	–	Each ind (opt.)	7.1

The subroutine allows the user to name each inductor, or alternatively default inductor names are used.

During netlist export of the main routine, which happens at the end of the program execution, the program will allow the choice of transistor particulars.

The flow chart of the complete power amplifier design program is illustrated in Fig. 8.17.

8.7.4 MATLAB Code

The MATLAB code for the power amplifier design program, complete with all additional procedures, is listed in this section. The main routine used try/catch command to catch unexpected program execution errors.

The MATLAB code listing or the main routine invoking all subroutines is given in Fig. 8.18. The code for the procedure that allows for the inductance value override is listed in Fig. 8.19. Figure 8.20 illustrates the code for the procedure that invokes the inductance search algorithm for each required inductor, and Fig. 8.21 shows the procedure that exports subcircuits of all designed spiral inductors. The procedure for export of the SPICE netlist of the complete power amplifier system is given in Fig. 8.22, and procedures that export the matching network in the same netlist file for ideal inductors and spiral inductors are given in Figs. 8.23 and 8.24 respectively.

8.8 Design Examples

In this section, we will illustrate how to use the program and its procedures to design several different power amplifiers. We will use tasks as we have done in Chap. 6. Unless otherwise specified, we will assume that the driving transistor (be it a power NMOS or an HBT) provides enough gain, and the drive signal characteristics will provide an adequate bias/swing combination.

Task 1 Design two integrated power amplifier configurations to deliver 50 mW of power into the appropriate load at 2.4 GHz, one operating from the 1-V supply and the other operating from the 1.5-V supply.

Solution: Since there is no specified output stage, one needs to first investigate which output stage is best for this purpose. Seeing that an integrated solution is required, and there is no linearity requirement, one can conclude that one can design two switch-mode classes. Thus, a Class-E configuration and a Class-F configuration are chosen. Naturally, in the case of the Class-E configuration, one uses the ZVS variant, and for the Class-F, to increase efficiency, one chooses the Class-F_{35} variant. Since there is no explicit requirement to design any inductors or matching network, one can use the Class-E and Class-F subroutines as standalone programs.

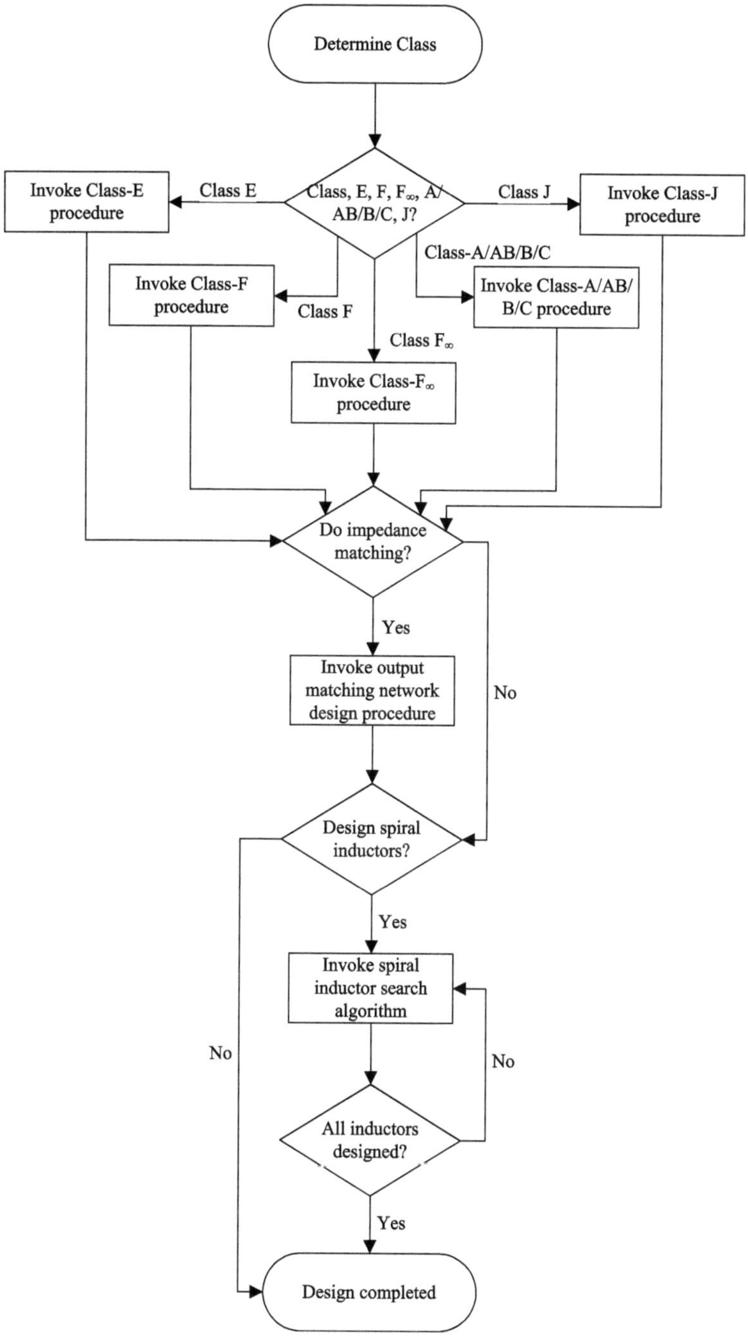

Fig. 8.17 Flow chart of the complete power amplifier design routine

```
try
    clear all; %Clear all variables from workspace
    fprintf('Switch-mode Power Amplifier Design Tool and Induc-
tance Calculator v3.0.0\n\n');
    fprintf('Developed by Mladen Božanic, as part of the book\n');
    fprintf('       POWER AMPLIFIERS FOR S-, C-, X- AND Ku
BANDS\n');
    fprintf('University of Johannesburg\n');

    choice = 0;
    inductor = zeros(1,10); %declare array size 10
    while (choice < 1 || choice > 5)
        choice = input('Please enter:\n1 for Class-E amplifier de-
sign, \n2 for Class-F amplifier design,\n3 for Class-F-QWT ampli-
fier design,\n4 for Class-A/AB/B/C amplifier design or\n5 for
Class-J amplifier design \n ');
        if (choice < 1 || choice > 5)
            fprintf('Wrong entry. Please try again.\n');
        end
    end
    switch choice
        case 1
           classE; %Class-E PA design program
        case 2
           classF; %Class-F PA design program
        case 3
           classFQWT; %Class-F with quarter-wave transformer PA
design program
        case 4
           classABC; %Class A/AB/B/C PA design program
        case 5
           classJ; %Class-J PA design program
    end%switch
    %Impedance matching
    wo = 2*pi*fo; %Calc. radian frequency
```

Fig. 8.18 MATLAB code of the main procedure of the power amplifier design program (paprog.m)

Class E To start the Class-E design, one runs `ClassE` from the MATLAB command window, and 50 mW is entered for output power, 1 V for voltage supply and 2400 MHz for frequency. Since the loaded Q-factor is not specified, one can choose it from experience. The authors decided to choose a value of 5. Since they had no process details, they could decide to ignore other program inputs.

The listing of the interaction with the program is shown in Fig. 8.25. With the aid of the program, the solution was determined to be $R_L = 11.5\ \Omega$, $L_f = 41.68$ nH, series inductance is $L = 3.83$ nH, series capacitance is $C_0 = 1.71$ pF and shunt capacitance is $C = 1.05$ pF. Furthermore, the expected DC current consumption was 49.97 mA and the peak collector voltage was 3.56 V.

```
if choice ~=3 %matching is not allowed for Class-F QWT
        cmat = input('\nDo you want to perform the impedance
matching to 50 ohm antenna impedance (y/n)? ', 's');
        if cmat == 'y' || cmat == 'Y';
            %Should be transparent except for the bandwidth and
matching
            %impedance
            BW = input('\nPlease enter the matching bandwidth
(MHz): '); %Input bandwidth
            BW = BW * 1e6;
            RS = RL; %Source resistance
            RL = 50; %Load resistance
            QL = fo/BW; %Quality factor
            XL = 0;
            if choice ~= 5
                XS = 0;
            else
                XS = XL; %For Class J
            end%if
            if(RS == RL)
                fprintf('Source and load resistances are the same.
No matching needed.\n');
                cmat = 'n'; %no matching
            else
                calcMatch;
            end%if
        end%if
        %cnet - flag - if selected, netlist will be performed
        if choice ~=5
            cnet = input('\nDo you want to export the netlist of
the PA (y/n)? ', 's');
        else
            cnet = 'n'; %no netlist export Class J
        end
else
        cmat = 'n'; %no matching for Class F-QWT
        cnet = 'n'; %no netlist export Class F-QWT
    end%if choice ~=3
    fprintf('\n');
    if cmat == 'y' || cmat == 'Y';
        %cind - flag - if selected, spiral inductors will be found
        cind = input('\nDo you want to attempt to find inductors
for the PA and the matching networks (y/n)? ', 's');
        if (cind == 'y' || cind == 'Y') || (cnet == 'y' || cnet ==
'Y');
            %cmn - choice of the matching network
            cmn = input('Please enter which matching network you
would like to use: \n 1 - L-network option 1 \n 2 - L-network op-
tion 2 \n 3 - L-network option 3 \n 4 - L-network option 4 \n 5 -
ind-cap-ind T-network  \n 6 - cap-ind-cap T network \n 7 - ind-
cap-ind Pi network \n 8 - cap-ind-cap Pi network: \n (quarter-wave
transformer not allowed) \n ');
```

Fig. 8.18 (continued)

```
  else
            cmn = 0;
        end
    else
        cmn = 0;
        %cind - flag - if selected, spiral inductors will be found
(no
        %matching case)
        cind = input('\nDo you want to attempt to find inductor(s)
for the PA (y/n)? ', 's');
    end

    if (cind == 'y' || cind == 'Y') || (cnet == 'y' || cnet ==
'Y');
        %Overrides any inductor values if needed
        overrideInductors;
    end
    fprintf('\n');
    if cind == 'y' || cind == 'Y';
        %Find inductors
        %Define constants
        b = 1.62e-3; a1 = -1.21; a2 = -0.147; a3 = 2.40; a4 =
1.78; a5 = -0.030;
        f = fo;
        omega = wo;
        %Setup the inductance search algorithm search paramters
        indSetup;
        cprocess = 0;
        if cnet == 'y' || cnet == 'Y';
            %Name of the PA used for netlist extraction
            paName = input('\nPlease enter the PA name (press "s"
to skip): ', 's');
            if paName(1) == 's' || paName(1) == 'S';
                %Default PA name
                paName = 'MyPA';
            end%if
            %RFC = 1;
            fileName = strcat(paName, '_net.spc');
            fid = fopen(fileName, 'w');
        end
        fprintf('\nINDUCTORS\n');
        %Find the inductors designed by the PA program
        findInductors;
    end
    %Export the netlist
    exportPA;
catch
    fprintf('MATLAB returns: \n%s\nError occurred while running
the program. Please try running the program again.\n\n', lasterr);
end%try
```

Fig. 8.18 (continued)

Class F For Class-F design, one runs `ClassF` from the MATLAB command window. With no specified base filter inductances, we choose 1 nH for the base filter and 0.5 nH for the third and fifth harmonic filters. The rest of the input

```
%Overrides any inductance values calculated by PA design program
cov = input('Do you want to override any of the inductance values
(y/n)? ', 's');
if cov == 'y' || cov == 'Y';
    fprintf('Old Lf = %.2f nH. ', Lf*1e9); %Feed inductor first
    iLf = input ('New Lf (-1 to skip): ');
    if iLf > 0;
        Lf = iLf/1e9;
    end
    if choice == 1;
        fprintf('Old series L = %.2f nH. ', L2*1e9);
        iL2 = input ('New L (-1 to skip): ');
        if iL2 > 0;
            L2 = iL2/1e9;
        end
    elseif choice == 2 || choice == 3;
        fprintf('Old L0 = %.2f nH. ', L0*1e9);
        iL0 = input ('New L0 (-1 to skip): ');
        if iL0 > 0;
            L0 = iL0/1e9;
        end
        if choice == 2
            fprintf('Old L3 = %.2f nH. ', L3*1e9);
            iL3 = input ('New L3 (-1 to skip): ');
            if iL3 > 0;
                L3 = iL3/1e9;
            end
        end
        if choice == 2 && harmonics == 2
           fprintf('Old L5 = %.2f nH. ', L5*1e9);
           iL5 = input ('New L5 (-1 to skip): ');
           if iL5 > 0;
               L5 = iL5/1e9;
           end
        end
    elseif choice == 4
        fprintf('Old resonant L = %.2f nH. ', L*1e9);
        iL = input ('New L (-1 to skip): ');
        if iL > 0;
            L = iL/1e9;
        end
    end
    if cmn == 1;
        fprintf('Old LL1 = %.2f nH. ', LL1*1e9);
        iLL1 = input ('New LL1 (-1 to skip): ');
        if iLL1 > 0;
            LL1 = iLL1/1e9;
        end
    elseif cmn == 2;
        fprintf('Old LL2 = %.2f nH. ', LL2*1e9);
        iLL2 = input ('New LL2 (-1 to skip): ');
        if iLL2 > 0;
            LL2 = iLL2/1e9;
        end
    elseif cmn == 3;
        fprintf('Old LL3 = %.2f nH. ', LL3*1e9);
        iLL3 = input ('New LL3 (-1 to skip): ');
        if iLL3 > 0;
            LL3 = iLL3/1e9;
        end
```

Fig. 8.19 MATLAB code for the inductance values override procedure (overrideInductors.m)

```
elseif cmn == 4;
        fprintf('Old LL4 = %.2f nH. ', LL4*1e9);
        iLL4 = input ('New LL4 (-1 to skip): ');
        if iLL4 > 0;
            LL4 = iLL4/1e9;
        end
    elseif cmn == 5;
        fprintf('Old LM1 = %.2f nH. ', LM1*1e9);
        iLM1 = input ('New LM1 (-1 to skip): ');
        if iLM1 > 0;
            LM1 = iLM1/1e9;
        end
        fprintf('Old LM2 = %.2f nH. ', LM2*1e9);
        iLM2 = input ('New LM2 (-1 to skip): ');
        if iLM2 > 0;
            LM2 = iLM2/1e9;
        end
    elseif cmn == 6;
        fprintf('Old LM3 = %.2f nH. ', LM3*1e9);
        iLM3 = input ('New LM3 (-1 to skip): ');
        if iLM3 > 0;
            LM3 = iLM3/1e9;
        end
    elseif cmn == 7;
        fprintf('Old LM4 = %.2f nH. ', LM4*1e9);
        iLM4 = input ('New LM4 (-1 to skip): ');
        if iLM4 > 0;
            LM4 = iLM4/1e9;
        end
        fprintf('Old LM5 = %.2f nH. ', LM5*1e9);
        iLM5 = input ('New LM5 (-1 to skip): ');
        if iLM5 > 0;
            LM5 = iLM5/1e9;
        end
    elseif cmn == 8;
        fprintf('Old LM6 = %.2f nH. ', LM6*1e9);
        iLM6 = input ('New LM6 (-1 to skip): ');
        if iLM6 > 0;
            LM6 = iLM6/1e9;
        end
    end
end
```

Fig. 8.19 (continued)

parameters are the same as for the Class-E power amplifier, and again we decide to ignore any other calculations. We also decide to use the default conduction angle setting.

The listing of the interaction with the program is shown in Fig. 8.26. With the aid of the program, we conclude that the solution is $R_L = 30.9\ \Omega$, $L_f = 20.49$ nH, base capacitance is $C_0 = 4.40$ pF, third harmonic capacitance is $C_3 = 0.98$ pF, fifth harmonic capacitance is $C_5 = 0.35$ pF and blocking capacitance is $C_B = 205$ pF. Further, the expected DC current consumption is 43.07 mA and peak collector voltage is 3 V.

Task 2 Design a power amplifier to deliver 1 W of power into a 50 Ω antenna from a 5-V supply at 15 GHz.

```
%Find spiral inductor geometries for all inductors
entProcParam; %Enter inductor process parameters
 Ls = Lf*1e9;
fprintf('\nInductor value: %.3f nH',Ls);
indSearch; %Inductance search algorithm
if cnet == 'y' || cnet == 'Y';
    inductorName = 'Lf';
    exportSubckt; %Exports the spiral inductor subcircuit
end
if choice == 1;
    Ls = L2*1e9;
    fprintf('\nInductor value: %.3f nH',Ls);
    indSearch; %Inductance search algorithm
    if cnet == 'y' || cnet == 'Y';
        inductorName = 'L2';
        exportSubckt; %Exports the spiral inductor subcircuit
    end
elseif choice == 2 || choice == 3;
    Ls = L0*1e9;
    fprintf('\nInductor value: %.3f nH',Ls);
    indSearch; %Inductance search algorithm
    if cnet == 'y' || cnet == 'Y';
        inductorName = 'L0';
        exportSubckt; %Exports the spiral inductor subcircuit
    end
    if choice == 2
        if L3 ~= L0;
            Ls = L3*1e9;
            fprintf('\nInductor value: %.3f nH',Ls);
            indSearch; %Inductance search algorithm
        end
        if cnet == 'y' || cnet == 'Y';
            inductorName = 'L3';
            exportSubckt; %Exports the spiral inductor subcircuit
        end
    end
    if choice == 2 && harmonics == 2;
        if L5 ~= L3 && L5 ~= L0;
            Ls = L5*1e9;
            fprintf('\nInductor value: %.3f nH',Ls);
            indSearch; %Inductance search algorithm
        end
        if cnet == 'y' || cnet == 'Y';
            inductorName = 'L5';
            exportSubckt; %Exports the spiral inductor subcircuit
        end
    end
end
if cmn == 1;
    Ls = LL1*1e9;
    fprintf('\nInductor value: %.3f nH',Ls);
    indSearch; %Inductance search algorithm
    if cnet == 'y' || cnet == 'Y';
        inductorName = 'LL1';
        exportSubckt; %Exports the spiral inductor subcircuit
    end
end
```

Fig. 8.20 MATLAB code for the procedure for search for geometries of all needed inductors (findInductors.m)

```
if cmn == 2;
    Ls = LL2*1e9;
    fprintf('\nInductor value: %.3f nH',Ls);
    indSearch; %Inductance search algorithm
    if cnet == 'y' || cnet == 'Y';
        inductorName = 'LL2';
        exportSubckt; %Exports the spiral inductor subcircuit
    end
end
if cmn == 3;
    Ls = LL3*1e9;
    fprintf('\nInductor value: %.3f nH',Ls);
    indSearch; %Inductance search algorithm
    if cnet == 'y' || cnet == 'Y';
        inductorName = 'LL3';
        exportSubckt; %Exports the spiral inductor subcircuit
    end
end
if cmn == 4;
    Ls = LL4*1e9;
    fprintf('\nInductor value: %.3f nH',Ls);
    indSearch; %Inductance search algorithm
    if cnet == 'y' || cnet == 'Y';
        inductorName = 'LL4';
        exportSubckt; %Exports the spiral inductor subcircuit
    end
end
if cmn == 5;
    Ls = LM1*1e9;
    fprintf('\nInductor value: %.3f nH',Ls);
    indSearch; %Inductance search algorithm
    if cnet == 'y' || cnet == 'Y';
        inductorName = 'LM1';
        exportSubckt; %Exports the spiral inductor subcircuit
    end
    Ls = LM2*1e9;
    fprintf('\nInductor value: %.3f nH',Ls);
    indSearch; %Inductance search algorithm
    if cnet == 'y' || cnet == 'Y';
        inductorName = 'LM2';
        exportSubckt; %Exports the spiral inductor subcircuit
    end
end
if cmn == 6;
    Ls = LM3*1e9;
    fprintf('\nInductor value: %.3f nH',Ls);
    indSearch; %Inductance search algorithm
    if cnet == 'y' || cnet == 'Y';
        inductorName = 'LM3';
        exportSubckt; %Exports the spiral inductor subcircuit
    end
end
if cmn == 7;
    Ls = LM4*1e9;
    fprintf('\nInductor value: %.3f nH',Ls);
    indSearch; %Inductance search algorithm
    if cnet == 'y' || cnet == 'Y';
        inductorName = 'LM4';
        exportSubckt; %Exports the spiral inductor subcircuit
    end
```

Fig. 8.20 (continued)

```
        Ls = LM5*1e9;
        fprintf('\nInductor value: %.3f nH',Ls);
        indSearch; %Inductance search algorithm
        if cnet == 'y' || cnet == 'Y';
            inductorName = 'LM5';
            exportSubckt; %Exports the spiral inductor subcircuit
        end
    end
    if cmn == 8;
        Ls = LM6*1e9;
        fprintf('\nInductor value: %.3f nH',Ls);
        indSearch; %Inductance search algorithm
        if cnet == 'y' || cnet == 'Y';
            inductorName = 'LM6';
            exportSubckt; %Exports the spiral inductor subcircuit
        end
    end
```

Fig. 8.20 (continued)

```
%Exports the spiral inductor subcircuit
try
   fprintf(fid, '.SUBCKT %s L1 L2 GND\n', inductorName);
    fprintf(fid, 'CS L1 L2 %.2ffF\n', floor(1e17 * Csstored + 0.5)
/ 100);
    fprintf(fid, 'LS N4 L1 %.2fn\n', floor(100 * Lclfstored + 0.5)
/ 100);
    fprintf(fid, 'RS N4 L2 %.2f\n', floor(100 * Rsstored + 0.5) /
100);
    fprintf(fid, 'Csi1 N2 GND %.2ffF\n', floor(1e17 * CSistored +
0.5) / 100);
    fprintf(fid, 'Csi2 N3 GND %.2ffF\n', floor(1e17 * CSistored +
0.5) / 100);
    fprintf(fid, 'Cox1 L1 N2 %.2ffF\n', floor(1e17 * Coxstored +
0.5) / 100);
    fprintf(fid, 'Cox2 L2 N3 %.2ffF\n', floor(1e17 * Coxstored +
0.5) / 100);
      fprintf(fid, 'Rsi1 N2 GND %.2f\n', floor(100 * RSistored +
0.5) / 100);
      fprintf(fid, 'Rsi2 N3 GND %.2f\n', floor(100 * RSistored +
0.5) / 100);
      fprintf(fid, '.ENDS\n\n');
  catch
      fprintf('MATLAB reports:\n%s\n', lasterr);
      fprintf('Netlist export failed.\n');
  end%try
```

Fig. 8.21 MATLAB code for the procedure for export of the spiral inductor subcircuit (exportSubckt.m)

Solution: In this task, the requirement is to deliver high power at a high frequency with a specified antenna. If we choose the Class-F_∞ power amplifier, we can automatically perform impedance matching, and implementing transmission lines is not a problem because this configuration is clearly a discrete implementation, with a

```
%This procedure exports the netlist of designed PA
if (cnet == 'y' || cnet == 'Y')
    cModel = 0;
    %Input the transistor choice
    while cModel ~= 1 && cModel ~= 2 && cModel ~= 3;
        fprintf('\n');
        cModel = input('Please enter 1 for HBT, 2 RF NMOS or 3 for
any other transistor: ');
    end
    if cModel == 1;
       modelName = input('Please enter the model name (e.g.
npn254h5): ', 's');
       if modelName(1) == 's' || modelName(1) == 'S';
           modelName = 'npn254';
       end
        wEmitters = input('Please enter the total emitter width
(um) (-1 to skip): ');
        if wEmitters < 0;
            wEmitters = 1;
        end
        transistor = strcat(['X',modelName,'_1 N_1 bias Gnd Gnd
',modelName,' area=', num2str(wEmitters)]);
    elseif cModel == 2;
        lnmos = input('Please enter the NMOS length (um): ');
        wnmos = input('Please enter the NMOS width (um): ');
        ng = input('Please enter the number of NMOS fingers: ');
        addparam = input('Please type in any additional SPICE pa-
rameters in format PARAM=param ("s" to skip): ', 's');
        if lnmos < 0;
            lnmos = 1;
        end
        if wnmos < 0;
            wnmos = 100;
        end
        if ng < 0;
            ng = 10;
        end
        if addparam(1) == 's'
            addparam = ' ';
        end
        transistor = strcat (['Xnmosrf_1 N_1 bias Gnd Gnd nmosrf
l=',num2str(lnmos),'u wtot=',num2str(wnmos),'u ng=',num2str(ng),'
',addparam]);
    else
        transistor = input('Please type the SPICE command for your
transistor: ','s');
    end
end
%Export the netlist
% Class E with ideal inductors
if (cnet == 'y' || cnet == 'Y') && (cind ~= 'y' && cind ~= 'Y') &&
choice == 1;
        paName = input('\nPlease enter the PA name (press "s" to
skip): ', 's');
        if paName(1) == 's' || paName(1) == 'S';
            paName = 'MyPA';
        end%if
        if Lf == 0;
            RFC = 100;
        else
            RFC = Lf*1e9;
        end
```

Fig. 8.22 MATLAB code for the procedure for netlist export (exportPA.m)

```
    fileName = strcat(paName, '_net.spc');
        fid = fopen(fileName, 'w');
        fprintf(fid, '.SUBCKT %s bias supply Gnd\n', paName);
        fprintf(fid, 'LRFC_2 supply N_1 %.2fn\n', RFC);
        if cmn == 0;
            fprintf(fid, 'LInductor_2 N_5 N_3 %.2fn\n', floor(1e11
* L2 + 0.5) / 100);
            fprintf(fid, 'RResistor_1 N_3 Gnd %.2f TC=0.0, 0.0\n',
floor(100 * RL + 0.5) / 100);
        else
            fprintf(fid, 'LInductor_2 N_5 N_M1 %.2fn\n',
floor(1e11 * L2 + 0.5) / 100);
            fprintf(fid, 'RResistor_1 N_M2 Gnd %.2f TC=0.0,
0.0\n', 50);
        end
        fprintf(fid, transistor);
        fprintf(fid, '\n');
        if Lf == 0;
            fprintf(fid, 'CCapacitor_1 N_1 Gnd %.2fp\n',
floor(1e14 * C1 + 0.5) / 100);
        else
            fprintf(fid, 'CCapacitor_1 N_1 Gnd %.2fp\n',
floor(1e14 * Cp + 0.5) / 100);
        end
        fprintf(fid, 'CCapacitor_2 N_1 N_5 %.2fp\n', floor(1e14 *
C2 + 0.5) / 100);
        matchIdeal; %Export matching inductors
        fprintf(fid, '.ENDS\n');
        fclose(fid);
        fprintf('File %s created successfully.\n\n', fileName);
    end
    %Class E with spiral inductors
    if (cnet == 'y' || cnet == 'Y') && (cind == 'y' || cind ==
'Y') && choice == 1;
        fprintf(fid, '.SUBCKT %s bias supply Gnd\n', paName);
        if Lf == 0;
            RFC = 100;
            fprintf(fid, 'LRFC_2 supply N_1 %.2fn\n', RFC);
        else
            fprintf(fid, 'XLf_1 supply N_1 Gnd Lf\n');
        end
        if cmn == 0;
            fprintf(fid, 'XL2_1 N_5 N_3 Gnd L2\n');
            fprintf(fid, 'RResistor_1 N_3 Gnd %.2f TC=0.0, 0.0\n',
floor(100 * RL + 0.5) / 100);
        else
            fprintf(fid, 'XL2_1 N_5 N_M1 Gnd L2\n');
            fprintf(fid, 'RResistor_1 N_M2 Gnd %.2f TC=0.0,
0.0\n', 50);
        end
        fprintf(fid, transistor);
        fprintf(fid, '\n');
        fprintf(fid, 'CCapacitor_1 N_1 Gnd %.2fp\n', floor(1e14 *
C1 + 0.5) / 100);
        fprintf(fid, 'CCapacitor_2 N_1 N_5 %.2fp\n', floor(1e14 *
C2 + 0.5) / 100);
        matchSpiral; %Export matching inductors
        fprintf(fid, '.ENDS\n');
        fclose(fid);
        fprintf('File %s created successfully.\n\n', fileName);
    end
```

Fig. 8.22 (continued)

```
%Class F with ideal inductors
    if (cnet == 'y' || cnet == 'Y') && (cind ~= 'y' && cind ~=
'Y') && choice == 2;
        paName = input('\nPlease enter the PA name (press "s" to
skip): ', 's');
        if paName(1) == 's' || paName(1) == 'S';
            paName = 'MyPA';
        end%if
        fileName = strcat(paName, '_net.spc');
        fid = fopen(fileName, 'w');
        RFC = Lf*1e9;
        fprintf(fid, '.SUBCKT %s bias supply Gnd\n', paName);
        fprintf(fid, 'LRFC_2 supply N_1 %.2fn\n', RFC);
        fprintf(fid, 'CCapacitor_3 N_1 N_4 %.2fp\n', 1);
        if cmn == 0;
            fprintf(fid, 'LInductor_2 N_2 Gnd %.2fn\n', floor(1e11
* L0 + 0.5) / 100);
            fprintf(fid, 'CCapacitor_1 N_2 Gnd %.2fp\n',
floor(1e14 * C0 + 0.5) / 100);
            if harmonics == 1;
                fprintf(fid, 'LInductor_1 N_4 N_2 %.2fn\n',
floor(1e11 * L3 + 0.5) / 100);
                fprintf(fid, 'CCapacitor_2 N_4 N_2 %.2fp\n',
floor(1e14 * C3 + 0.5) / 100);
            end
            if harmonics == 2;
                fprintf(fid, 'LInductor_1 N_4 N_3 %.2fn\n',
floor(1e11 * L3 + 0.5) / 100);
                fprintf(fid, 'CCapacitor_2 N_4 N_3 %.2fp\n',
floor(1e14 * C3 + 0.5) / 100);
                fprintf(fid, 'LInductor_3 N_3 N_2 %.2fn\n',
floor(1e11 * L5 + 0.5) / 100);
                fprintf(fid, 'CCapacitor_4 N_3 N_2 %.2fp\n',
floor(1e14 * C5 + 0.5) / 100);
            end
            fprintf(fid, 'RResistor_1 N_2 Gnd %.2f TC=0.0, 0.0\n',
floor(100 * RL + 0.5) / 100);
        else
            fprintf(fid, 'LInductor_2 N_M1 Gnd %.2fn\n',
floor(1e11 * L0 + 0.5) / 100);
            fprintf(fid, 'CCapacitor_1 N_M1 Gnd %.2fp\n',
floor(1e14 * C0 + 0.5) / 100);
            if harmonics == 1;
                fprintf(fid, 'LInductor_1 N_4 N_M1 %.2fn\n',
floor(1e11 * L3 + 0.5) / 100);
                fprintf(fid, 'CCapacitor_2 N_4 N_M1 %.2fp\n',
floor(1e14 * C3 + 0.5) / 100);
            end
            if harmonics == 2;
                fprintf(fid, 'LInductor_1 N_4 N_3 %.2fn\n',
floor(1e11 * L3 + 0.5) / 100);
                fprintf(fid, 'CCapacitor_2 N_4 N_3 %.2fp\n',
floor(1e14 * C3 + 0.5) / 100);
                fprintf(fid, 'LInductor_3 N_3 N_M1 %.2fn\n',
floor(1e11 * L5 + 0.5) / 100);
                fprintf(fid, 'CCapacitor_4 N_3 N_M1 %.2fp\n',
floor(1e14 * C5 + 0.5) / 100);
            end
```

Fig. 8.22 (continued)

```
   fprintf(fid, 'RResistor_1 N_M2 Gnd %.2f TC=0.0, 0.0\n', 50);
        end

        fprintf(fid, transistor);
        fprintf(fid, '\n');
        matchIdeal; %Export matching inductors
        fprintf(fid, '.ENDS\n');
        fclose(fid);
        fprintf('File %s created successfully.\n\n', fileName);
    end
    %Class F with spiral inductors
    if (cnet == 'y' || cnet == 'Y') && (cind == 'y' || cind ==
'Y') && choice == 2;
        fprintf(fid, '.SUBCKT %s bias supply Gnd\n', paName);
        RFC = Lf*1e9;
        fprintf(fid, 'LRFC_2 supply N_1 %.2fn\n', RFC);
        fprintf(fid, 'CCapacitor_3 N_1 N_4 %.2fp\n', 1);
        if cmn == 0;
           fprintf(fid, 'XL0_1 N_2 Gnd Gnd L0\n');
           fprintf(fid, 'CCapacitor_1 N_2 Gnd\n');
           if harmonics == 1;
                fprintf(fid, 'XL3_1 N_4 N_2 Gnd L3\n');
                fprintf(fid, 'CCapacitor_2 N_4 N_2 %.2fp\n',
floor(1e14 * C3 + 0.5) / 100);
           end
           if harmonics == 2;
                fprintf(fid, 'XL3_1 N_4 N_3 Gnd L3\n');
                fprintf(fid, 'CCapacitor_2 N_4 N_3 %.2fp\n',
floor(1e14 * C3 + 0.5) / 100);
                fprintf(fid, 'XL5_1 N_3 N_2 Gnd L5\n');
                fprintf(fid, 'CCapacitor_4 N_3 N_2 %.2fp\n',
floor(1e14 * C5 + 0.5) / 100);
           end
           fprintf(fid, 'RResistor_1 N_2 Gnd %.2f TC=0.0, 0.0\n',
floor(100 * RL + 0.5) / 100);
        else
           fprintf(fid, 'XL0_1 N_M1 Gnd Gnd L0\n');
           fprintf(fid, 'CCapacitor_1 N_M1 Gnd %.2fp\n',
floor(1e14 * C0 + 0.5) / 100);
           if harmonics == 1;
                fprintf(fid, 'XL3_1 N_4 N_M1 Gnd L3\n');
                fprintf(fid, 'CCapacitor_2 N_4 N_M1 %.2fp\n',
floor(1e14 * C3 + 0.5) / 100);
           end
           if harmonics == 2;
                fprintf(fid, 'XL3_1 N_4 N_3 Gnd L3\n');
                fprintf(fid, 'CCapacitor_2 N_4 N_3 %.2fp\n',
floor(1e14 * C3 + 0.5) / 100);
                fprintf(fid, 'XL5_1 N_3 N_M1 Gnd L5\n');
                fprintf(fid, 'CCapacitor_4 N_3 N_M1 %.2fp\n',
floor(1e14 * C5 + 0.5) / 100);
           end
           fprintf(fid, 'RResistor_1 N_M2 Gnd %.2f TC=0.0,
0.0\n', 50);
        end
        fprintf(fid, transistor);
        fprintf(fid, '\n');
        matchSpiral; %Export matching inductors
```

Fig. 8.22 (continued)

```
  fprintf(fid, '.ENDS\n');
        fclose(fid);
        fprintf('File %s created successfully.\n\n', fileName);
    end
    %Class ABC with ideal inductors
    if (cnet == 'y' || cnet == 'Y') && (cind ~= 'y' && cind ~=
'Y') && choice == 3;
        paName = input('\nPlease enter the PA name (press "s" to
skip): ', 's');
        if paName(1) == 's' || paName(1) == 'S';
            paName = 'MyPA';
        end%if
        fileName = strcat(paName, '_net.spc');
        fid = fopen(fileName, 'w');
        RFC = Lf*1e9;
        fprintf(fid, '.SUBCKT %s bias supply Gnd\n', paName);
        fprintf(fid, 'LRFC_2 supply N_1 %.2fn\n', RFC);
        if cmn == 0;
           fprintf(fid, 'CCapacitor_3 N_1 N_2 %.2fp\n', 1);
           fprintf(fid, 'LInductor_2 N_2 Gnd %.2fn\n', floor(1e11
* L0 + 0.5) / 100);
           fprintf(fid, 'CCapacitor_1 N_2 Gnd %.2fp\n',
floor(1e14 * C0 + 0.5) / 100);
           fprintf(fid, 'RResistor_1 N_2 Gnd %.2f TC=0.0, 0.0\n',
floor(100 * RL + 0.5) / 100);
        else
           fprintf(fid, 'CCapacitor_3 N_1 N_M1 %.2fp\n', 1);
           fprintf(fid, 'LInductor_2 N_M1 Gnd %.2fn\n',
floor(1e11 * L0 + 0.5) / 100);
           fprintf(fid, 'CCapacitor_1 N_M1 Gnd %.2fp\n',
floor(1e14 * C0 + 0.5) / 100);
           fprintf(fid, 'RResistor_1 N_M2 Gnd %.2f TC=0.0,
0.0\n', 50);
        end
        fprintf(fid, transistor);
        fprintf(fid, '\n');
        matchIdeal; %Export matching inductors
        fprintf(fid, '.ENDS\n');
        fclose(fid);
        fprintf('File %s created successfully.\n\n', fileName);
    end
    %Class ABC with spiral inductors
    if (cnet == 'y' || cnet == 'Y') && (cind == 'y' || cind ==
'Y') && choice == 3;
        fprintf(fid, '.SUBCKT %s bias supply Gnd\n', paName);
        RFC = Lf*1e9;
        fprintf(fid, 'LRFC_2 supply N_1 %.2fn\n', RFC);
        if cmn == 0;
           fprintf(fid, 'CCapacitor_3 N_1 N_2 %.2fp\n', 1);
           fprintf(fid, 'XL0_1 N_2 Gnd Gnd L0\n');
           fprintf(fid, 'CCapacitor_1 N_2 Gnd\n');
           fprintf(fid, 'RResistor_1 N_2 Gnd %.2f TC=0.0, 0.0\n',
floor(100 * RL + 0.5) / 100);
        else
           fprintf(fid, 'CCapacitor_3 N_1 N_M1 %.2fp\n', 1);
           fprintf(fid, 'XL0_1 N_M1 Gnd Gnd L0\n');
           fprintf(fid, 'CCapacitor_1 N_M1 Gnd %.2fp\n',
floor(1e14 * C0 + 0.5) / 100);
```

Fig. 8.22 (continued)

```
fprintf(fid, 'RResistor_1 N_M2 Gnd %.2f TC=0.0, 0.0\n', 50);
     end
     fprintf(fid, transistor);
     fprintf(fid, '\n');
     matchSpiral; %Export matching inductors
     fprintf(fid, '.ENDS\n');
     fclose(fid);
     fprintf('File %s created successfully.\n\n', fileName);
 end
```

Fig. 8.22 (continued)

frequency that allows for reasonable transmission line lengths. Again, there is no reason to run the full program. So we run `ClassFQWT` from the MATLAB command line, enter 1000 mW, 5 V, 15,000 MHz, 50 Ω and choose a 0.5 nH base filtering inductor. Since the substrate for the transmission line design is not specified, one can use the default value used by the program ($\varepsilon_r = 3$).

The listing of the interaction with the program is shown in Fig. 8.27. From this figure, the resulting power amplifier requires a quarter-wave transmission line 5 mm long with characteristic impedance of 31.83 Ω, which can be achieved with the *w*/*h* ratio of 4.35. The capacitance of the base capacitor is $C_0 = 0.23$ pF, that of the feed inductor is Lf = 2.15 nH and the blocking capacitor is CB = 21.5 pF. The expected current drawn from the supply is 200 mA.

Task 3 Design a linear power amplifier for an application at 5 GHz. Let $P_O = 1.5$ *mW, VDD/VCC = 10 V, and let the amplifier have a bandwidth of BW = 1 GHz.*

Solution: The fact that linearity is a requirement steers one towards designing Class-AB amplifiers. However, to retain a fair amount of efficiency, one can choose a conduction angle that is a compromise between linearity and efficiency. In this case we opt for $\theta = 120°$. We then invoke the Class-A/AB/B/C power amplifier design program by typing `ClassABC` into the MATLAB command line, and enter all required parameters.

The listing of the interaction with the program is shown in Fig. 8.28. From this figure, the solution calls for a 0.21 nH inductor and 2.39 pF capacitor for the resonant tank. The amplifier draws 227 mA from the supply and has the expected efficiency figure of roughly 66 %.

Task 4 Design a power amplifier utilizing a minimum number of components that can deliver 500 mW at 8 GHz from a supply of 7 V. The power amplifier should reach 65 % collector efficiency in theory. Assume that the matching can be performed elsewhere.

Solution: The requirement for a minimum number of components calls for the use of the Class-J output stage, which, in addition to the driving transistor, requires only one capacitor and an RFC. To perform the design, one can run `ClassJ` from the MATLAB main window. Although the matching for this stage is not a requirement,

```
if RL > RS
    if cmn == 1;
        fprintf(fid, 'CCapacitor_L1 N_M1 N_M2 %.2fp\n',
floor(1e14 * CL1 + 0.5) / 100);
        fprintf(fid, 'LInductor_L1 Gnd N_M2 %.2fn\n', floor(1e11
* Ll1 + 0.5) / 100);
    elseif cmn == 2;
        fprintf(fid, 'CCapacitor_L2 N_M1 N_M2 %.2fp\n',
floor(1e14 * CL2 + 0.5) / 100);
        fprintf(fid, 'LInductor_L2 Gnd N_M2 %.2fn\n', floor(1e11
* LL2 + 0.5) / 100);
    elseif cmn == 3;
        fprintf(fid, 'LInductor_L3 N_M1 N_M2 %.2fn\n', floor(1e11
* LL3 + 0.5) / 100);
        fprintf(fid, 'CCapacitor_L3 Gnd N_M2 %.2fp\n', floor(1e14
* CL3 + 0.5) / 100);
    elseif cmn == 4;
        fprintf(fid, 'LInductor_L4 N_M1 N_M2 %.2fn\n', floor(1e11
* LL4 + 0.5) / 100);
        fprintf(fid, 'CCapacitor_L4 Gnd N_M2 %.2fp\n', floor(1e14
* CL4 + 0.5) / 100);
    elseif cmn == 5;
        fprintf(fid, 'LInductor_M1 N_M1 N_M %.2fn\n', floor(1e12
* LM1 + 0.5) / 1000);
        fprintf(fid, 'LInductor_M2 N_M N_M2 %.2fn\n', floor(1e12
* LM2 + 0.5) / 1000);
        fprintf(fid, 'CCapacitor_M1 Gnd N_M %.2fp\n', floor(1e15
* CM1 + 0.5) / 1000);
    elseif cmn == 6
        fprintf(fid, 'CCapacitor_M1 N_M1 N_M %.2fp\n', floor(1e15
* CM2 + 0.5) / 1000);
        fprintf(fid, 'CCapacitor_M2 N_M N_M2 %.2fp\n', floor(1e15
* CM3 + 0.5) / 1000);
        fprintf(fid, 'LInductor_M1 Gnd N_M %.2fn\n', floor(1e12 *
LM3 + 0.5) / 1000);
   elseif cmn == 7;
        fprintf(fid, 'LInductor_M1 Gnd N_M1 %.2fn\n', floor(1e12
* LM4 + 0.5) / 1000);
        fprintf(fid, 'LInductor_M2 Gnd N_M2 %.2fn\n', floor(1e12
* LM5 + 0.5) / 1000);
        fprintf(fid, 'CCapacitor_M1 N_M2 N_M1 %.2fp\n',
floor(1e15 * CM4 + 0.5) / 1000);
    elseif cmn == 8;
        fprintf(fid, 'CCapacitor_M1 Gnd N_M1 %.2fp\n', floor(1e15
* CM5 + 0.5) / 1000);
```

Fig. 8.23 MATLAB code for the procedure for export of the part of the netlist that involves matching when ideal inductors are used (matchIdeal.m)

one can attempt to simplify the matching network by designing this stage to work with a typical antenna load of 50 Ω. Note that by definition a complex load will be required, thus one can assume that one can place a capacitor or an inductor next to the load to achieve this. Thus one enters 50 Ω for required load impedance, 65 % for required efficiency and 500 mW and 7 V for the required power and supply voltage. The program also requires one to choose the initial phase angle, which is chosen as 20°.

```
        fprintf(fid, 'CCapacitor_M2 Gnd N_M2 %.2fp\n', floor(1e15
* CM6 + 0.5) / 1000);
        fprintf(fid, 'LInductor_M1 N_M2 N_M1 %.2fn\n', floor(1e12
* LM6 + 0.5) / 1000); end
else
    if cmn == 1;
        fprintf(fid, 'LInductor_L1 Gnd N_M1 %.2fn\n', floor(1e11
* LL1 + 0.5) / 100);
        fprintf(fid, 'CCapacitor_L1 N_M1 N_M2 %.2fp\n',
floor(1e14 * CL1 + 0.5) / 100);
    elseif cmn == 2;
        fprintf(fid, 'LInductor_L2 Gnd N_M1 %.2fn\n', floor(1e11
* LL2 + 0.5) / 100);
        fprintf(fid, 'CCapacitor_L2 N_M1 N_M2 %.2fp\n',
floor(1e14 * CL2 + 0.5) / 100);
    elseif cmn == 3;
        fprintf(fid, 'CCapacitor_L3 Gnd N_M1 %.2fp\n', floor(1e14
* CL3 + 0.5) / 100);
        fprintf(fid, 'LInductor_L3 N_M1 N_M2 %.2fn\n', floor(1e11
* LL3 + 0.5) / 100);
    elseif cmn == 4;
        fprintf(fid, 'CCapacitor_L4 Gnd N_M1 %.2fp\n', floor(1e14
* CL4 + 0.5) / 100);
        fprintf(fid, 'LInductor_L4 N_M1 N_M2 %.2fn\n', floor(1e11
* LL4 + 0.5) / 100);
    elseif cmn == 5;
        fprintf(fid, 'LInductor_M1 N_M2 N_M %.2fn\n', floor(1e12
* LM1 + 0.5) / 1000);
        fprintf(fid, 'LInductor_M2 N_M N_M1 %.2fn\n', floor(1e12
* LM2 + 0.5) / 1000);
        fprintf(fid, 'CCapacitor_M1 Gnd N_M %.2fp\n', floor(1e15
* CM1 + 0.5) / 1000);
    elseif cmn == 6
        fprintf(fid, 'CCapacitor_M1 N_M2 N_M %.2fp\n', floor(1e15
* CM2 + 0.5) / 1000);
        fprintf(fid, 'CCapacitor_M2 N_M N_M1 %.2fp\n', floor(1e15
* CM3 + 0.5) / 1000);
        fprintf(fid, 'LInductor_M1 Gnd N_M %.2fn\n', floor(1e12 *
LM3 + 0.5) / 1000);
    elseif cmn == 7;
        fprintf(fid, 'LInductor_M1 Gnd N_M2 %.2fn\n', floor(1e12
* LM4 + 0.5) / 1000);
        fprintf(fid, 'LInductor_M2 Gnd N_M1 %.2fn\n', floor(1e12
* LM5 + 0.5) / 1000);
        fprintf(fid, 'CCapacitor_M1 N_M2 N_M1 %.2fp\n',
floor(1e15 * CM4 + 0.5) / 1000);
    elseif cmn == 8;
        fprintf(fid, 'CCapacitor_M1 Gnd N_M2 %.2fp\n', floor(1e15
* CM5 + 0.5) / 1000);
        fprintf(fid, 'CCapacitor_M2 Gnd N_M1 %.2fp\n', floor(1e15
* CM6 + 0.5) / 1000);
        fprintf(fid, 'LInductor_M1 N_M2 N_M1 %.2fn\n', floor(1e12
* LM6 + 0.5) / 1000);
    end
end
```

Fig. 8.23 (continued)

```
if RL > RS
    if cmn == 1;
        fprintf(fid, 'CCapacitor_L1 N_M1 N_M2 %.2fp\n',
floor(1e14 * CL1 + 0.5) / 100);
        fprintf(fid, 'XLL1_1 Gnd N_M2 Gnd LL1\n');
    elseif cmn == 2;
        fprintf(fid, 'CCapacitor_L2 N_M1 N_M2 %.2fp\n',
floor(1e14 * CL2 + 0.5) / 100);
        fprintf(fid, 'XLL2_1 Gnd N_M2 Gnd LL2\n');
    elseif cmn == 3;
        fprintf(fid, 'XLL3_1 N_M1 N_M2 Gnd LL3\n');
        fprintf(fid, 'CCapacitor_L3 Gnd N_M2 %.2fp\n', floor(1e14
* CL3 + 0.5) / 100);
    elseif cmn == 4;
        fprintf(fid, 'XLL4_1 N_M1 N_M2 Gnd LL4\n');
        fprintf(fid, 'CCapacitor_L4 Gnd N_M2 %.2fp\n', floor(1e14
* CL4 + 0.5) / 100);
    elseif cmn == 5;
        fprintf(fid, 'XLM1_1 N_M1 N_M Gnd LM1\n');
        fprintf(fid, 'XLM2_1 N_M N_M2 Gnd LM2\n');
        fprintf(fid, 'CCapacitor_M1 Gnd N_M %.2fp\n', floor(1e15
* CM1 + 0.5) / 1000);
    elseif cmn == 6
        fprintf(fid, 'CCapacitor_M1 N_M1 N_M %.2fp\n', floor(1e15
* CM2 + 0.5) / 1000);
        fprintf(fid, 'CCapacitor_M2 N_M N_M2 %.2fp\n', floor(1e15
* CM3 + 0.5) / 1000);
        fprintf(fid, 'XLM3_1 Gnd N_M Gnd LM3\n');
    elseif cmn == 7;
        fprintf(fid, 'XLM4_1 Gnd N_M1 Gnd LM4\n');
        fprintf(fid, 'XLM5_1 Gnd N_M2 Gnd LM5\n');
        fprintf(fid, 'CCapacitor_M1 N_M2 N_M1 %.2fp\n',
floor(1e15 * CM4 + 0.5) / 1000);
    elseif cmn == 8;
        fprintf(fid, 'CCapacitor_M1 Gnd N_M1 %.2fp\n', floor(1e15
* CM5 + 0.5) / 1000);
        fprintf(fid, 'CCapacitor_M2 Gnd N_M2 %.2fp\n', floor(1e15
* CM6 + 0.5) / 1000);
        fprintf(fid, 'XLM6_1 N_M1 N_M2 Gnd LM6\n');
    end
else
    if cmn == 1;
        fprintf(fid, 'XLL1_1 Gnd N_M1 Gnd LL1\n');
        fprintf(fid, 'CCapacitor_L1 N_M1 N_M2 %.2fp\n',
floor(1e14 * CL1 + 0.5) / 100);
    elseif cmn == 2;
        fprintf(fid, 'XLL2_1 Gnd N_M1 Gnd LL2\n');
        fprintf(fid, 'CCapacitor_L2 N_M1 N_M2 %.2fp\n',
floor(1e14 * CL2 + 0.5) / 100);
    elseif cmn == 3;
        fprintf(fid, 'CCapacitor_L3 Gnd N_M1 %.2fp\n', floor(1e14
* CL3 + 0.5) / 100);
        fprintf(fid, 'XLL3_1 N_M1 N_M2 Gnd LL3\n');
    elseif cmn == 4;
        fprintf(fid, 'CCapacitor_L4 Gnd N_M1 %.2fp\n', floor(1e14
* CL4 + 0.5) / 100);
        fprintf(fid, 'XLL4_1 N_M1 N_M2 Gnd LL4\n');
```

Fig. 8.24 MATLAB code for the procedure for export of the part of the netlist that involves matching when spiral inductors are used (matchSpiral.m)

```
    elseif cmn == 5;
        fprintf(fid, 'XLM1_1 N_M1 N_M Gnd LM2\n');
        fprintf(fid, 'XLM2_1 N_M N_M2 Gnd LM1\n');
        fprintf(fid, 'CCapacitor_M1 Gnd N_M %.2fp\n', floor(1e15
* CM1 + 0.5) / 1000);
    elseif cmn == 6
        fprintf(fid, 'CCapacitor_M1 N_M2 N_M %.2fp\n', floor(1e15
* CM2 + 0.5) / 1000);
        fprintf(fid, 'CCapacitor_M2 N_M N_M1 %.2fp\n', floor(1e15
* CM3 + 0.5) / 1000);
        fprintf(fid, 'XLM3_1 Gnd N_M Gnd LM3\n');
    elseif cmn == 7;
        fprintf(fid, 'XLM4_1 Gnd N_M2 Gnd LM4\n');
        fprintf(fid, 'XLM5_1 Gnd N_M1 Gnd LM5\n');
        fprintf(fid, 'CCapacitor_M1 N_M2 N_M1 %.2fp\n',
floor(1e15 * CM4 + 0.5) / 1000);
    elseif cmn == 8;
        fprintf(fid, 'CCapacitor_M1 Gnd N_M2 %.2fp\n', floor(1e15
* CM5 + 0.5) / 1000);
        fprintf(fid, 'CCapacitor_M2 Gnd N_M1 %.2fp\n', floor(1e15
* CM6 + 0.5) / 1000);
        fprintf(fid, 'XLM6_1 N_M1 N_M2 Gnd LM6\n');
    end
end
```

Fig. 8.24 (continued)

```
>> classE
Please enter the frequency of operation, fo (MHz): 2400
Please enter the loaded quality factor, QL: 5
Please enter the required output power, Po (mW): 50
Please enter the value of the regulated power supply (Vcc/Vdd): 1
Please enter VCEsat/VDSsat (-1 to skip): -1
Please enter the highest allowed inductance (nH) (-1 to skip): -1
Please enter the lowest allowed inductance (nH) (-1 to skip): -1
Please  enter  collector-emitter  breakdown  voltage  (V)  (-1  to
skip): -1
Please enter the value of Cce/Cds (pF) (-1 to skip): -1

RESULTS:
R = 11.54 Ohm
L = 3.83 nH
Lf = 41.68 nH
C0 = 1.71 pF
C = 1.05 pF
Cp = 1.05 pF
Idc = 49.97 mA
vcp = 3.56 V
isp = 142.93 mA
```

Fig. 8.25 Interaction with the Class-E power amplifier design program in task 1

```
>>classF
Please enter 1 for 3-harmonic design or 2 for 5-harmonic design:
2
Please enter the frequency of operation, fo (MHz): 2400
Please enter the required output power, Po (mW): 50
Please enter the value of the regulated power supply (Vcc/Vdd):
1.5
Please enter collector-emitter breakdown voltage (V) (-1 to
skip): -1
Please enter the inductance for the base filter (nH): 1
Please enter the inductance for the third harmonic filter (nH):
0.5
Please enter the inductance for the fifth harmonic filter (nH):
0.5
Please enter the theta (deg) (-1 to skip): -1

RESULTS:
R = 30.90 Ohm
L0 = 1.00 nH
C0 = 4.400 pF
L3 = 0.50 nH
C3 = 0.980 pF
Lf = 20.49 nH
CB = 204.890 pF
L5 = 0.50 nH
C5 = 0.350 pF
Idc = 43.07 mA
vm = 1.76 V
VCEM = 3.00 V
```

Fig. 8.26 Interaction with the Class-F power amplifier design program in task 1

The listing of the interaction with the program is shown in Fig. 8.29. From this figure, the required components are $C = 1.95$ pF and $X_L = 150\ \Omega$. It is interesting at this point to show the shape of the output voltage as well. This is done in Fig. 8.30a.

Note that this is only one solution to the problem, where different choices of R_L and ϕ will result in different solutions (designs). Designs for $R_L = 10$, 50, 100 and 250 Ω and different choices of ϕ are shown in Table 8.19, with the voltage waveforms shown in Fig. 8.30.

Task 5 Design at least three matching networks to match a 10 Ω source to a 50 Ω load at 5 GHz. If required, assume a matching bandwidth of 500 MHz and a matching network for an integrated solution.

Solution: In the matching network subroutine that was developed, a choice of matching using nine different matching networks was presented. Thus one can use one L-shape network, one T-network and one Π-network. We start by running `match` from the MATLAB command line and entering the required parameters.

The listing of the interaction with the program is shown in Fig. 8.31, where in the case of L-matching networks we have only two possible solutions, since both matching impedances are real. The other two solutions are the same as the first two,

```
>> classFQWT
Please enter the frequency of operation, fo (MHz): 15000
Please enter the required output power, Po (mW): 1000
Please enter the value of the regulated power supply (Vcc/Vdd): 5
Please enter collector-emiiter breakdown voltage (V) (-1 to
skip): -1
Please enter the inductance for the base filter (nH): 0.5
Please enter antenna impedance (Ohms) (-1 to skip): 50
Please enter relative dielectric permittivity (-1 to skip): -1

RESULTS:
Z0 = 31.83 Ohm
r = w/h = 4.35
length of QWT = 5.00 mm
Ri = 20.26 Ohm
L0 = 0.50 nH
C0 = 0.23 pF
Lf = 2.15 nH
CB = 21.50 pF
Idc = 200.00 mA
vm = 6.37 V
VCEM = 10.00 V
```

Fig. 8.27 Interaction with the Class-F_{∞} power amplifier design program in task 2

```
>> classABC
Please enter the frequency of operation, fo (MHz): 5000
Please enter the required output power, Po (mW): 1500
Please enter the value of the regulated power supply (Vcc/Vdd):
10
Please enter collector-emitter breakdown voltage (V) (-1 to
skip): -1
Please enter the bandwidth of the resonant tank (MHz): 1000
Please enter the theta (deg) (-1 to skip): 120

RESULTS:
R = 33.33 Ohm
L = 0.21 nH
C = 4.77 pF
Lf = 10.61 nH
CB = 106.10 pF
Idc = 227.10 mA
Vm = 10.00 V
VCEM = 20.00 V
Efficiency = 66.05 percent
```

Fig. 8.28 Interaction with the Class-AB power amplifier design program in task 3

but the movement is on the opposite side on the Smith chart. The minus sign in this case indicates that the reactances are of the opposite type (capacitive vs. inductive).

We make our choice of matching networks based on the practicality of the implementation of spiral inductors on-chip. Thus, as a solution to this task, we choose one L-network (option 1) and two T-networks.

```
>> classJ
Please enter the frequency of operation, fo (MHz): 8000
Please enter the required output power, Po (mW): 500
Please enter the value of the regulated power supply (Vcc/Vdd): 7
Please enter the required impedance (Ohm): 50
Please enter the required efficiency (%): 65
Please choose a value of phi (deg): 20
RL = 50.00 Ohm
XL = 147.58 Ohm
C = 1.95 pF
Lf = 9.95 nH
voff = 0.06 V
I1 = 141.42 mA
Idc = 109.89 mA
ICEM = 345.23 mA
```

Fig. 8.29 Interaction with the Class-J power amplifier design program in task 4

Task 6 Design a matching network to match the output impedance required in the power amplifier of task 4 to the 50 Ω antenna.

Solution: For each designed Class-J amplifier from Task 4, there are a number of solutions. Consider the matching case where $Z_L = (25 - j58.4)\ \Omega$. We run `match` again, recalling that the extension to the program was developed to match complex impedances only using L-networks. We can provide the frequency (8000 MHz), and the program requires the bandwidth, which in this case is not applicable, thus we can provide a dummy value. For the required matching impedances, we obtain four solutions, but have to disregard two solutions: option 1 and option 3 have components of different types, but the absence of the minus sign indicates they should be of the same type.

The listing of the interaction with the program is shown in Fig. 8.32, where we only include L-network solutions.

Task 7 Repeat Task 1 but design an optimum matching network. Also, design spiral inductors for each required inductor and suggest the length of the bond wire that can be used for RFC ($a = 0.2$ mm). Export the netlists for both systems. Assume that you have access to the amsAG S35 process and that thick metal inductors can be used with $s = 2\ \mu m$.

Solution: Now that we have to design the output stage together with matching networks and inductors, we can resort to running the program that takes care of complete system integration. Recall that in Task 1 we designed one Class-E power amplifier and one Class-F power amplifier.

Class-E We run `paProg` and start by choosing option 1. Then we enter all input parameters as we did in Task 1. The next step is to confirm that we want to perform the matching and that we want to export the netlist into a file. We choose the matching bandwidth of 1 GHz. We then decide to use a capacitor-inductor-capacitor narrowband Π-network for matching in the design because it contains

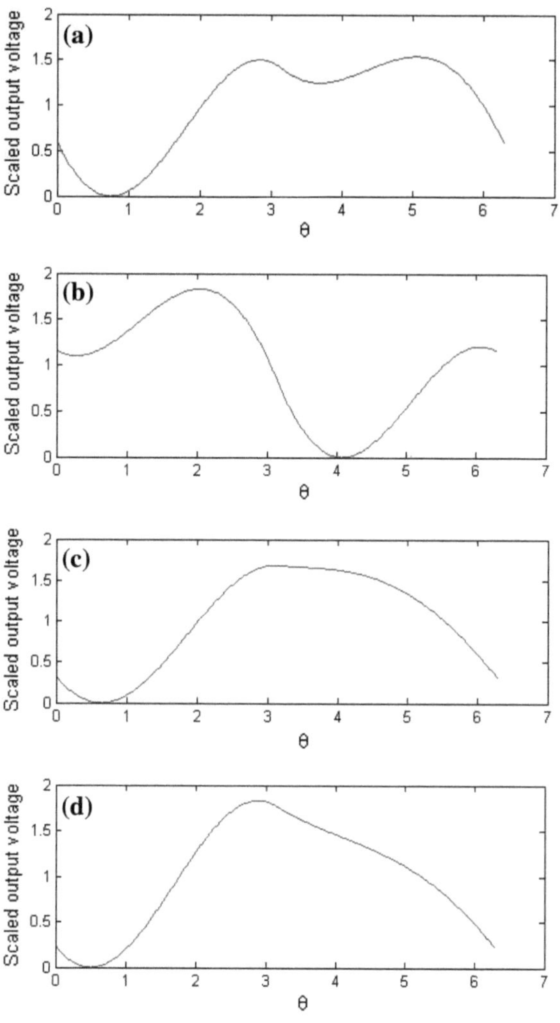

Fig. 8.30 Scaled output voltage waveforms of four variations of Class-J power amplifiers from Table 8.19: **a** Config. 1, **b** Config. 2, **c** Config. 3 and **d** Config. 4

Table 8.19 Design parameters of four different Class-J power amplifier configurations

Amplifier no.	R_L (Ω)	ϕ (°)	X_L (Ω)	C (pF)	I_1 (mA)
1	50	20	148	1.95	141
2	25	−20	−58.4	1.93	200
3	100	70	41.3	4.04	100
4	250	35	363	3.12	63.3

only one inductor, minimizing the floorspace, and proceed to override inductors. Since in this case we can trust the inductor calculation, we only modify the RFC value to 40 nH and follow the program execution to the end without modifying any default process parameters. Finally, we choose an npn254h5 transistor with emitter

```
>> match
Please enter the value for the source resistance RS (ohm): 10
Please enter the value for the source reactance XS (ohm) (0 to
skip): 0
Please enter the value for the load resistance RL (ohm): 50
Please enter the value for the load reactance XL (ohm) (0 to
skip): 0
Please enter the operating frequency (MHz): 5000
Please enter the bandwidth (MHz): 500

L network:
First component in series with source and second in shunt with
load
Option 1:
CL1 = -1.590 pF
LL1 = 0.800 nH
Option 2:
CL2 = 1.590 pF
LL2 = -0.800 nH
Option 3:
LL3 = -0.640 nH
CL3 = 1.270 pF
Option 4:
LL4 = 0.640 nH
CL4 = -1.270 pF

ind-cap-ind T network:
L1 = 3.183 nH (side of the smaller resistance)
C1 = 0.453 pF
L2 = 6.974 nH (side of the larger resistance)
cap-ind-cap T network:
C1 = 0.318 pF (side of the smaller resistance)
L1 = 2.235 nH
C2 = 0.145 pF (side of the larger resistance)

ind-cap-ind Pi network:
L1 = 0.073 nH (side of the smaller resistance)
C1 = 4.471 pF
L2 = 0.159 nH (side of the larger resistance)
cap-ind-cap Pi network:
C1 = 13.948 pF (side of the smaller resistance)
L1 = 0.227 nH
C2 = 6.366 pF (side of the larger resistance)

Quarter-wave transformer:

Please enter relative dielectric permittivity  (-1 to skip): -1

RESULTS:
Z0 = 22.36 Ohm
r = w/h = 6.13
```

Fig. 8.31 Interaction with the matching design program in task 5

length of 96 μm. We choose the name PA_task7_classE as the power amplifier name for netlist extraction purposes, and name the inductors M40383N and M40154N.

```
>> match
Please enter the value for the source resistance RS (ohm): 25
Please enter the value for the source reactance XS (ohm) (0 to
skip): -58.4
Please enter the value for the load resistance RL (ohm): 50
Please enter the value for the load reactance XL (ohm) (0 to
skip): 0
Please enter the operating frequency (MHz): 8000
Please enter the bandwidth (MHz): 800

L network:
First component in series with source and second in shunt with
load
Option 1:
CL1 = 0.600 pF
LL1 = 0.990 nH
Option 2:
CL2 = 0.240 pF
LL2 = -0.990 nH
Option 3:
LL3 = 0.660 nH
CL3 = 0.400 pF
Option 4:
LL4 = 1.660 nH
CL4 = -0.400 pF
```

Fig. 8.32 Interaction with the matching design program in task 6

The listing of the interaction with the program is shown in Fig. 8.33, and the output netlist is shown in Fig. 8.34. Note that the program attempts to find a spiral inductor for the 40 nH inductor but it is not successful, which generates an error in the netlist export. This is fine, given the fact that we need to design a bond wire for the RFC inductor. Thus we need to make a manual modification to the netlist to include the bond wire, which we treat as an ideal inductor of 40 nH.

A summary of all parameters (input, output, and inductor) of the designed Class-E power amplifier is shown in Table 8.20.

Class-F We run `paProg` and repeat the process while entering input parameters for the Class-F_{35} design. We again choose the capacitor-inductor-capacitor Π-matching network (BW = 500 MHz in this case), and name the power amplifier PA_task7_classF. Furthermore, we use M40100N, M40500N and M40118N as inductor names of 1, 0.5 and 1.18 nH inductors respectively. The listing of the interaction with the program is shown in Fig. 8.35. A summary of all parameters (input, output, and inductor) of the designed Class-F power amplifier is shown in Table 8.21, and the exported netlist is shown in Fig. 8.36, where we have again edited the netlist manually to include the 40 nH bondwire inductor.

Bond wire: The last step is to run the design of the bondwire inductor for RFC. This seems like an adequate choice because the exact inductance is not paramount, but the quality factor will remain high. Thus we run the `bondwire` program

```
>> paprog
Switch-mode Power Amplifier Design Tool and Inductance Calculator
v3.0.0

Developed by Mladen Božanic, as part of the book
       POWER AMPLIFIERS FOR S-, C-, X- AND Ku BANDS
University of Johannesburg
Please enter:
1 for Class-E amplifier design,
2 for Class-F amplifier design,
3 for Class-F-QWT amplifier design,
4 for Class-A/AB/B/C amplifier design or
5 for Class-J amplifier design
 1
Please enter the frequency of operation, fo (MHz): 2400
Please enter the loaded quality factor, QL: 5
Please enter the required output power, Po (mW): 50
Please enter the value of the regulated power supply (Vcc/Vdd): 1
Please enter VCEsat/VDSsat (-1 to skip): -1
Please enter the highest allowed inductance (nH) (-1 to skip): -1
Please enter the lowest allowed inductance (nH) (-1 to skip): -1
Please enter collector-emitter breakdown voltage (V) (-1 to
skip): -1
Please enter the value of Cce/Cds (pF) (-1 to skip): -1

RESULTS:
R = 11.54 Ohm
L = 3.83 nH
Lf = 41.68 nH
C0 = 1.71 pF
C = 1.05 pF
Cp = 1.05 pF
Idc = 49.97 mA
vcp = 3.56 V
isp = 142.93 mA

Do you want to perform the impedance matching to 50 ohm antenna
impedance (y/n)? y

Please enter the matching bandwidth (MHz): 1000

L network:
First component in series with source and second in shunt with
load
Option 1:
CL1 = -3.150 pF
LL1 = 1.820 nH
Option 2:
CL2 = 3.150 pF
LL2 = -1.820 nH
Option 3:
LL3 = -1.400 nH
CL3 = 2.420 pF
Option 4:
LL4 = 1.400 nH
CL4 = -2.420 pF
```

Fig. 8.33 Interaction with complete power amplifier design program in task 7 for Class-E power amplifier variation

```
ind-cap-ind T network:
L1 = 1.837 nH (side of the smaller resistance)
C1 = 2.676 pF
L2 = 2.482 nH (side of the larger resistance)
cap-ind-cap T network:
C1 = 2.394 pF (side of the smaller resistance)
L1 = 1.643 nH
C2 = 1.772 pF (side of the larger resistance)

ind-cap-ind Pi network:
L1 = 1.022 nH (side of the smaller resistance)
C1 = 2.848 pF
L2 = 1.382 nH (side of the larger resistance)
cap-ind-cap Pi network:
C1 = 4.301 pF (side of the smaller resistance)
L1 = 1.544 nH
C2 = 3.183 pF (side of the larger resistance)

Quarter-wave transformer:

Please enter relative dielectric permittivity (-1 to skip): -1

RESULTS:
Z0 = 24.02 Ohm
r = w/h = 5.75

Do you want to export the netlist of the PA (y/n)? y

Do you want to attempt to find inductors for the PA and the
matching networks (y/n)? y
Please enter which matching network you would like to use:
 1 - L-network option 1
 2 - L-network option 2
 3 - L-network option 3
 4 - L-network option 4
 5 - ind-cap-ind T-network
 6 - cap-ind-cap T network
 7 - ind-cap-ind Pi network
 8 - cap-ind-cap Pi network:
 (quarter-wave transformer not allowed)
 8
Do you want to override any of the inductance values (y/n)? y
Old Lf = 41.68 nH. New Lf (-1 to skip): 40
Old series L = 3.83 nH. New L (-1 to skip): -1
Old LM6 = 1.54 nH. New LM6 (-1 to skip): -1

Please enter the tolerance for the inductance value (%) (-1 to
skip): -1
Please enter the search grid resolution (um) (-1 to skip): -1
Do you want to change any of the default geometry parameters
(y/n)? n
Using the default values of 30.0, 500.0, 2.0 and 2.0 for dinmin,
doutmax, s and wmin respectively.

Please enter the PA name (press "s" to skip): PA_task7_classE
```

Fig. 8.33 (continued)

```
INDUCTORS
Do you want to change any of the default process parameters
(y/n)? n

Inductor value: 40.000 nH
Looking for the geometry for highest quality factor Q...

Could not find a geometry for 40.00 nH
limited to dinmin and doutmax with Q greater than 1 at 2400.00
MHz.
MATLAB reports:
Undefined function or variable 'Lclfstored'.
Netlist export failed.

Inductor value: 3.826 nH
Looking for the geometry for highest quality factor Q...

Ls = 3.84 nH
Lslf = 3.46 nH
Q = 4.84
fo = 6.86 GHz
w = 17.00 um
s = 2.00 um
din = 165.00 um
dout = 275.00 um
n = 3

Inductor value: 1.544 nH
Looking for the geometry for highest quality factor Q...

Ls = 1.55 nH
Lslf = 1.43 nH
Q = 7.43
fo = 8.35 GHz
w = 43.00 um
s = 2.00 um
din = 149.00 um
dout = 325.00 um
n = 2

Please enter 1 for HBT, 2 RF NMOS or 3 for any other transistor:
1
Please enter the model name (e.g. npn254h5): npn254h5
Please enter the total emitter width (um) (-1 to skip): 96
File PA_task7_classE_net.spc created successfully.
```

Fig. 8.33 (continued)

designed in Chap. 7 from the MATLAB command line and follow the prompts. The resulting bond wire length is 38.5 mm, which is somewhat impractical. Our other option is to use two bond wires with a narrow spacing, demanding the length of 28.2 mm, which is acceptable. Placing a few more bonds in parallel would bring the wire length down even further. The listing of the interaction with the program is given in Fig. 8.37.

```
.SUBCKT Lf L1 L2 GND
CS L1 L2 0.00fF
.SUBCKT L2 L1 L2 GND
CS L1 L2 30.69fF
LS N4 L1 3.46n
RS N4 L2 6.02
Csi1 N2 GND 81.75fF
Csi2 N3 GND 81.75fF
Cox1 L1 N2 230.68fF
Cox2 L2 N3 230.68fF
Rsi1 N2 GND 273.26
Rsi2 N3 GND 273.26
.ENDS

.SUBCKT LM6 L1 L2 GND
CS L1 L2 130.91fF
LS N4 L1 1.43n
RS N4 L2 1.79
Csi1 N2 GND 73.99fF
Csi2 N3 GND 73.99fF
Cox1 L1 N2 370.08fF
Cox2 L2 N3 370.08fF
Rsi1 N2 GND 300.89
Rsi2 N3 GND 300.89
.ENDS

.SUBCKT PA_task7_classE bias supply Gnd
XLf_1 supply N_1 Gnd Lf
XL2_1 N_5 N_M1 Gnd L2
RResistor_1 N_M2 Gnd 50.00 TC=0.0, 0.0
Xnpn254h5_1 N_1 bias Gnd Gnd npn254h5 area=96
CCapacitor_1 N_1 Gnd 1.05p
CCapacitor_2 N_1 N_5 1.71p
CCapacitor_M1 Gnd N_M1 4.30p
CCapacitor_M2 Gnd N_M2 3.18p
XLM6_1 N_M1 N_M2 Gnd LM6
.ENDS
```

Fig. 8.34 Exported netlist of the Class-E power amplifier design

We can import the netlists of both power amplifiers into a schematic editor. If we use Cadence Schematic Editor [3] and rearrange the netlist for higher readability, and add some simulation sources, schematics of Fig. 8.38a, b are obtained for Class-E and Class-F power amplifiers respectively.

In this chapter we have not treated exporting of the layout of inductors into GDSII format, one of the topics of Chap. 9. Therefore we will reconsider this task in Chap. 9 to illustrate the layouts of designed inductors.

Task 8 Repeat task 4 but add a matching network and choice of appropriate inductor from the Coilcraft library introduced in Chap. 5. Alternatively, calculate the number of windings needed to implement a wire-wound inductor with a non-magnetic core.

Table 8.20 Chosen and calculated parameters for the design of the Class-E power amplifier

Parameter	Value	Unit
Centre frequency (f_o)	2.4	GHz
Loaded quality factor (Q_L)	5	–
Aimed output power (P_{out})	50	mW
Supply voltage (V_{CC})	1	V
Matching bandwidth (BW)	1	GHz
RFC (L_1)	40	nH
Required load resistance (R_L)	11.5	Ω
Series inductance (L_2)	3.83	nH
Series capacitance (C_2)	1.71	pF
Shunt capacitance (C_1)	1.05	pF
DC current (I_{DC})	49.97	mA
Peak collector voltage (v_{Cp})	3.56	V
Peak collector current (i_{sp})	143	mA
Cap-ind-cap Π-network matching capacitance 1 (C_{M1})	4.301	pF
Cap-ind-cap Π-network matching inductance 1 (L_{M1})	1.544	nH
Cap-ind-cap Π-network matching capacitance 2 (C_{M2})	3.183	nH
Inductance at 2.4 GHz (L_S) (M40383N)	3.84	nH
Inductance at DC (L_{SLF}) (M40383N)	3.46	nH
Q-factor (Q) (M40383N)	4.84	–
Resonant frequency (f_R) (M40383N)	6.86	GHz
Turn width (w) (M40383N)	17.0	μm
Turn spacing (s) (M40383N)	2.00	μm
Inner diameter (d_{in}) (M40383N)	165	μm
Outer diameter (d_{out}) (M40383N)	275	μm
Number of turns (n) (M40154N)	3	–
Inductance at 2.4 GHz (L_S) (M40154N)	1.55	nH
Inductance at DC (L_{SLF}) (M40154N)	1.43	nH
Q-factor (Q) (M40154N)	7.43	–
Resonant frequency (f_R) (M40154N)	8.35	GHz
Turn width (w) (M40154N)	43.0	μm
Turn spacing (s) (M40154N)	2.00	μm
Inner diameter (d_{in}) (M40154N)	149	μm
Outer diameter (d_{out}) (M40154N)	325	μm
Number of turns (n) (M40154N)	2	–

Solution: We handled the first two parts of this task separately in Tasks 4 and 6. Now we can redo it, but by invoking `paProg`. We follow the prompts and the interaction with the program looks as in the listing in Fig. 8.37. The matching inductor has a value of roughly 1 nH. By inspecting the Coilcraft library, we find a

```
>> paprog
Switch-mode Power Amplifier Design Tool and Inductance Calculator
v3.0.0

Developed by Mladen Božanic, as part of the book
       POWER AMPLIFIERS FOR S-, C-, X- AND Ku BANDS
University of Johannesburg
Please enter:
1 for Class-E amplifier design,
2 for Class-F amplifier design,
3 for Class-F-QWT amplifier design,
4 for Class-A/AB/B/C amplifier design or
5 for Class-J amplifier design
 2
Please enter 1 for 3-harmonic design or 2 for 5-harmonic design:
2
Please enter the frequency of operation, fo (MHz): 2400
Please enter the required output power, Po (mW): 50
Please enter the value of the regulated power supply (Vcc/Vdd):
1.5
Please enter collector-emitter breakdown voltage (V) (-1 to
skip): -1
Please enter the inductance for the base filter (nH): 1
Please enter the inductance for the third harmonic filter (nH):
0.5
Please enter the inductance for the fifth harmonic filter (nH):
0.5
Please enter the theta (deg) (-1 to skip): -1

RESULTS:
R = 30.90 Ohm
L0 = 1.00 nH
C0 = 4.400 pF
L3 = 0.50 nH
C3 = 0.980 pF
Lf = 20.49 nH
CB = 204.890 pF
L5 = 0.50 nH
C5 = 0.350 pF
Idc = 43.07 mA
vm = 1.76 V
VCEM = 3.00 V

Do you want to perform the impedance matching to 50 ohm antenna
impedance (y/n)? y

Please enter the matching bandwidth (MHz): 500

L network:
First component in series with source and second in shunt with
load
Option 1:
CL1 = -2.730 pF
LL1 = 4.220 nH
Option 2:
CL2 = 2.730 pF
LL2 = -4.220 nH
```

Fig. 8.35 Interaction with complete power amplifier design program in task 7 for Class-F power amplifier variation

```
Option 3:
LL3 = -1.610 nH
CL3 = 1.040 pF
Option 4:
LL4 = 1.610 nH
CL4 = -1.040 pF

ind-cap-ind T network:
L1 = 9.835 nH (side of the smaller resistance)
C1 = 0.761 pF
L2 = 12.342 nH (side of the larger resistance)
cap-ind-cap T network:
C1 = 0.447 pF (side of the smaller resistance)
L1 = 5.780 nH
C2 = 0.356 pF (side of the larger resistance)

ind-cap-ind Pi network:
L1 = 0.550 nH (side of the smaller resistance)
C1 = 3.741 pF
L2 = 0.691 nH (side of the larger resistance)
cap-ind-cap Pi network:
C1 = 7.989 pF (side of the smaller resistance)
L1 = 1.175 nH
C2 = 6.366 pF (side of the larger resistance)

Quarter-wave transformer:

Please enter relative dielectric permittivity  (-1 to skip): -1

RESULTS:
Z0 = 39.30 Ohm
r = w/h = 3.43

Do you want to export the netlist of the PA (y/n)? y

Do you want to attempt to find inductors for the PA and the
matching networks (y/n)? y
Please enter which matching network you would like to use:
 1 - L-network option 1
 2 - L-network option 2
 3 - L-network option 3
 4 - L-network option 4
 5 - ind-cap-ind T-network
 6 - cap-ind-cap T network
 7 - ind-cap-ind Pi network
 8 - cap-ind-cap Pi network:
 (quarter-wave transformer not allowed)
 8
Do you want to override any of the inductance values (y/n)? y
Old Lf = 20.49 nH. New Lf (-1 to skip): 40
Old L0 = 1.00 nH. New L0 (-1 to skip): -1
Old L3 = 0.50 nH. New L3 (-1 to skip): -1
Old L5 = 0.50 nH. New L5 (-1 to skip): -1
Old LM6 = 1.18 nH. New LM6 (-1 to skip): -1

Please enter the tolerance for the inductance value (%) (-1 to
skip): -1
```

Fig. 8.35 (continued)

```
Please enter the search grid resolution (um) (-1 to skip): -1
Do you want to change any of the default geometry parameters
(y/n)? n
Using the default values of 30.0, 500.0, 2.0 and 2.0 for dinmin,
doutmax, s and wmin respectively.

Please enter the PA name (press "s" to skip): PA_task7_classF

INDUCTORS
Do you want to change any of the default process parameters
(y/n)? n

Inductor value: 40.000 nH
Looking for the geometry for highest quality factor Q...

Could not find a geometry for 40.00 nH
limited to dinmin and doutmax with Q greater than 1 at 2400.00
MHz.
MATLAB reports:
Undefined function or variable 'Lclfstored'.
Netlist export failed.

Inductor value: 1.000 nH
Looking for the geometry for highest quality factor Q...
Ls = 1.00 nH
Lslf = 0.95 nH
Q = 8.17
fo = 9.72 GHz
w = 50.00 um
s = 2.00 um
din = 89.00 um
dout = 293.00 um
n = 2

Inductor value: 0.500 nH
Looking for the geometry for highest quality factor Q...

Ls = 0.50 nH
Lslf = 0.49 nH
Q = 7.12
fo = 14.99 GHz
w = 47.00 um
s = 2.00 um
din = 30.00 um
dout = 222.00 um
n = 2

Inductor value: 1.175 nH
Looking for the geometry for highest quality factor Q...

Ls = 1.18 nH
Lslf = 1.10 nH
Q = 8.02
fo = 8.87 GHz
w = 50.00 um
s = 2.00 um
din = 108.00 um
```

Fig. 8.35 (continued)

```
dout = 312.00 um
n = 2

Please enter 1 for HBT, 2 RF NMOS or 3 for any other transistor:
1
Please enter the model name (e.g. npn254h5): npn254h5
Please enter the total emitter width (um) (-1 to skip): 96
File PA_task7_classF_net.spc created successfully.
```

Fig. 8.35 (continued)

Table 8.21 Chosen and calculated parameters for the design of the Class-F power amplifier

Parameter	Value	Unit
Centre frequency (f_o)	2.4	GHz
Aimed output power (P_{out})	50	mW
Supply voltage (V_{CC})	1.5	V
RFC (L_1)	40	nH
Required load resistance (R_L)	30.9	Ω
Base filter inductance (L_O)	1.00	nH
Base filter capacitance (C_O)	4.40	pF
Third-harmonic filter inductance (L_3)	0.50	nH
Third-harmonic filter capacitance (C_3)	0.98	pF
Fifth-harmonic filter inductance (L_5)	0.50	nH
Fifth-harmonic filter capacitance (C_5)	0.35	pF
DC current (I_{DC})	43.1	mA
Peak collector voltage (v_{Cp})	3.00	V
Cap-ind-cap Π-network matching capacitance 1 (C_{M1})	7.989	pF
Cap-ind-cap Π-network matching inductance 1 (L_{M1})	1.175	nH
Cap-ind-cap Π-network matching capacitance 2 (C_{M2})	6.366	nH
Inductance at 2.4 GHz (L_S) (M40100N)	1.00	nH
Inductance at DC (L_{SLF}) (M40100N)	0.94	nH
Q-factor (Q) (M40100N)	8.17	–
Resonant frequency (f_R) (M40100N)	9.72	GHz
Turn width (w) (M40100N)	48.0	μm
Turn spacing (s) (M40100N)	2.00	μm
Inner diameter (d_{in}) (M40100N)	89.0	μm
Outer diameter (d_{out}) (M40100N)	293	μm
Number of turns (n) (M40100N)	2	–
Inductance at 2.4 GHz (L_S) (M400500N)	0.50	nH
Inductance at DC (L_{SLF}) (M400500N)	0.49	nH
Q-factor (Q) (M400500N)	7.12	–
Resonant frequency (f_R) (M400500N)	15.0	GHz

(continued)

Table 8.21 (continued)

Parameter	Value	Unit
Turn width (w) (M400500N)	47.0	μm
Turn spacing (s) (M400500N)	2.00	μm
Inner diameter (d_{in}) (M400500N)	30.0	μm
Outer diameter (d_{out}) (M400500N)	222	μm
Number of turns (n) (M400500N)	2	–
Inductance at 2.4 GHz (L_S) (M40119N)	1.18	nH
Inductance at DC (L_{SLF}) (M40119N)	1.10	nH
Q-factor (Q) (M40119N)	8.02	–
Resonant frequency (f_R) (M40119N)	8.87	GHz
Turn width (w) (M40119N)	50.0	μm
Turn spacing (s) (M40119N)	2.00	μm
Inner diameter (d_{in}) (M40119N)	108	μm
Outer diameter (d_{out}) (M40119N)	312	μm
Number of turns (n) (M40119N)	2	–

suitable choice in the 0201DS-1N2 ceramic chip inductor with a nominal value of 1.2 nH. The inductor is rated at 3 GHz, but with an SFR of 17.9 GHz.

To implement the wire-wound inductor, we run the `toroid` subroutine developed in Chap. 7, as illustrated in Fig. 8.39, with default length of 1.5 cm and default area of 1.5 cm^2. The number of turns required is 11. We can also design an RFC of say 100 nH. The required number of turns is 109.

8.9 Summary

The aim of this chapter was to illustrate to the readers how analytical design equations can be used to develop their own EDA programs for power amplifier design in a programming or scripting package such as MATLAB. We started by developing separate subroutines for each power amplifier stage we deemed interesting, after which we took on developing subroutines for impedance matching. When dealing with Class-F power amplifiers, we also developed a procedure for the design of the quarter-wave transmission line. That gave us a solid foundation to combine all different subroutines into a comprehensive standalone program for power amplifier design. The program allows users to select an output stage (a choice between Class-E, Class-F, Class-A/AB/B/C and Class-J), decide on the impedance matching network, and decide if they want to design integrated spiral inductors (if applicable). We also included a capability to export the netlist of the fully designed system in SPICE format.

We illustrated the use of the program and all of its subroutines by means of several examples. The examples were chosen to illustrate the full capability of the

```
.SUBCKT L0 L1 L2 GND
CS L1 L2 177.00fF
LS N4 L1 0.95n
RS N4 L2 1.29
Csi1 N2 GND 63.96fF
Csi2 N3 GND 63.96fF
Cox1 L1 N2 353.05fF
Cox2 L2 N3 353.05fF
Rsi1 N2 GND 347.79
Rsi2 N3 GND 347.79
.ENDS
.SUBCKT L3 L1 L2 GND
CS L1 L2 156.40fF
LS N4 L1 0.49n
RS N4 L2 0.95
Csi1 N2 GND 43.75fF
Csi2 N3 GND 43.75fF
Cox1 L1 N2 231.92fF
Cox2 L2 N3 231.92fF
Rsi1 N2 GND 508.58
Rsi2 N3 GND 508.58
.ENDS
.SUBCKT L5 L1 L2 GND
CS L1 L2 156.40fF
LS N4 L1 0.49n
RS N4 L2 0.95
Csi1 N2 GND 43.75fF
Csi2 N3 GND 43.75fF
Cox1 L1 N2 231.92fF
Cox2 L2 N3 231.92fF
Rsi1 N2 GND 508.58
Rsi2 N3 GND 508.58
.ENDS
.SUBCKT LM6 L1 L2 GND
CS L1 L2 177.00fF
LS N4 L1 1.10n
RS N4 L2 1.40
Csi1 N2 GND 69.56fF
Csi2 N3 GND 69.56fF
Cox1 L1 N2 383.96fF
Cox2 L2 N3 383.96fF
Rsi1 N2 GND 319.79
Rsi2 N3 GND 319.79
.ENDS

.SUBCKT PA_task7_classF bias supply Gnd
LRFC_2 supply N_1 40.00n
CCapacitor_3 N_1 N_4 1.00p
XL0_1 N_M1 Gnd Gnd L0
CCapacitor_1 N_M1 Gnd 4.40p
XL3_1 N_4 N_3 Gnd L3
CCapacitor_2 N_4 N_3 0.98p
XL5_1 N_3 N_M1 Gnd L5
CCapacitor_4 N_3 N_M1 0.35p
RResistor_1 N_M2 Gnd 50.00 TC=0.0, 0.0
Xnpn254h5_1 N_1 bias Gnd Gnd npn254h5 area=96
CCapacitor_M1 Gnd N_M1 7.99p
CCapacitor_M2 Gnd N_M2 6.37p
XLM6_1 N_M1 N_M2 Gnd LM6
.ENDS
```

Fig. 8.36 Exported netlist of the Class-F power amplifier design

```
>> bondwire

Please enter 1 for analysis and 2 for synthesis: 2
Please enter bond wire radius (mm): 0.2
Please enter spacing between two wires (mm) (-1 to skip): 0.2
Please enter wire resistivity (ohm.m) (-1 to skip): -1
Using the default value of 2.82e-008 ohm.m for resistivity.
Please enter the wanted inductance value (nH): 40
Please enter the operating frequency (GHz): 2400

One-wire option:
----------------------

ell = 38.46 mm
Lc = 40.00 nH
Rc = 15.82 Ohm
Q = 38122

Two-wire option:
----------------------

ell = 28.24 mm
Lc = 40.00 nH
Rc = 5.81 Ohm
Q = 103833
```

Fig. 8.37 Interaction with the bond wire design program of task 7

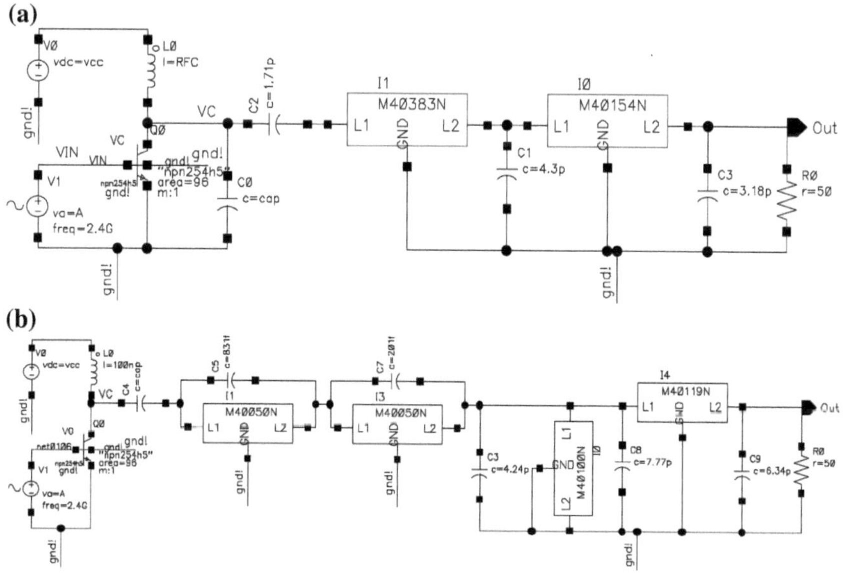

Fig. 8.38 Imported Class-E and Class-F power amplifier schematics

```
>> toroid

Please enter 1 to use permeability formula and 2 to use induct-
ance-index formula: 1

Please enter the initial permeability of the core (-1 to skip): 1

Please enter the effective length of the core in cm (-1 to skip):
-1
Using the default value of 1.50 cm for effective length.

Please enter the effective area of the core in cm2 (-1 to skip):
-1
Using the default value of 1.50 cm2 for effective area.
Please enter the wanted inductance value (nH): 1

N = 11
```

Fig. 8.39 Interaction with the toroid inductor program of task 8

program and its procedures. We also attempted to make every procedure or subroutine standalone so that they could be invoked directly, therefore several examples first showed how to use separate subroutines. The last few examples showed how to use the complete program. A final few examples also demonstrated the use of the bond wire and toroid inductor subroutines, left over from Chap. 7.

In Chap. 9 we will proceed to tackle some practical aspects of power amplifier design that need to be taken into consideration when using the EDA programs developed in this chapter, or any programs and procedures developed by the readers based on the techniques presented in this book.

References

1. Grebennikov A, Sokal NO, Franco MJ. Switchmode RF and microwave power amplifiers. 2nd ed. Burlington: Elsevier; 2012.
2. Sokal NA, Sokal AD. Class-E-A new class of high-efficiency tuned single-ended switching power amplifiers. IEEE J Solid-State Circuits. 1975;10(3):168–76.
3. Cadence Design Systems. Virtuoso schematic composed user guide. San Jose: Cadence Design Systems; 2007.

Chapter 9
Practical Considerations of Integrated and Discrete Power Amplifier Solutions

This chapter deals with some practical considerations for both integrated and discrete power amplifier solutions. The discussion includes practical considerations common to all types of designs, as well as considerations typically associated with just one type: layout of the integrated circuits, packaging of the integrated circuits, and packaging of discrete devices with miniature external components (system-on-package). Some of the topics mentioned in this chapter have already been discussed throughout this book but in less detail. The idea of this chapter is to make the reader aware that even with a streamlined design approach by means of design programs, unwanted real-life design problems will creep through. Some of the problems can be rectified by modifying some design specifications; for example, some can be improved on by increasing the complexity of design models, or by simply understanding the problem via simulations or experimentation.

We also find it fitting to share another MATLAB procedure in this chapter with the readers of this book. It is hoped that the procedure for inductor layout export into GDS format will help ease the pain of integrated power amplifier design. Finally, in this chapter we suggest how to execute a design of a practical power amplifier.

9.1 Practical Considerations Common to Integrated, Discrete and Hybrid (System-on-Package) Solutions

Some considerations are common to more than one type of power amplifier implementation.

9.1.1 *Gain of the Power Transistor*

Typically, the power gain of the transistor needs to be known in order to understand what kind of output power level can be reached in practice. It was seen in Chaps. 3

M. Božanić and S. Sinha, *Power Amplifiers for the S-, C-, X- and Ku-bands*,
Signals and Communication Technology, DOI 10.1007/978-3-319-28376-0_9

and 4 that the biasing and the input signal level are crucial to the operation of each power amplifier stage. Typically, the signal shape and level will be dictated by the previous amplification stage, which is superimposed onto the biasing to enable the particular class of operation. This means that the power transistor needs to be able to amplify with enough gain to reach the required output power level at a specified frequency, given the input signal shape. The overall gain is directly related to the transitional frequency f_T described in Chap. 2.

In integrated solutions, the designer has some control over the transistor gain, which is not the case for discrete solutions. For the MOSFET transistor, the gain of the transistor is proportional to the width-to-length ratio and can be controlled fairly closely. For bipolar counterparts, the power amplifier needs to be designed such that the forward gain (β) of the transistor is adequate, be it an integrated or a discrete solution. Specifically, when working with HBTs, the designer has some control over the gain by tuning the emitter length. If even more gain is needed, several power amplifier transistors can be connected in parallel or power-combining techniques can be used.

As stated earlier, one of the metrics for describing the power amplifier is the output power capability. Various output stages have various output power capabilities; therefore the amount of gain in each power amplification stage that is required to reach the same amount of power differs. With some output classes, the amount of amplification can be tuned by, among others, modifying the conduction angle value.

It thus makes sense that before attempting the design of a power amplifier, with the aid of the procedures described or otherwise, the designers should evaluate whether the required output power level can be reached. The stability analysis on the final power amplifier design, either by simulations or experimentally, as discussed in Chaps. 1 and 2, must not be omitted.

9.1.2 Component Loss

Another important aspect to consider when designing power amplifiers is the component loss. The theory used to develop the procedures in Chap. 8 assumes that there is no loss in various power amplifier and system components. In practice, various components (e.g. the resonant tank or the transmission) will have some loss. Thus the overall efficiency of the output stage will be less than what was predicted by calculations. An additional efficiency figure, the component efficiency, can be introduced into the picture. The overall power amplifier efficiency is then a multiple of all component efficiencies and the drain or collector efficiency.

With correct modeling of components, accurate simulations can be performed on the final system to determine the efficiency and power loss mechanisms. Before attempting any simulations, it is paramount that all the components are modeled correctly. Modeling of active devices is not one of the topics of this book, but the reader is referred, for example, to [1]. When developing more accurate models of passive components, specialized tools for simulation of passives can be utilized to gain confidence in the model. An example of this approach was shown earlier in this book when investigating the spiral inductor model with the aid of an EM simulator.

As described before, inductors are typically the components that are expected to contribute greatly to system loss and they must be modeled such that their finite Q-factor is catered for. With increasing frequencies, it is possible that the loss of other passives (e.g. capacitors) will need to be modeled as well.

For integrated scenarios incorporating spiral inductors, we have already demonstrated how the modeling problem is typically overcome by using one of the inductor models detailed in Chap. 5. This was done in Chaps. 7 and 8 we demonstrated that we can automatically import the SPICE netlist of the inductor. The same can be done for any component that exhibits a loss with a known loss mechanism. With netlists and schematics generated in such a manner, accurate simulations can be executed.

9.1.3 Feasibility of Component Values

Some of the component values calculated based on the specifications for the system can become difficult to reach in practice. Therefore, care needs to be taken to evaluate each component and decide whether it can be used in a given implementation, or whether the design specifications need to be reevaluated.

In discrete implementations, a problem that may arise is that the values obtained by program calculations may be different from standard component values. In integrated implementations, a typical problem that could arise is that the design may lead to excessively large capacitors and/or inductors, which cannot be integrated as easily. A possible solution is to adopt the approach taken when developing the procedure for the design of Class-E power amplifiers in Chap. 8 where the limits on the inductance value were used to recommend the practical range of the loaded quality factors. In Chaps. 6 and 8 we also discussed the fact that various matching networks can be used to perform the same matching task and that the ability to choose different matching networks for the particular frequency and application introduces another degree of freedom on the component size. Matching networks can also help eliminate blocking capacitors, for example, which will decrease the overall component count. We have already used this approach in some examples in Chap. 8.

During the design, it may turn out that the component values dictate whether an integrated, discrete or combined implementation is required. Typically, a power amplifier could have most of its circuitry on-chip, but if a large inductor is necessary, it might require the particular inductor to be implemented outside the chip. On the other hand, the requirement for larger output power could necessitate larger supply voltages, which could result in the need for a discrete implementation where an integrated implementation may have been considered initially. Thus, the choice among integrated, SOP and discrete implementation has to be made not only from the standpoint of given specifications, but also from the standpoint of passives.

9.1.4 Influence of the Frequency and the Wavelength

The relation between frequency and wavelength was discussed previously in Chap. 2 when the frequency spectrum was discussed. It was also mentioned several times in this book that the wavelength directly influences the feasibility of the implementation of transmission lines, and that in the frequency range this book deals with, the transmission lines are only practical off-chip. In general in this book we have tried to stay away from intense transmission line theory. We have limited our scope to the quarter-wave transformer used for matching and for the design of the Class-F power amplifier and we have demonstrated that it should be used only if frequency and space allow for it.

Furthermore, we cannot stress enough how important the modeling of all components (and especially passives) becomes with increased frequency. Every wire and every track has a tendency to turn into a transmission line and to introduce some parasitics. The inductive behavior of the wire, for example, was treated in Chap. 5 and it will be revised later in this chapter. The best practice is to minimize any wire and track lengths. Another possible approach is to use the parasitics in the design by absorbing them in wanted passive calculations. Still, all simulations have to be performed with parasitics in mind, especially with increasing frequencies.

It is also important to remember that impedance matching is necessary for maximum power transfer. Because of the fact that with decreasing wavelength, every track or a wire turns into a transmission line, great care needs to be taken either to keep any tracks to about 1/10 of the wavelength to avoid matching, or to use matched transmission lines to carry the signals.

The concept of the maximum operating frequency at which the power amplifier has some gain was discussed in Chap. 1. Thus readers are reminded here to check first if their power amplifier specifications (power and supply voltage) and the choice of active device allow for the system to be designed for the required frequency band.

9.1.5 Coupling

Analog and RF circuitry are also subject to unwanted signal coupling, adding to the noise figure. The signals can couple from adjacent signal routing lines, through passives electromagnetically, or via the substrate [2]. The problem with coupling becomes prominent with increased circuit density, thus mostly affecting integrated and SOP implementations. Therefore, when performing the design and considering component placement, it should be done with coupling in mind.

9.2 Integrated Circuit Considerations

Integrated design is prone to many more problems than discrete design. In this section, considerations typical to IC implementations are discussed in some detail.

9.2.1 Design Rule Checks and Technology Considerations

The most important concept to grasp here is that when designing integrated power amplifiers using the procedures described in Chap. 8, it has to done with the specific technology and its design rules in mind. This is important for both active and passive devices. For active devices, the maximum size of the transistor could impose a restriction on the active side. For passive components, it is important to stick to the DRC rules. Typically, the minimum metal spacing would determine the pitch between each turn of the spiral, or how closely together the capacitor fingers can be placed. The layout also has to be done with heating in mind. Different transistors suffer from different heating issues, but there are layout techniques that can be used to minimize the effect.

Some other layout considerations (especially in the case of inductors), have already been discussed in Chap. 5 [3].

9.2.2 Extraction of a Spiral Inductor Layout into a GDSII File

When placing inductors onto the substrate, it may prove tedious to draw the inductor layout by hand in the layout editor. Although a small variation, in say the width of the spiral in one turn or the spacing on one of the turns, might not greatly influence the behavior of the inductor, a much more plausible scenario is that spacing DRC rules could be broken. A much simpler way is to develop a procedure to automatically export the inductor layout into a file that can be used in the final power amplifier system layout. This also increases the speed at which inductors can be designed for various power amplifier designs. Thus, as an example, in this section we include a procedure for the design of square spiral inductors as part of the complete power amplifier design program discussed in Chap. 8.

Several layout formats can be used for the extraction of inductors. However, the current industry standard is the GDSII format, which works well with most commercial IC design tools. The GDSII format (or simply GDS format) is a stream format, and the extraction of inductor layout into a GDS file requires tedious number format and number-to-string conversions and string manipulation. We have developed a main routine (convertGDS.m) and subroutines (str2hex.m, coordinatesSpiral.m, coordinatesUnderpass.m, coordinatesVia.m, coordinatesP1.m, coordinatesP2.m and coordinatesName.m), all listed in Figs. 9.1, 9.2, 9.3, 9.4, 9.5, 9.6, 9.7 and 9.8. These routines were implemented by using the GDS format definitions and guidelines listed in [4].

In Chap. 8 we designed several spiral inductors when dealing with Task 7. Here we export the layouts of these spiral inductors to GDS format to illustrate the use of the developed GDS export subroutine. In order to do the export, we call the main inductor design procedure from Chap. 7 by typing `indcalc` in the main MATLAB window for each inductor. We then pass the geometry parameters, after which we allow the inductors to be exported. This process is illustrated in Fig. 9.9. for M40383N inductor. Finally, we use free layout viewing software named Layout Editor to open the designed inductors, and show them in Fig. 9.10.

Note that the export of the layout of the inductor into older text-based CIF layout file format was also devised but it will not be discussed.

9.2.3 Bond-Pad Considerations

In order to connect the internal IC layout to the bond wire, which then connects to the package on the other side, bonding pads need to be placed on the layout. A bonding pad typically consists of several layers of metals and vias for integrity purposes. Therefore the size and the shape of the bonding pad will introduce parasitics that will have to be included in the design and the simulation of the complete power amplifier system.

9.2.4 Bond-Wire Considerations

The inductive behavior of bond wires was discussed and described mathematically in previous chapters. In low-frequency designs, bond wires are typically modeled as short circuits (i.e. can be ignored). With increasing frequencies bond wires should be modeled and included in simulations. The best practice for high power design is to utilize the bond wire parasitics in calculations. One possible approach is to design the matching network such that part of the inductance value is contributed by the bond wire. However, this can only be done if the system can allow for the increase on the tolerance of the inductance value that the bond wire will introduce.

```
%This procedure exports the layout of an inductor into a stream
based (GDSII) file
ccif = input('Do you want to change any GDS parameters (y/n)? ',
's');
if ccif == 'y' || ccif == 'Y';
    %Change default GDS parameters
    scaling = input('Please enter GDS scaling factor (-1 to
skip): ');
    if scaling <= 0;
        scaling = 1000;
    end%if
    viaSpacing = input('Please enter via spacing (-1 to skip):
');
    if viaSpacing <= 0;
        viaSpacing = 1;
    end%if
    viaWidth = input('Please enter via width (-1 to skip): ');
    if viaWidth <= 0;
        viaWidth = 0.5;
    end%if
else
    %Default GDS parameters - Change for particular process
    scaling = 1000;
    viaSpacing = 1;
    viaWidth = 0.5;
end%if
%Calculate usable values
dout2 = round(doutstored*scaling); zero = 0;
w2 = round(wstored * scaling); s2 = round(sstored * scaling);
n = nstored;
spacing = viaSpacing * scaling;
width = viaWidth * scaling;
%Specify the inductor name and open the file
cellName = input('Please enter the cell (inductor) name (press
"s" to skip): ', 's');
if cellName(1) == 's' || cellName(1) == 'S';
    cellName = 'mycell';
end%if
fileName = strcat(cellName, '_lay.gds');
%Stream export
fprintf('Exporting GDS. Please wait...\n');
%Header
gdsHexVector = '0006';
gdsHexVector = strvcat(gdsHexVector, '0002');
gdsHexVector = strvcat(gdsHexVector, '0003');
%Bgnlib
gdsHexVector = strvcat(gdsHexVector, '001C');
gdsHexVector = strvcat(gdsHexVector, '0102');
dateInt = fix(clock); %date and time
dateInt(1) = dateInt(1) - 1900;
dateHex = dec2hex(dateInt,4);
gdsHexVector = strvcat(gdsHexVector, dateHex); %modified
gdsHexVector = strvcat(gdsHexVector, dateHex); %accessed
%STRNAME
textString = 'Inductor';
str2hex; %Convert text into hex number
gdsHexVector = strvcat(gdsHexVector, hexLength);
```

Fig. 9.1 Main inductor GDS export procedure (convertGDS.m)

```
gdsHexVector = strvcat(gdsHexVector, '0206');
gdsHexVector = strvcat(gdsHexVector, hexString2);
%units
gdsHexVector = strvcat(gdsHexVector, '0014');
gdsHexVector = strvcat(gdsHexVector, '0305');
gdsHexVector = strvcat(gdsHexVector, '3E41');
gdsHexVector = strvcat(gdsHexVector, '8937');
gdsHexVector = strvcat(gdsHexVector, '4BC6');
gdsHexVector = strvcat(gdsHexVector, 'A7EF');
gdsHexVector = strvcat(gdsHexVector, '3944');
gdsHexVector = strvcat(gdsHexVector, 'B82F');
gdsHexVector = strvcat(gdsHexVector, 'A09B');
gdsHexVector = strvcat(gdsHexVector, '5A51');
%BGNSTR
gdsHexVector = strvcat(gdsHexVector, '001C');
gdsHexVector = strvcat(gdsHexVector, '0502');
gdsHexVector = strvcat(gdsHexVector, dateHex); %modified
gdsHexVector = strvcat(gdsHexVector, dateHex); %accessed
%STRNAME
textString = 'Inductor';
str2hex;
gdsHexVector = strvcat(gdsHexVector, hexLength);
gdsHexVector = strvcat(gdsHexVector, '0606');
gdsHexVector = strvcat(gdsHexVector, hexString2);
%Boundary object - Spiral
gdsHexVector = strvcat(gdsHexVector, '0004'); %Boundary
gdsHexVector = strvcat(gdsHexVector, '0800');
gdsHexVector = strvcat(gdsHexVector, '0006'); %Layer 39 (metal 3)
gdsHexVector = strvcat(gdsHexVector, '0D02');
gdsHexVector = strvcat(gdsHexVector, '0027'); %metal3
gdsHexVector = strvcat(gdsHexVector, '0006'); %Datatype 0
gdsHexVector = strvcat(gdsHexVector, '0E02');
gdsHexVector = strvcat(gdsHexVector, '0000');
coordinatesSpiral; %Calculate the coordinates of the spiral
gdsHexVector = strvcat(gdsHexVector, hexRecLength); %XY
gdsHexVector = strvcat(gdsHexVector, '1003');
gdsHexVector = strvcat(gdsHexVector, hexCoordinate2);
gdsHexVector = strvcat(gdsHexVector, '0004');
gdsHexVector = strvcat(gdsHexVector, '1100');
%Boundary object - Underpass
gdsHexVector = strvcat(gdsHexVector, '0004'); %Boundary
gdsHexVector = strvcat(gdsHexVector, '0800');
gdsHexVector = strvcat(gdsHexVector, '0006'); %Layer 37 (metal 2)
gdsHexVector = strvcat(gdsHexVector, '0D02');
gdsHexVector = strvcat(gdsHexVector, '0025'); %metal 2
gdsHexVector = strvcat(gdsHexVector, '0006'); %Datatype 0
gdsHexVector = strvcat(gdsHexVector, '0E02');
gdsHexVector = strvcat(gdsHexVector, '0000');
coordinatesUnderpass; %Calculate the coordinates of the underpass
gdsHexVector = strvcat(gdsHexVector, hexRecLength); %XY
gdsHexVector = strvcat(gdsHexVector, '1003');
gdsHexVector = strvcat(gdsHexVector, hexCoordinate2);
gdsHexVector = strvcat(gdsHexVector, '0004');
gdsHexVector = strvcat(gdsHexVector, '1100');
%Boundary object - Vias
viaSpacing = 0.9; viaWidth = 0.5;
spacing = viaSpacing * scaling;
```

Fig. 9.1 (continued)

```
width = viaWidth * scaling;
x = floor(w2 / (spacing + width));
y = floor((2*w2 + s2) / (spacing + width));
for i=0:x-1;
   for j=0:y-1;
        gdsHexVector = strvcat(gdsHexVector, '0004'); %Boundary
        gdsHexVector = strvcat(gdsHexVector, '0800');
        gdsHexVector = strvcat(gdsHexVector, '0006'); %Layer 38
(via2)
        gdsHexVector = strvcat(gdsHexVector, '0D02');
        gdsHexVector = strvcat(gdsHexVector, '0026'); %via2
        gdsHexVector = strvcat(gdsHexVector, '0006'); %Datatype 0
        gdsHexVector = strvcat(gdsHexVector, '0E02');
        gdsHexVector = strvcat(gdsHexVector, '0000');
        coordinatesVia; %Calculate the coordinates of the vias
        gdsHexVector = strvcat(gdsHexVector, hexRecLength); %XY
        gdsHexVector = strvcat(gdsHexVector, '1003');
        gdsHexVector = strvcat(gdsHexVector, hexCoordinate2);
        gdsHexVector = strvcat(gdsHexVector, '0004');
        gdsHexVector = strvcat(gdsHexVector, '1100');
   end%for
end%for
%Text - P1
gdsHexVector = strvcat(gdsHexVector, '0004'); %Text
gdsHexVector = strvcat(gdsHexVector, '0C00');
gdsHexVector = strvcat(gdsHexVector, '0006'); %Layer 61 (text)
gdsHexVector = strvcat(gdsHexVector, '0D02');
gdsHexVector = strvcat(gdsHexVector, '003D');
gdsHexVector = strvcat(gdsHexVector, '0006'); %texttype 0
gdsHexVector = strvcat(gdsHexVector, '1602');
gdsHexVector = strvcat(gdsHexVector, '0000');
gdsHexVector = strvcat(gdsHexVector, '0006'); %presentation mid-
dle vertical centre horizontal
gdsHexVector = strvcat(gdsHexVector, '1701');
gdsHexVector = strvcat(gdsHexVector, '0005');
%gdsHexVector = strvcat(gdsHexVector, '0006'); %absolute magnifi-
cation; absolute angle;
gdsHexVector = strvcat(gdsHexVector, '000C'); %Text magnified 2
times
gdsHexVector = strvcat(gdsHexVector, '1B05');
gdsHexVector = strvcat(gdsHexVector, '41A0');
gdsHexVector = strvcat(gdsHexVector, '0000');
gdsHexVector = strvcat(gdsHexVector, '0000');
gdsHexVector = strvcat(gdsHexVector, '0000');
coordinatesP1; %Calculate the coordinates of point P1
gdsHexVector = strvcat(gdsHexVector, hexRecLength); %XY
gdsHexVector = strvcat(gdsHexVector, '1003');
gdsHexVector = strvcat(gdsHexVector, hexCoordinate2);
textString = 'P1';
str2hex;
gdsHexVector = strvcat(gdsHexVector, hexLength);
gdsHexVector = strvcat(gdsHexVector, '1906');
gdsHexVector = strvcat(gdsHexVector, hexString2);
gdsHexVector = strvcat(gdsHexVector, '0004');
gdsHexVector = strvcat(gdsHexVector, '1100');
%Text - P2
gdsHexVector = strvcat(gdsHexVector, '0004'); %Text
```

Fig. 9.1 (continued)

```
gdsHexVector = strvcat(gdsHexVector, '0C00');
gdsHexVector = strvcat(gdsHexVector, '0006'); %Layer 61 (text)
gdsHexVector = strvcat(gdsHexVector, '0D02');
gdsHexVector = strvcat(gdsHexVector, '003D');
gdsHexVector = strvcat(gdsHexVector, '0006'); %texttype 0
gdsHexVector = strvcat(gdsHexVector, '1602');
gdsHexVector = strvcat(gdsHexVector, '0000');
gdsHexVector = strvcat(gdsHexVector, '0006'); %presentation mid-
dle vertical centre horizontal
gdsHexVector = strvcat(gdsHexVector, '1701');
gdsHexVector = strvcat(gdsHexVector, '0005');
gdsHexVector = strvcat(gdsHexVector, '000C'); %Text magnified 2
times
gdsHexVector = strvcat(gdsHexVector, '1B05');
gdsHexVector = strvcat(gdsHexVector, '41A0');
gdsHexVector = strvcat(gdsHexVector, '0000');
gdsHexVector = strvcat(gdsHexVector, '0000');
gdsHexVector = strvcat(gdsHexVector, '0000');
coordinatesP2; %Calculate the coordinates of point P2
gdsHexVector = strvcat(gdsHexVector, hexRecLength); %XY
gdsHexVector = strvcat(gdsHexVector, '1003');
gdsHexVector = strvcat(gdsHexVector, hexCoordinate2);
textString = 'P2';
str2hex;
gdsHexVector = strvcat(gdsHexVector, hexLength);
gdsHexVector = strvcat(gdsHexVector, '1906');
gdsHexVector = strvcat(gdsHexVector, hexString2);
gdsHexVector = strvcat(gdsHexVector, '0004');
gdsHexVector = strvcat(gdsHexVector, '1100');
%Text - Inductor
gdsHexVector = strvcat(gdsHexVector, '0004'); %Text
gdsHexVector = strvcat(gdsHexVector, '0C00');
gdsHexVector = strvcat(gdsHexVector, '0006'); %Layer 61 (text)
gdsHexVector = strvcat(gdsHexVector, '0D02');
gdsHexVector = strvcat(gdsHexVector, '003D');
gdsHexVector = strvcat(gdsHexVector, '0006'); %texttype 0
gdsHexVector = strvcat(gdsHexVector, '1602');
gdsHexVector = strvcat(gdsHexVector, '0000');
gdsHexVector = strvcat(gdsHexVector, '0006'); %presentation mid-
dle vertical centre horizontal
gdsHexVector = strvcat(gdsHexVector, '1701');
gdsHexVector = strvcat(gdsHexVector, '0005');
gdsHexVector = strvcat(gdsHexVector, '000C'); %Text magnified 3
times
gdsHexVector = strvcat(gdsHexVector, '1B05');
gdsHexVector = strvcat(gdsHexVector, '41A0');
gdsHexVector = strvcat(gdsHexVector, '0000');
gdsHexVector = strvcat(gdsHexVector, '0000');
gdsHexVector = strvcat(gdsHexVector, '0000');
coordinatesName; %Calculate the coordinates of the text
gdsHexVector = strvcat(gdsHexVector, hexRecLength); %XY
gdsHexVector = strvcat(gdsHexVector, '1003');
gdsHexVector = strvcat(gdsHexVector, hexCoordinate2);
textString = cellName;
str2hex;
gdsHexVector = strvcat(gdsHexVector, hexLength);
gdsHexVector = strvcat(gdsHexVector, '1906');
```

Fig. 9.1 (continued)

```
gdsHexVector = strvcat(gdsHexVector, hexString2);
gdsHexVector = strvcat(gdsHexVector, '0004');
gdsHexVector = strvcat(gdsHexVector, '1100');
%EndStr
gdsHexVector = strvcat(gdsHexVector, '0004');
gdsHexVector = strvcat(gdsHexVector, '0700');
%EndLib
gdsHexVector = strvcat(gdsHexVector, '0004');
gdsHexVector = strvcat(gdsHexVector, '0400');
gdsDecVector = hex2dec(gdsHexVector);
%Write Gds File
fid = fopen(fileName,'wb','b');
fwrite(fid,gdsDecVector,'uint16');
fclose all;
fprintf('Export complete.\n');
```

Fig. 9.1 (continued)

```
%This procedure converts any string into a hexadecimal number for
GDS extraxtion
clear hexString2;
numString = uint8(textString);
strLength = length(numString);
if mod(strLength, 2) == 1;
    strLength = strLength + 1;
    numString(strLength) = 32;
end
hexString = dec2hex(numString);
for k = 1:strLength/2;
    hexString2(k,:) = strcat(hexString(2*k-1,:), hex-
String(2*k,:));
end
hexLength = dec2hex(strLength + 4,4);
```

Fig. 9.2 str2hex.m procedure

9.2.5 Package Lead Considerations

Packaging of ICs is needed to protect the die from external influences. Leads in the lead frame provide the physical framework and electrical connections from the IC to the PCB or another system outside the IC [5]. Thus package leads should also be taken into consideration when doing the design or performing simulations [6].

Leads are typically made of copper, aluminum or various alloys (e.g. alloy42), all having high conductivity. Therefore, leads mostly exhibit inductive behavior, but this obviously depends on the type of package used. The package models are now becoming commonly available to designers because of the ever-increasing need for accurate simulation and modeling [7].

```
%This procedure calculates the coordinates for the spiral for GDS
%extraction of the spiral inductor
clear decCoordinate;
clear hexCoordinate;
clear hexCoordinate2;
 %inner points
 %first turn
decCoordinate(1)= zero;
decCoordinate(2)= zero;
decCoordinate(3)= w2;
decCoordinate(4)= zero;
decCoordinate(5)= w2;
decCoordinate(6)= dout2 - w2;
decCoordinate(7)= dout2 - w2;
decCoordinate(8)= dout2 - w2;
decCoordinate(9)= dout2 - w2;
decCoordinate(10)= w2;
%inner turns
for turn = 2:n-1;
    coordIndex = 8*(turn-2)+10;
    decCoordinate(coordIndex + 1) = turn*w2 + (turn-1)*s2;
    decCoordinate(coordIndex + 2) = (turn-1)*w2 + (turn-2)*s2;
    decCoordinate(coordIndex + 3) = turn*w2 + (turn-1)*s2;
    decCoordinate(coordIndex + 4) = dout2 - (turn-1)*s2 -
turn*w2;
    decCoordinate(coordIndex + 5) = dout2 - (turn-1)*s2 -
turn*w2;
    decCoordinate(coordIndex + 6) = dout2 - (turn-1)*s2 -
turn*w2;
    decCoordinate(coordIndex + 7) = dout2 - (turn-1)*s2 -
turn*w2;
    decCoordinate(coordIndex + 8) = turn*w2 + (turn-1)*s2;
end%for
%last turn
turn = n;
coordIndex = 8*(n-2)+10;
decCoordinate(coordIndex + 1) = turn*w2 + (turn-1)*s2;
decCoordinate(coordIndex + 2) = (turn-1)*w2 + (turn-2)*s2;
decCoordinate(coordIndex + 3) = turn*w2 + (turn-1)*s2;
decCoordinate(coordIndex + 4) = dout2 - (turn-1)*s2 - turn*w2;
decCoordinate(coordIndex + 5) = dout2 - (turn-1)*s2 - turn*w2;
decCoordinate(coordIndex + 6) = dout2 - (turn-1)*s2 - turn*w2;
decCoordinate(coordIndex + 7) = dout2 - (turn-1)*s2 - turn*w2;
decCoordinate(coordIndex + 8) = (turn-1)*w2 + (turn-1)*s2;
decCoordinate(coordIndex + 9) = dout2 - (turn-1)*s2 - (turn-
1)*w2;
decCoordinate(coordIndex + 10) = (turn-1)*w2 + (turn-1)*s2;
decCoordinate(coordIndex + 11) = dout2 - (turn-1)*s2 - (turn-
1)*w2;
decCoordinate(coordIndex + 12) = dout2 - (turn-1)*w2 - (turn-
1)*s2;
decCoordinate(coordIndex + 13) = (turn-1)*w2 + (turn-1)*s2;
decCoordinate(coordIndex + 14) = dout2 - (turn-1)*w2 - (turn-
1)*s2;
decCoordinate(coordIndex + 15) = (turn-1)*w2 + (turn-1)*s2;
decCoordinate(coordIndex + 16) = (turn-2)*w2 + (turn-2)*s2;
```

Fig. 9.3 coordinatesSpiral.m procedure

```
%inner turns
for turn = n-1:-1:2;
    coordIndex = 8*(n-2)+26 + 8*((n-1)-turn);
    decCoordinate(coordIndex + 1) = dout2 - (turn-1)*s2 - (turn-
1)*w2;
    decCoordinate(coordIndex + 2) = (turn-1)*w2 + (turn-1)*s2;
    decCoordinate(coordIndex + 3) = dout2 - (turn-1)*s2 - (turn-
1)*w2;
    decCoordinate(coordIndex + 4) = dout2 - (turn-1)*w2 - (turn-
1)*s2;
    decCoordinate(coordIndex + 5) = (turn-1)*s2 + (turn-1)*w2;
    decCoordinate(coordIndex + 6) = dout2 - (turn-1)*w2 - (turn-
1)*s2;
    decCoordinate(coordIndex + 7) = (turn-1)*s2 + (turn-1)*w2;
    decCoordinate(coordIndex + 8) = (turn-2)*w2 + (turn-2)*s2;
end%for
%first turn
coordIndex = 16*(n-2)+26;
decCoordinate(coordIndex + 1) = dout2;
decCoordinate(coordIndex + 2) = zero;
decCoordinate(coordIndex + 3) = dout2;
decCoordinate(coordIndex + 4) = dout2;
decCoordinate(coordIndex + 5) = zero;
decCoordinate(coordIndex + 6) = dout2;
decCoordinate(coordIndex + 7) = zero;
decCoordinate(coordIndex + 8) = zero;

%Convert everything into hex numbers
hexCoordinate = dec2hex(decCoordinate,8);
coorLength = length(decCoordinate);
for count = 1:coorLength;
    hexCoordinate2(2*count-1,1:4) = hexCoordinate(count,1:4);
    hexCoordinate2(2*count,  1:4) = hexCoordinate(count,5:8);
end
decRecLength = 2*length(hexCoordinate2) + 4;
hexRecLength = dec2hex(decRecLength, 4);
```

Fig. 9.3 (continued)

One of the possible solutions to compensate for the power (and efficiency) loss that is expected because of packaging and otherwise is to design for somewhat higher output power than required [8]. In fact, this approach can be used to compensate for bond-pad or bond-wire parasitics if needed. This will obviously not compensate for the distortion from the ideal waveforms that may arise owing to packaging [9].

9.3 Systems-on-Package Considerations

SOP was mentioned in Chap. 2 in the discussion of discrete substrates. Further, in the earlier chapters of this book we kept mentioning that hybrid power amplifier systems are possible, which are supposed to combine the advantages of both

```
%This procedure calculates the coordinates for the underpass for
GDS extraction of the spiral inductor
clear decCoordinate;
clear hexCoordinate;
clear hexCoordinate2;
decCoordinate(1) = dout2 - n*w2 - (n - 1)*s2;
decCoordinate(2) = zero;
decCoordinate(3) = dout2 - n*w2 - (n - 1)*s2 + w2;
decCoordinate(4) = zero;
decCoordinate(5) = dout2 - n*w2 - (n - 1)*s2 + w2;
decCoordinate(6) = w2*(n + 1) + s2*n;
decCoordinate(7) = dout2 - n*w2 - (n - 1)*s2;
decCoordinate(8) = w2*(n + 1) + s2*n;
decCoordinate(9) = dout2 - n*w2 - (n - 1)*s2;
decCoordinate(10) = zero;
%Convert everything into hex numbers
hexCoordinate = dec2hex(decCoordinate,8);
coorLength = length(decCoordinate);
for count = 1:coorLength;
    hexCoordinate2(2*count-1,1:4) = hexCoordinate(count,1:4);
    hexCoordinate2(2*count,  1:4) = hexCoordinate(count,5:8);
end
decRecLength = 2*length(hexCoordinate2) + 4;
hexRecLength = dec2hex(decRecLength, 4);
```

Fig. 9.4 coordinatesUnderpass.m procedure

```
%This procedure calculates the coordinates for the vias for GDS
%extraction of the spiral inductor
clear decCoordinate;
clear hexCoordinate;
clear hexCoordinate2;
basex = dout2 - n*w2 - (n - 1)*s2 + (spacing + width)*i + spac-
ing/2;
basey = (n-1)*w2 + (n-1)*s2 + (spacing + width)*j + spacing/2;
decCoordinate(1) = basex;
decCoordinate(2) = basey;
decCoordinate(3) = basex + width;
decCoordinate(4) = basey;
decCoordinate(5) = basex + width;
decCoordinate(6) = basey + width;
decCoordinate(7) = basex;
decCoordinate(8) = basey + width;
decCoordinate(9) = basex;
decCoordinate(10) = basey;

%Convert everything into hex numbers
hexCoordinate = dec2hex(decCoordinate,8);
coorLength = length(decCoordinate);
for count = 1:coorLength;
    hexCoordinate2(2*count-1,1:4) = hexCoordinate(count,1:4);
    hexCoordinate2(2*count,  1:4) = hexCoordinate(count,5:8);
end
decRecLength = 2*length(hexCoordinate2) + 4;
hexRecLength = dec2hex(decRecLength, 4);
```

Fig. 9.5 coordinatesVia.m procedure

```
%This procedure calculates the coordinates for node P1 for GDS
%extraction of the spiral inductor
clear decCoordinate;
clear hexCoordinate;
clear hexCoordinate2;
decCoordinate(1) = w2/2;
decCoordinate(2) = zero;

%Convert everything into hex numbers
hexCoordinate = dec2hex(decCoordinate,8);
coorLength = length(decCoordinate);
for count = 1:coorLength;
    hexCoordinate2(2*count-1,1:4) = hexCoordinate(count,1:4);
    hexCoordinate2(2*count,   1:4) = hexCoordinate(count,5:8);
end
decRecLength = 2*length(hexCoordinate2) + 4;
hexRecLength = dec2hex(decRecLength, 4);
```

Fig. 9.6 coordinatesP1.m procedure

```
%This procedure calculates the coordinates for node P2 for GDS
%extraction of the spiral inductor
clear decCoordinate;
clear hexCoordinate;
clear hexCoordinate2;
decCoordinate(1) = dout2 - (n-0.5)*w2 - (n-1)*s2;
decCoordinate(2) = zero;

%Convert everything into hex numbers
hexCoordinate = dec2hex(decCoordinate,8);
coorLength = length(decCoordinate);
for count = 1:coorLength;
    hexCoordinate2(2*count-1,1:4) = hexCoordinate(count,1:4);
    hexCoordinate2(2*count,   1:4) = hexCoordinate(count,5:8);
end
decRecLength = 2*length(hexCoordinate2) + 4;
hexRecLength = dec2hex(decRecLength, 4);
```

Fig. 9.7 coordinatesP2.m procedure

discrete and integrated worlds. These are more formally known as systems on package or SOP. The idea behind SOP is that many system components can be designed in the same packaging, with components placed closer together, allowing for shorter interconnects, which in turn can decrease parasitic effects and result in overall improvement of performance [2, 5].

A very similar concept to SOP is systems-in-package or SIP, where the former typically implies a mini-PCB assembly with components on different substrates, and the latter implies a common substrate for all components. A multi-chip package is a special case of SIP and SOP, where the package contains several ICs and their interconnects only. With SOP, a compact microsystem can be realized without any

```
%This procedure calculates the coordinates for the inductor name
for GDS
%extraction of the spiral inductor
clear decCoordinate;
clear hexCoordinate;
clear hexCoordinate2;
decCoordinate(1) = dout2/2;
decCoordinate(2) = dout2/2;

%Convert everything into hex numbers
hexCoordinate = dec2hex(decCoordinate,8);
coorLength = length(decCoordinate);
for count = 1:coorLength;
    hexCoordinate2(2*count-1,1:4) = hexCoordinate(count,1:4);
    hexCoordinate2(2*count,  1:4) = hexCoordinate(count,5:8);
end
decRecLength = 2*length(hexCoordinate2) + 4;
hexRecLength = dec2hex(decRecLength, 4);
```

Fig. 9.8 coordinatesName.m procedure

material or process limits and with increased flexibility [10]. Since all the components are inside the package, the package has a lower influence on the performance of the system. SOP is becoming increasingly popular in mobile applications.

In RF systems, in addition to an unpackaged integrated die with various systems, typical components that can be combined are antennas, capacitors, inductors, resistors, baluns, filters and MEMS components fabricated on substrates different from silicon. These components can be fabricated in the same way their discrete counterparts are fabricated, but the size of the whole system is decreased. As all of these components are used with typical power amplifiers, SOC implementations could be an important consideration to a power amplifier designer as well. In addition, at frequencies that form the topic of this book, shorter distances in SOP may allow for relaxed matching considerations, but at the same time there may be more space for transmission-line packaging. The advantages of SOP were shown in Table 2.5, but they do not stop there, because additional subsystems, such as antennas and heat sinks, can also be included in the package. Circuits other than power amplifiers can benefit from even more possibilities, such as optoelectronic circuitry.

Some of the challenges for power amplifier SOP designs are the necessity for accurate modeling and optimization, choice of substrate for passive fabrication, crosstalk, and many other concerns previously discussed common to all power amplifier implementations. The reliability of the whole package also needs to be ensured, which becomes increasingly difficult, because of the number of subsystems in one package.

```
>>indcalc2_2

Welcome to Inductor Calculator v2.2 for Matlab. Please follow the
prompts.

Please enter 1 to specify geometry or 2 to specify inductance: 1
Please enter the dout diameter (dout) (um): 275
Please enter the din diameter (din) (um): 165
Please enter the turn width (w) (um): 17
Please enter the number of turns: 3
Please enter the operating frequency (MHz): 2400
Do you want to change any of the default process parameters
(y/n)? n

Ls = 3.84 nH
Lslf = 3.46 nH
Q = 4.84
fo = 6.86 GHz
w = 17.00 um
s = 2.00 um
din = 165.00 um
dout = 275.00 um
n = 3

Do you want to export the netlist of this inductor into a SPC
file (y/n)? y
Please enter the inductor name (press "s" to skip): M40383N
File M40383N_net.spc created successfully.

Do you want to export the layout of this inductor into a CIF or
GDS file (c - CIF, g - GDS, n - none)? g
Do you want to change any GDS parameters (y/n)? n
Please enter the cell (inductor) name (press "s" to skip):
M40383N
Exporting GDS. Please wait...
Export complete.
```

Fig. 9.9 Execution of the indcalc.m procedure illustrating the utilization of layout export concept

9.4 From Theoretical Design Using Custom EDA to Practical Design

From the discussions carried out in this chapter and earlier in this book, it is clear that several steps need to be taken in the power amplifier design process until the final design is implemented. Although this may be obvious, it makes sense to outline the flow in a formal manner. Some readers will find some similarity with the power amplifier design procedure listed in [11]. Thus, the full design procedure can be carried out as follows:

1. Synthesize the design from transient equations, using custom-developed CAD programs similar to the ones described in Chap. 8.

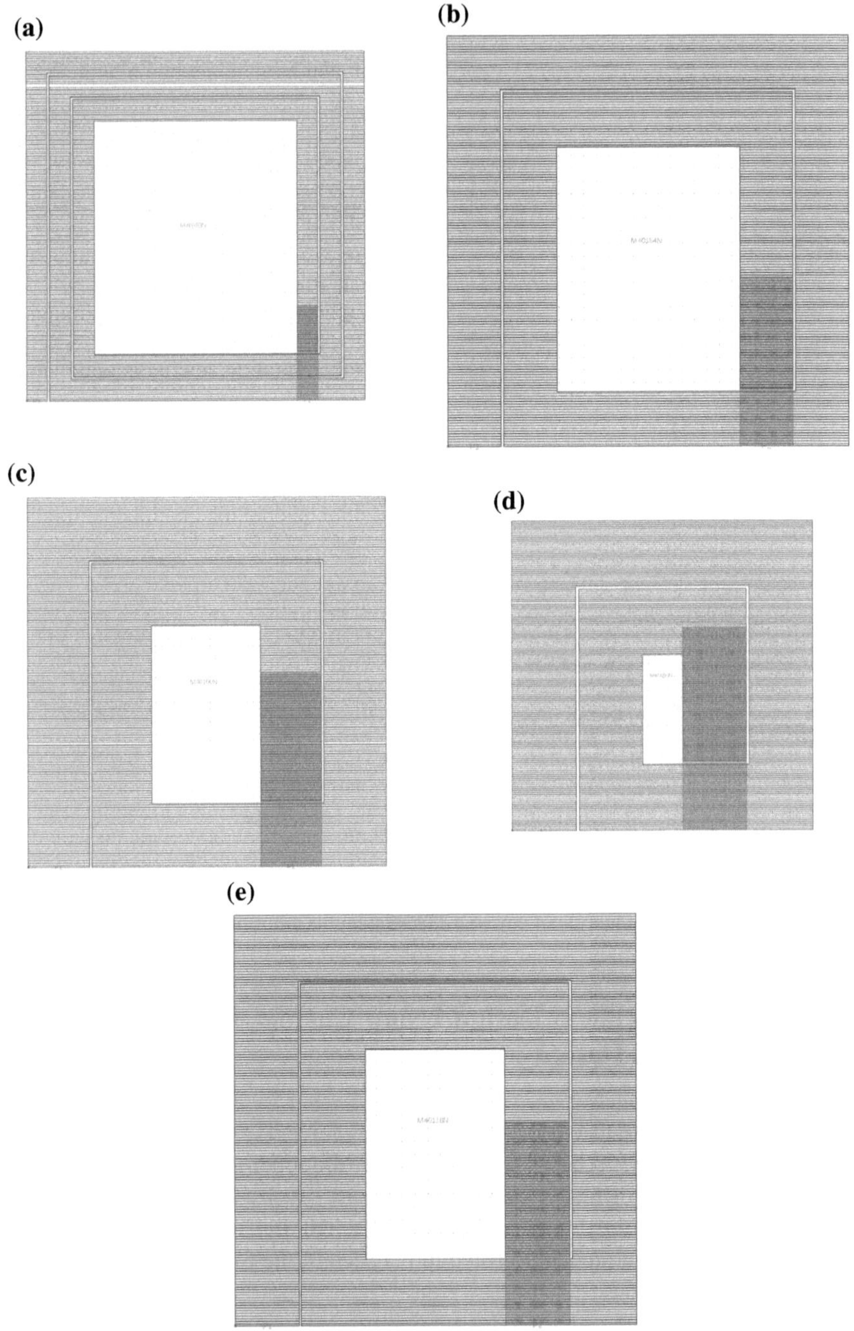

Fig. 9.10 Layouts of inductors designed in task 7 of Chap. 8: **a** M40383N inductor, **b** M40154N inductor, **c** M40100N, **d** M40050N and **e** M40118N

2. Validate the circuit in time and frequency domain by means of simulations, using idealized components to verify that the analytic solution is correct. Then modify the design if necessary; repeat the simulations until satisfied with the performance.
3. Perform accurate modeling of all components, connections, signal tracks, packaging, etc. Then repeat the simulations with these more advanced models. Try to absorb the parasitic in the design components, or attempt to compensate for the losses by modifying the design specifications.
4. Perform the final simulation and execute the physical design.

A final simulation test bench may look like the one depicted in Fig. 9.11 for an integrated implementation. The block marked PA is the power amplifier, with internal circuitry designed using the procedures of Chap. 8, with spiral inductors modeled and synthesized as well. The RFC is added externally, where we have also modeled the RFC resistance. Bond pads and bond wires are modeled as inductors, and we have also added some bond pad capacitance. Finally, we added the bias circuit, the source resistance and the load (antenna) resistance.

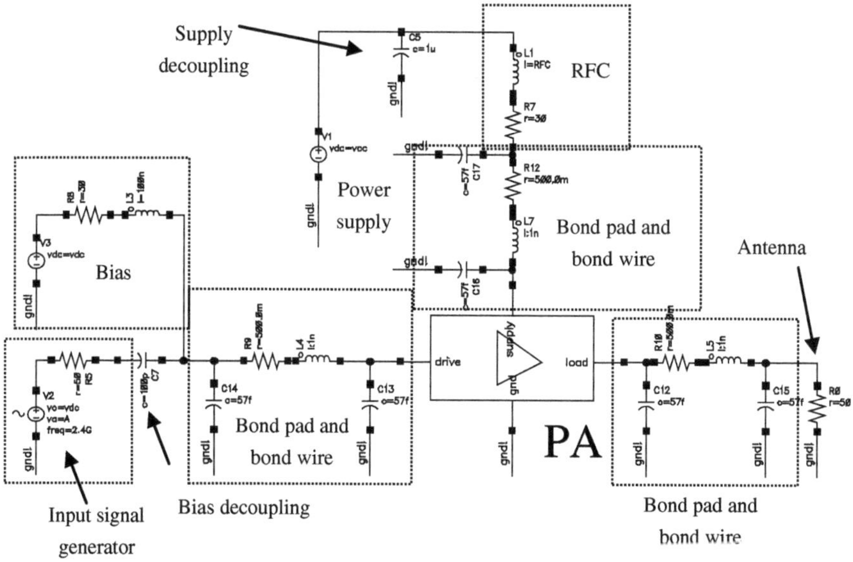

Fig. 9.11 The power amplifier in relation to bond-pad and bond-wire capacitances, bond-wire inductances, RF inductance and resistance and the antenna

9.5 Summary

In this chapter, we introduced some practical issues applicable to discrete, integrated and SOP power amplifier implementations. We stressed that irrespective of the exact implementation, designers need to be careful to ensure that the active device can provide enough gain at the frequency of interest, as well as to budget for the component loss. The feasibility of passives was re-evaluated from the perspective of component values. We also discussed the influence of the operating frequency on the whole system, on the impedance matching, and on the type of implementation. Integrated implementations, although able to reach lower power, allow for compact designs, but introduce additional considerations, because of DRC and layout issues and the influence of packaging and bonding. The SOP implementation, which can be understood as a compromise between discrete and integrated implementations, is considered good alternative to traditional designs, but, as expected, comes with practical considerations of its own.

Finally, we have introduced a formal procedure for power amplifier design, where custom EDA solutions are utilized to replace the initial paper design. However, another major step is to perform sufficient modeling of all subsystems, tracks, leads and packaging if applicable, followed by extensive simulations and re-evaluation of the initial design parameters, where typically, one can redesign for higher power than required to compensate for various losses. The resulting procedure is still iterative, but sufficiently simplified from the situation where all steps are performed manually.

In the following chapter, the last chapter of this book, we will look at future trends and present some concluding remarks.

References

1. Raghavan A, Srirattana N, Laskar J. Modeling and design techniques for RF power amplifiers. 1st ed. Hoboken: Wiley; 2008.
2. Tummala RR, Swaminathan M. System-on-package: miniaturization of the entire system. 1st ed. New York: McGraw-Hill Professional; 2008.
3. Hastings A. The art of analog layout. 2nd ed. Upper Saddle River: Prentice Hall; 2006.
4. Cadence Design Systems. Design data translator's reference. San Jose: Cadence Design Systems; 2007.
5. Greig WJ. Integrated circuit packaging, assembly and interconnections. 1st ed. New York: Springer; 2007.
6. Grebennikov A, Sokal NO, Franco MJ. Switchmode RF and microwave power amplifiers. 2nd ed. Burlington: Elsevier; 2012.
7. Canning T, Tasker PJ, Cripps SC. Continuous mode power amplifier design using harmonic clipping contours: theory and practice. IEEE Trans Microw Theory Tech. 2014;62(1):100–10.
8. Trabelsi H, Barraj I. A 3–5 GHz FSK-UWB transmitter for wireless personal healthcare applications. AEU-Int J Electron Commun. 2015;69(1):262–73.

9. Ji L, Xu Z, Zhou J, Zhai J. Highly efficient 10 W GaN class F power amplifier using DPD. Microwave J. 2013;56(10):120–30.
10. Chao TY, Li CH, Chen YC, Cheng YT, Kuo CN. An interconnecting technology for RF MEMS heterogeneous chip integration. IEEE Trans Electron Devices. 2010;57(4):928–38.
11. Ozalas M. A synthesis-based approach to quickly and easily design a class E amplifier. Microwave J. 2015;58(6):80–6.

Chapter 10
Future Directions and Final Remarks

The previous chapter dealt with some practical considerations for both integrated and discrete power amplifier solutions, including discrete implementation, SOC and SOP. In this chapter the focus is on future directions in the field of power amplification and using custom EDA solutions to aid the power amplifier design task.

The focus of this book was amplification up to and including the Ku-band, which spans up to 18 GHz. On the other hand, the authors also focused on amplification of frequencies above 2 GHz, because they felt that the field of power amplifications below 2 GHz had been fairly well explored. Furthermore, the influence of parasitics is less prominent in these bands and there is no need for very accurate models at these frequencies. That only leaves frequencies above 18 GHz as a possible focus of future work when considering EDA for power amplification. Seeing that mm-waves and THz frequencies are becoming increasingly popular research fields, considerations important for both will be discussed in some detail in this chapter.

In addition, developing custom EDA is not limited only to the field of power amplifiers. Some of the topics and ideas shared in this book are also applicable to other systems, RF and microwave, or otherwise. This chapter will be used to suggest some other areas for research as well.

10.1 Power Amplifiers Utilizing Transmission Lines

As frequency increases, the length of the transmission lines required for implementing passive components both on-chip and off-chip decreases due to decreased wavelength. Furthermore, the Q-factor of a transmission line is directly proportional to the square root of the frequency of operation. Thus at mm-wave frequency range, the Q-factor of a transmission line is enhanced where for passive components, it decreases. Up to now, this book has only considered including a quarter-wavelength transmission line for the implementation of the Class-F power amplifier and for matching. However, other transmission lines are also practical and possible. In fact, transmission lines (typically microstrip lines on chip or SOP) are the only choice not only because of their higher Q-factor and lower insertion loss

M. Božanić and S. Sinha, *Power Amplifiers for the S-, C-, X- and Ku-bands*,
Signals and Communication Technology, DOI 10.1007/978-3-319-28376-0_10

and lower influence of parasitics, but also because of the prospect for high data rates, and in essence, because they are easier to implement [1].

From the EDA perspective, this means that the algorithms presented in previous chapters need to be expanded such that after all required circuit elements have been computed, the sizes of equivalent transmission line segments need to be computed as well [2]. The concept will be illustrated with two Class-E and Class-F configurations. Each power amplifier configuration needs to be analyzed separately and a methodology has to be established. In these examples, the approach to finding the solution will be suggested, but the exact implementation in MATLAB will not be given.

10.1.1 Class-E Power Amplifier Utilizing Transmission Lines

As a first example, consider the Class-E power amplifier with a capacitor-inductor-capacitor matching network designed in Chap. 8 and shown in Fig. 8.38a. If the series inductor is replaced with inductance $L + L_0$, then the general schematics can be shown as in Fig. 10.1. If we allow C_0 and L_0 to resonate each other out, and we can assume that C is of similar size as the parasitic output capacitance of the transistor, the only components that need to be designed are inductors L, L_M and shunt capacitors C_{M1} and C_{M2}. Inductors can be designed as microstrip lines of length

$$l \leq \lambda/8 \tag{10.1}$$

while satisfying

$$L_i = \frac{Z_0}{2\pi f}\tan\theta, \tag{10.2}$$

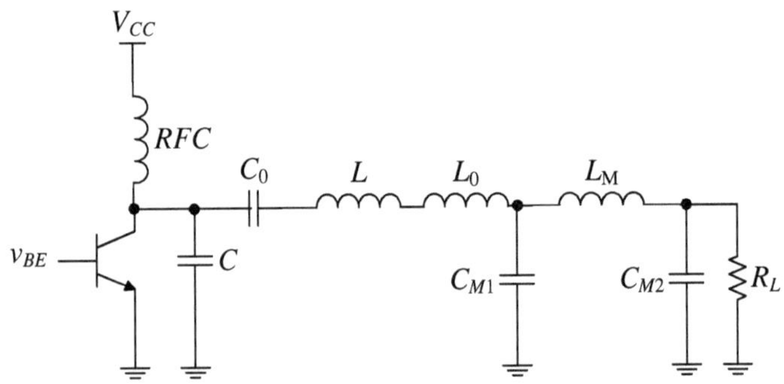

Fig. 10.1 Schematic of the Class-E ZVS power amplifier designed in Chap. 8

where f is the operating frequency, λ is the wavelength, Z_0 is the characteristic impedance of the transmission line and

$$\theta = \beta l = \frac{2\pi}{\lambda} l \tag{10.3}$$

is the electrical length of the line. Shunt capacitors can be implemented with a combination of open-circuit stubs satisfying

$$C_i = \frac{\tan \theta}{2\pi f Z_0}, \tag{10.4}$$

where θ is also described by (10.1) and (10.3).

For both inductors and capacitors, the extension to the design problem (that can be solved in our case in MATLAB), consists in the task of finding the electrical length θ and characteristic impedance Z_0 for each transmission line segment that satisfies Eq. (10.1) and either (10.2) or (10.4). From the characteristic impedance, the width and the height of the transmission line can both be calculated as before for the quarter-wavelength transformer—that is, using the narrow and wide line equations presented in Chap. 5.

Furthermore, the output matching network also serves to suppress harmonics at the load, and better harmonic suppression is achieved by using more than one stub for each shunt capacitor. An implementation where two stubs are used for each capacitor is shown in Fig. 10.2. It is worth mentioning that the Class-E power amplifier implemented using transmission lines, unlike the Class-F power amplifier with a quarter-wave transformer, cannot reach efficiency of 100 %, but the efficiency can be increased with an increased number of harmonic terminations.

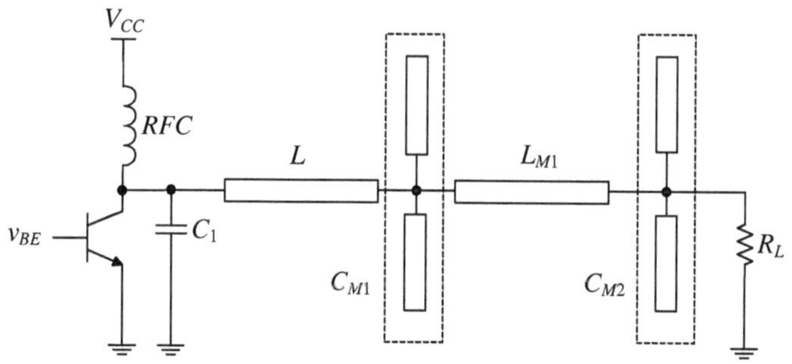

Fig. 10.2 Class-E ZVS power amplifier implemented with microstrip lines

10.1.2 Class-F Power Amplifier Utilizing Transmission Lines

To implement the Class-F power amplifier using transmission lines only (as opposed to the configuration with a quarter-wave transformer and a resonant tank or Class-F$_\infty$ amplifier presented in Chap. 8), we need a slightly different approach. One possible implementation is to start with the same Class-F$_\infty$ circuit (Fig. 10.3), and first note that with increasing frequencies, we cannot ignore the output capacitance of the transistor anymore. Thus, we choose to replace the RFC with a transmission line that will resonate out the output capacitance C. Furthermore, we replace the resonant tank with another transmission line that serves to suppress the harmonics at the third harmonics. Finally, the fundamental quarter-wave transmission line is moved to sit between the supply and the collector of the transistor.

The resulting circuit is shown in Fig. 10.4, where l_1 marks the fundamental quarter-wave line, l_2 marks the line that resonates with the output capacitance of the transistor, and l_3 is the line for harmonic suppression. IMN in this circuit denotes an arbitrary impedance matching network. The electrical lengths of l_1, l_2 and l_3 are

$$\theta_1 = \frac{\pi}{2}, \tag{10.5}$$

$$\theta_2 = \frac{1}{3}\arctan\left(\frac{1}{3Z_0 2\pi f_0 C_{out}}\right), \tag{10.6}$$

and

$$\theta_3 = \frac{\pi}{6}, \tag{10.7}$$

respectively.

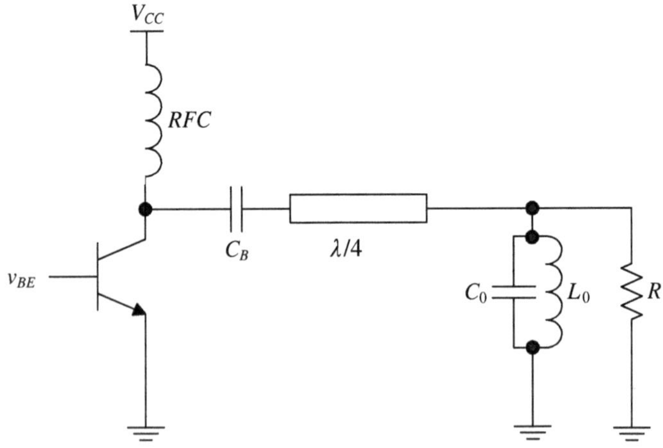

Fig. 10.3 Class-F$_\infty$ power amplifier circuit using some discrete components

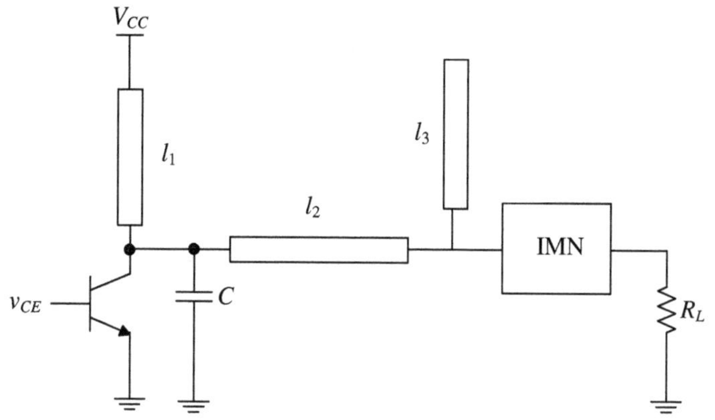

Fig. 10.4 Class-F$_\infty$ power amplifier circuit with all components replaced with transmission lines

The idea behind automating the design of this output stage is to once more design microstrip lines for correct θ and Z_0, which translates to designing for microstrip line length, width and height, similar to what has been done in Chap. 8 for quarter-wave lines.

10.2 Near-Terahertz Frequencies

The focus of the previous section was on the proposed methodology for the EDA power amplifier design for frequencies above 18 GHz and more specifically, mm-wave frequencies. The main assumption was that transistors capable of delivering a substantial amount of gain and power exist and that the main emphasis was on designing passives for those applications. This assumption has in general been true up to now.

Near-THz frequency applications are becoming popular areas of research because of the possibility of high-bandwidth communication, and opportunities for applications in medical equipment, automotive radars and optoelectronics [3, 4]. However, as designs near THz frequencies, the focus of design shifts back towards active devices. The main challenge in implementing relatively low-cost integrated and SOC circuits lies in fabricating transistors with f_T of 0.3 THz, 0.5 THz or even 0.7 THz.

Constant efforts in both lateral and vertical scaling of SiGe HBTs resulted in transistors with acceptable f_T (0.7 THz), but not without problems. Two main issues resulting from such scaling are the need for ever more accurate transistor modeling and thermal considerations. More advanced transistor models (such as the extension of the HICUM model) are now needed. On the other hand, thermal issues arise from high currents and current densities, use of materials with low thermal conductivities and the need for a buried oxide layer to ensure full dielectric isolation. HBTs at near THz frequencies are known to have thermal resistances of several thousand K/W.

Once the challenges with near-THz HBTs are overcome, passives at these frequencies can be considered, followed by efforts to extend the EDA ideas presented in this book to automate the design of power amplifiers.

10.3 Other Design Automation Ideas and Ideas for Expansion of Devised Programs

In Chaps. 7 and 8 several fully functional power amplifier design programs were presented. As stressed before, many design tasks were undertaken only for illustration purposes. However, this does not mean that those are the only output stages which the design can be automated. An attempt to automate the design of any output stage or variation thereof, mentioned in preceding chapters, can be made. This includes various other Class-E and Class-F configurations omitted in Chap. 8 but discussed in Chap. 4, various hybrids and other configurations. Here, both lumped designs attempted throughout this book and transmission-line designs suggested in this chapter are an option.

Power-combining techniques, including the Doherty technique and Chireix outphasing, were also mentioned earlier. With proper analysis of the theory behind their operation, automation of the design of these stages can also be considered.

Any passive components of which the design needs to be automated can be used in any other circuitry, not only power amplifier circuits. For example, any transmission lines can be used for impedance matching, or any other microwave or mm-wave circuits that might require transmission lines. Inductors designed using routines suggested in Chap. 7 can be used in other circuits requiring inductors, such as LNAs or oscillators.

In addition, any procedures developed throughout the book can be used in systems other than power amplifier systems, if applicable. For example, the automated design procedures for matching networks can be used to match any circuitry that requires impedance matching.

Finally, what the authors tried to accomplish in this book was to point out the methodology and the reasoning for incorporating computer-aided design. In summary, the methodology consists of:

- Identifying the circuit topology that needs to be designed,
- Identifying the parameters or components that need to designed,
- Finding the correct equations that allow the designer to compute design parameters,
- Obtaining the parameters used by the equations either from the specifications or otherwise (e.g. material parameters),

and only when all of the above has been completed:

- Structuring and developing the subroutine or a procedure to execute the design task using a programming language that is available to the reader or that he or she understands best.

This method worked very well for power amplifiers because of well-established equations resulting from years of research, but in principle, can be extended to any field of electronics or RF electronics.

10.4 Final Remarks

This brings us to the conclusion of this book. In Chap. 1 we presented the justification for the use of custom EDA in power amplifier design and in later chapters we shared some ideas on what stages of design can be automated to speed up the design process. The aim was not to simply present algorithms that can be used as is, but rather to illustrate by means of examples how every designer can use existing knowledge to improve power amplifier design flow. The algorithms are meant to be modified and updated in on-the-fly fashion, and can be customized to fit the particular problem. As also seen in this chapter, another idea was that the methodology presented throughout the book can first be extended to other frequencies and bands using methods used in those bands, and later even in other fields of electronics or microelectronics.

With ever increasing computing power and simpler programming languages becoming available, some of which are even free or inexpensive, much of the manual work can be automated. As seen throughout the book, some manual work will still have to be done, but ideally it can be limited to research to decide on the applicable topology and to find or devise describing equations, then developing the programs in an available programming or scripting language, before modeling applicable blocks and setting up extensive simulations.

References

1. Grebennikov A, Sokal NO, Franco MJ. Switchmode RF and microwave power amplifiers. 2nd ed. Burlington: Elsevier; 2012.
2. Božanić M, Sinha S. Switch-mode power amplifier design method. Microwave Opt Technol Lett. 2011;53(12):2724–8.
3. d'Alessandro V, Sasso G, Rinaldi N, Aufinger K. Influence of scaling and emitter layout on the thermal behavior of toward-THz SiGe: C HBTs. IEEE Trans Electr Dev. 2014;61(10):3386–94.
4. Pawlak A, Schroter M. An improved transfer current model for RF and mm-wave SiGe (C) heterojunction bipolar transistors. IEEE Trans Electr Dev. 2014;61(8):2612–8.

MIX
Papier aus verantwortungsvollen Quellen
Paper from responsible sources
FSC® C105338

If you have any concerns about our products,
you can contact us on
ProductSafety@springernature.com

In case Publisher is established outside the EU,
the EU authorized representative is:
Springer Nature Customer Service Center GmbH
Europaplatz 3, 69115 Heidelberg, Germany

Printed by Libri Plureos GmbH
in Hamburg, Germany